Basic Trigonometric Graphs

The graph of $y = c + a \sin b(x - d)$ or $y = c + a \cos b(x - d)$ has amplitude $|a|$, period $2\pi/b$, a vertical translation c units $c > 0$ or $|c|$ units down if $c < 0$, and a phase shift d units to the right if $d > 0$ or $|d|$ units to the left if $d < 0$. Throughout, we assume $b > 0$. (To find d, write the argument as $b(x - d)$.) The graph of $y = a \tan bx$ or $y = a \cot bx$ has period π/b.

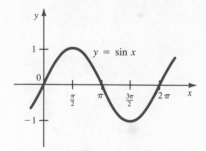

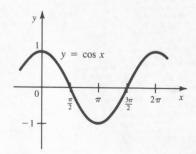

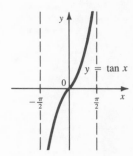

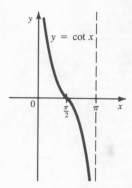

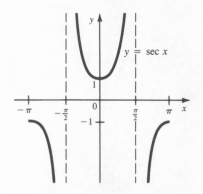

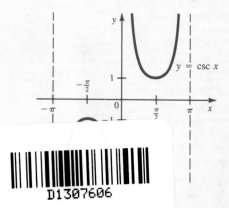

Fundamentals of Trigonometry

Charles D. Miller

Margaret L. Lial

American River College

Scott, Foresman and Company • Glenview, Illinois • London, England

Library of Congress Cataloging-in-Publication Data

Miller, Charles David, 1942–
  Fundamentals of trigonometry.

  Includes excerpts from Algebra and trigonometry /
Margaret L. Lial, Charles D. Miller. 4th ed. 1986.
  Includes index.
  1. Trigonometry. I. Lial, Margaret L. II. Lial,
Margaret L. Algebra and trigonometry. III. Title.
QA531.M63   1987   516.2′4   86–15447
ISBN 0-673-15868-3

Preface

Fundamentals of Trigonometry is designed for a one-semester or one-quarter course that will prepare students either for calculus or for further work in electronics and other technical fields. Applications for both groups of students are given throughout the text.

We have written the book assuming a background in algebra. A course in geometry is a desirable prerequisite, but many students reach trigonometry with little or no background in geometry. For this reason, the *Instructor's Guide* contains a geometry review unit, which can be reproduced and made available for students if desired.

Content Highlights

To prepare for the study of trigonometry, Chapter 1 includes a **review of algebraic ideas,** emphasizing graphs, functions, and composite and inverse functions. It may not be necessary to cover this material in some courses.

The **trigonometric functions are introduced via the unit circle** in Chapter 2; later in the chapter the domains of the trigonometric functions are extended to include angle measures in addition to arc measures.

Thorough coverage of the **graphs** of the trigonometric functions, **inverse trigonometric functions,** and **trigonometric identities and equations** follows. **Applications** of right-triangle trigonometry and oblique-triangle trigonometry and vectors are given in Chapter 5. The **trigonometric forms of complex numbers** and **polar equations** are covered in Chapter 6. Chapters on the **exponential and logarithmic functions** and **analytic geometry** are included and may be covered according to the needs of the course.

The book features a great **many applications problems** from the fields of engineering, physics, biology, astronomy, navigation, and demography. These are spread throughout the text rather than concentrated in one chapter, and are designed to be optional, so instructors may choose which ones to assign.

Key Features   The **format** of the book has been carefully designed to facilitate learning. **Definitions and rules are set off in boxes with marginal headings,** to help students with review and study, and also to make it easy for instructors to use the text. **Second color has been used pedagogically** to clarify explanations of techniques. **Numerous figures and graphs** illustrate examples and exercises.

The textbook features an **abundance of examples,** which present major points through detailed steps and explanations. The symbol ■ makes it clear when an example ends and the discussion continues. Sections are designed to be covered in one class period.

Without making the text a calculator instruction book, we do tell students when a calculator would be appropriate and how to use it. Flexibility has not been sacrificed; table use and evaluation of problems is taught as well as calculator use, and either approach can be used in the classroom.

Exercises   The range of difficulty in the exercise sets affords students ample practice with drill problems. Then they are eased gradually through problems of increasing difficulty to problems that will challenge outstanding students. More than 3400 exercises, including approximately 3120 drill problems and 280 word problems, are provided in the text. There are exercise sets for each section plus review exercises at the end of each chapter (about 600 review problems in all). Answers to odd-numbered exercises are located at the back of the book.

Instructor's Guide   The text is supplemented by an *Instructor's Guide* that gives a geometry review unit, answers to the even-numbered exercises in the text, and an extensive test bank that can be used to prepare examinations or to provide additional problems for students to work.

Acknowledgments   A great many people helped with this revision through their suggestions and comments. In particular, we would like to thank those who reviewed the manuscript: Charles Applebaum, Bowling Green State University; Lee H. Armstrong, University of Central Florida; Alan A. Bishop, Western Illinois University; August J. Garver, University of Missouri, Rolla; James Hodge, College of Lake County; John Hornsby, University of New Orleans; and Brenda Marshall, Parkland College. We also would like to thank the following people, who helped us check the answers: James Arnold, University of Wisconsin–Milwaukee; Lewis Blake III, Duke University; Louis F. Hoelzle, Bucks County Community College; and Marjorie Seachrist.

Our appreciation also goes to the staff at Scott, Foresman, who did an excellent job in working with us toward publication. In particular, we want to thank Bill Poole, Linda Youngman, and Sarah Joseph.

Charles D. Miller

Margaret L. Lial

Contents

Prerequisites for Trigonometry

The foundations of trigonometry go back at least three thousand years. The ancient Egyptians, Babylonians, and Greeks developed trigonometry to find the lengths of the sides of triangles and the measures of their angles. In Egypt trigonometry was used to reestablish land boundaries after the annual flood of the Nile River. In Babylonia it was used in astronomy. The very word *trigonometry* comes from Greek words for triangle *(trigon)* and measurement *(metry)*. Today trigonometry is used in electronics, surveying, and other engineering areas, and is necessary for further courses in mathematics, such as calculus.

1.1 The Number Line and Absolute Value

The relationships among the various sets of numbers are shown in Figure 1. All the numbers shown are real numbers. (The number π shown in Figure 1 is the ratio of

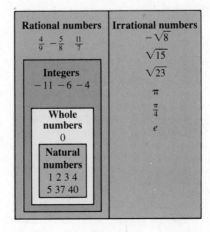

Figure 1

the circumference of a circle to its diameter; π is approximately 3.14159. Also, e is an irrational number discussed in Chapter 7; e is approximately 2.7182818.)

It is often important to know which of two given real numbers is the smaller. Deciding which is smaller is sometimes easier with a **number line,** a geometric representation of the set of real numbers. Set up a number line as follows: Draw a line and choose any point on the line to represent 0. (See Figure 2.) Then choose any point to the right of 0 and label it 1. The distance from 0 to 1 sets up a unit measure that can be used to locate other points to the right of 1, which are labeled 2, 3, 4, 5, and so on, and points to the left of 0, labeled -1, -2, -3, -4, and so on.

Points representing rational numbers, such as 2/3, 16/9, $-11/7$, and so on, can be located by dividing the intervals between integers. For example, 16/9 (or 1 7/9) can be found by dividing the interval from 1 to 2 into 9 equal parts (using methods given in geometry), and then choosing the correct point. Numbers such as $\sqrt{2}$, $\sqrt{3}$, $\sqrt{5}$, and so on, can be located by other geometric constructions. Points corresponding to other irrational numbers can be found to any desired degree of accuracy by using decimal approximations for the numbers.

Figure 2 shows a number line with the points corresponding to several different numbers marked on it. A number that corresponds to a particular point on a line is called the **coordinate** of the point. For example, the leftmost marked point in Figure 2 has coordinate -4. The correspondence between points on a line and the real numbers is called a **coordinate system** for the line. (From now on, the phrase "the point on a number line with coordinate a" will be abbreviated as "the point with coordinate a," or simply "the point a.")

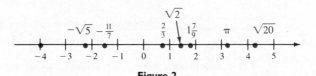

Figure 2

Example I Locate the elements of the set $\{-2/3, 0, \sqrt{2}, \sqrt{5}, \pi, 4\}$ on a number line.

The number π is irrational, with $\pi \approx 3.14159$ ($\approx$ means "approximately equal to"). From a calculator, $\sqrt{2} \approx 1.414$ and $\sqrt{5} \approx 2.236$. Using this information, place the given points on a number line as shown in Figure 3.

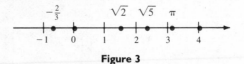

Figure 3

Suppose a and b are two real numbers. If the difference $a - b$ is positive, then *a* **is greater than** b, written $a > b$. If the difference $a - b$ is negative, then *a* **is less than** b, written $a < b$.

These algebraic statements can be given a geometric interpretation. If $a - b$ is positive, so that $a > b$, then a would be to the *right* of b on a number line. Also, if

$a < b$, then a would have to be to the *left* of b. Both the algebraic and geometric statements are summarized below.

Inequality Statements

Statement	Algebraic form	Geometric form
$a > b$	$a - b$ is positive	a is to the right of b
$a < b$	$a - b$ is negative	a is to the left of b

Example 2 (a) In Figure 3 above, $-2/3$ is to the left of $\sqrt{2}$, so that $-2/3 < \sqrt{2}$. Also, $\sqrt{2}$ is to the right of $-2/3$, giving $\sqrt{2} > -2/3$.

(b) The difference $-3 - (-8)$ is positive, showing that $-3 > -8$. The difference $-8 - (-3)$ is negative, so that $-8 < -3$. ■

The following variations on $<$ and $>$ are often used.

Symbol	Meaning
$\leq$	is less than or equal to
$\geq$	is greater than or equal to
$\not<$	is not less than
$\not>$	is not greater than

Statements involving these symbols, as well as $<$ and $>$, are called **inequalities.**

Example 3 (a) $8 \leq 10$ (since $8 < 10$)   (b) $8 \leq 8$ (since $8 = 8$)

(c) $-9 \geq -14$ (since $-9 > -14$)   (d) $-8 \not> -2$ (since $-8 < -2$)

(e) $4 \not< 2$ (since $4 > 2$) ■

The expression $a < b < c$ says that b is *between* a and c, since $a < b < c$ means $a < b$ and $b < c$. Also, $a \leq b \leq c$ means $a \leq b$ and $b \leq c$. When writing these "between" statements, make sure that both inequality symbols point in the same direction. For example, both $2 < 7 < 11$ and $5 > -1 > -6$ are true statements, but a statement such as $3 < 5 > -1$ is meaningless.

The following **properties of order** give the basic properties of $<$ and $>$.

Properties of Order

For all real numbers a, b, and c,

Transitive property	If $a < b$ and $b < c$, then $a < c$.
Addition property	If $a < b$, then $a + c < b + c$.
Multiplication property	If $a < b$, and $c > 0$, then $ac < bc$.
	If $a < b$, and $c < 0$, then $ac > bc$.
Trichotomy property	Given the real numbers a and b, either $a < b$, $a > b$, or $a = b$.

The distance on the number line from a number to 0 is called the **absolute value** of that number. The absolute value of the number a is written $|a|$. For example, the distance on the number line from 9 to 0 is 9, as is the distance from -9 to 0 (see Figure 4), so $|9| = 9$ and $|-9| = 9$.

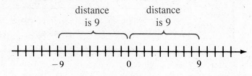

Figure 4

Example 4   (a) $|-4| = 4$

(b) $|2\pi| = 2\pi$

(c) $-|8| = -(8) = -8$

(d) $-|-2| = -(2) = -2$ ■

The definition of absolute value can be stated as follows:

Absolute Value

For every real number a,

$$|a| = \begin{cases} a \text{ if } a \geq 0 \\ -a \text{ if } a < 0. \end{cases}$$

The second part of this definition requires some thought. If a is a negative number, that is, if $a < 0$, then $-a$ is positive. Thus, for a *negative a*,

$$|a| = -a.$$

For example, if $a = -5$, then $|a| = |-5| = -(-5) = 5$.

Example 5   Write each of the following without absolute value bars.

(a) $|\sqrt{5} - 2|$
From Figure 3, $\sqrt{5} > 2$, making $\sqrt{5} - 2 > 0$, and $|\sqrt{5} - 2| = \sqrt{5} - 2$.

(b) $|\pi - 4|$
Since $\pi < 4$, then $\pi - 4 < 0$, and $|\pi - 4| = -(\pi - 4) = -\pi + 4 = 4 - \pi$.

(c) $|m - 2|$ if $m < 2$
If $m < 2$, then $m - 2 < 0$, so $|m - 2| = -(m - 2) = 2 - m$.   ■

The definition of absolute value can be used to prove the following properties of absolute value. (See the exercises below.)

Properties of
Absolute Value

For all real numbers a and b,

$$|a| \geq 0 \qquad\qquad |a| \cdot |b| = |ab|$$

$$|-a| = |a| \qquad\qquad \left|\frac{a}{b}\right| = \frac{|a|}{|b|} \qquad b \neq 0$$

$$|a + b| \leq |a| + |b| \quad \text{(the triangle inequality).}$$

Example 6   Prove that for all real numbers a, $|a| \geq 0$.

If $a \geq 0$, then by the definition of absolute value, $|a| = a$, which is assumed to be greater than or equal to 0. Thus, if $a \geq 0$, then $|a| \geq 0$. If $a < 0$, then $|a| = -a$, which is positive. By these results, if either $a \geq 0$ or $a < 0$, then $|a| \geq 0$, so that $|a| \geq 0$ for every real number a.   ■

The number line of Figure 5 shows the point A, with coordinate -3, and the point B, with coordinate 5. The distance between points A and B is 8 units, which can be found by subtracting the smaller coordinate from the larger. If $d(A, B)$ represents the distance between points A and B, then

$$d(A, B) = 5 - (-3) = 8.$$

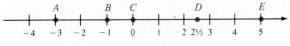

Figure 5

To avoid worrying about which coordinate is smaller, use absolute value as in the following definition.

Distance

Suppose points A and B have coordinates a and b respectively. The distance between A and B, written $d(A, B)$, is

$$d(A, B) = |a - b|.$$

Example 7   Let points A, B, C, D, and E have coordinates as shown on the number line of Figure 6. Find the indicated distances.

Figure 6

(a) $d(B, E)$

Since B has coordinate -1 and E has coordinate 5,

$$d(B, E) = d(-1, 5) = |-1 - 5| = 6.$$

(b) $d(D, A) = \left| 2\frac{1}{2} - (-3) \right| = 5\frac{1}{2}$

(c) $d(B, C) = |-1 - 0| = 1$

(d) $d(E, E) = |5 - 5| = 0$ ∎

I.I Exercises

Write the following numbers in numerical order, from smallest to largest. Use a calculator as necessary.

1. $-9, -2, 3, -4, 8$

2. $7, -6, 0, -2, -3$

3. $|-8|, -|9|, -|-6|$

4. $-|-9|, -|7|, -|-2|$

5. $\sqrt{8}, -4, -\sqrt{3}, -2, -5, \sqrt{6}, 3$

6. $\sqrt{2}, -1, 4, 3, \sqrt{8}, -\sqrt{6}, \sqrt{7}$

7. $3/4, \sqrt{2}, 7/5, 8/5, 22/15$

8. $-9/8, -3, -\sqrt{3}, -\sqrt{5}, -9/5, -8/5$

9. $|-8 + 2|, -|3|, -|-2|, -|-2| + (-3), -|-8| - |-6|$

10. $-2 -|-4|, -3 + |2|, -4, -5 + |-3|$

Let $x = -4$ and $y = 2$. Evaluate each of the following.

11. $|2x|$

12. $|-3y|$

13. $|x - y|$

14. $|2x + 5y|$

15. $|3x + 4y|$

16. $|-5y + x|$

17. $|-4x + y| - |y|$

18. $|-8y + x| - |x|$

19. $\dfrac{|x| + 2|y|}{5 + x}$

20. $\dfrac{2|x| - |y + 2|}{|x| \cdot |y|}$

21. $\dfrac{x|-2 + y + |x||}{|x + 3|}$

22. $\dfrac{(y - 3)|5 - |x| + 2|y + 4||}{2 - |y - 1|}$

Write an equivalent expression for each of the following without using absolute value bars.

23. $|-6|$

24. $-|-2|$

25. $-|-8| + |-2|$

26. $3 - |-4|$

27. $|8 - \sqrt{50}|$

28. $|2 - \sqrt{3}|$

29. $|\sqrt{7} - 5|$

30. $|\sqrt{2} - 3|$

31. $|\pi - 3|$

32. $|\pi - 5|$

33. $|x - 4|$, if $x > 4$

34. $|y - 3|$, if $y < 3$

35. $|2k - 8|$, if $k < 4$

36. $|3r - 15|$, if $r > 5$

37. $|7m - 56|$, if $m < 4$

38. $|2k - 7|$, if $k > 4$

39. $|-8 - 4m|$, if $m > -2$

40. $|6 - 5r|$, if $r < -2$

41. $|x - y|$, if $x < y$

42. $|x - y|$, if $x > y$

43. $|3 + x^2|$

44. $|x^2 + 4|$

45. $|-1 - p^2|$

46. $|-r^4 - 16|$

47. $|\pi - 5| + 1$

48. $|2 - \pi| + 5$

49. $|\sqrt{7} - 2| + 1$

50. $|3 - \sqrt{11}| + 2$

51. $|m - 3| + |m - 4|$, if $3 < m < 4$

52. $|z - 6| - |z - 5|$, if $5 < z < 6$

In the following exercises, the coordinates of four points are given. Find **(a)** $d(A, B)$; **(b)** $d(B, C)$; **(c)** $d(B, D)$; **(d)** $d(D, A)$; **(e)** $d(A, B) + d(B, C)$.

53. $A, -4; B, -3; C, -2; D, 10$

54. $A, -8; B, -7; C, 11; D, 5$

55. $A, -3; B, -5; C, -12; D, -3$

56. $A, 0; B, 6; C, 9; D, -1$

Justify each of the following statements by giving the correct property from this section. Assume that all variables represent real numbers and that no denominators are zero.

57. If $2k < 8$, then $k < 4$.

58. If $x + 8 < 15$, then $x < 7$.

59. If $-4x < 24$, then $x > -6$.

60. If $x < 5$ and $5 < m$, then $x < m$.

61. If $m > 0$, then $9m > 0$.

62. If $k > 0$, then $8 + k > 8$.

63. $|8 + m| \leq |8| + |m|$

64. $|k - m| \leq |k| + |-m|$

65. $|8| \cdot |-4| = |-32|$

66. $|12 + 11r| \geq 0$

67. $\left|\dfrac{-12}{5}\right| = \dfrac{|-12|}{|5|}$

68. $\left|\dfrac{6}{5}\right| = \dfrac{|6|}{|5|}$

69. If p is a real number, then $p < 5$, $p > 5$, or $p = 5$.

70. If z is a real number, then $z > -2$, $z < -2$, or $z = -2$.

Under what conditions are the following statements true?

71. $|x| = |y|$

72. $|x + y| = |x| + |y|$

73. $|x + y| = |x| - |y|$

74. $|x| \leq 0$

75. $|x - y| = |x| - |y|$

76. $||x + y|| = |x + y|$

Evaluate each of the following for real numbers x and y, if no denominators are equal to 0.

77. $\dfrac{|x|}{x}$

78. $\left|\dfrac{x}{|x|}\right|$

79. $\left|\dfrac{x - y}{y - x}\right|$

80. $\left|\dfrac{x + y}{-x - y}\right|$

Prove each of the following properties. Assume a and b are real numbers, and A and B are points on a number line with A having coordinate a and B having coordinate b.

81. $|-a| = |a|$

82. $|a - b| = |b - a|$

83. $|a| \cdot |b| = |ab|$

84. $\left|\dfrac{a}{b}\right| = \dfrac{|a|}{|b|}$, if $b \neq 0$

85. $-|a| \leq a \leq |a|$

86. $|a + b| \leq |a| + |b|$

87. $d(A, B) = d(B, A)$

88. $d(A, B) \geq 0$

89. $d(A, 0) = |a|$

90. $d(A, A) = 0$

91. Suppose $x^2 \leq 81$. Must it then be true that $x \leq 9$?

92. Suppose $x^2 \geq 81$. Must it then be true that $x \geq 9$?

93. Let x be a nonzero real number. Under what conditions is $1/x < x$?

94. Let x be a nonzero real number. When is $x < x^2$?

1.2 A Two-Dimensional Coordinate System

Section 1.1 showed a correspondence between real numbers and points on a number line, a correspondence set up by establishing a coordinate system for the line. This idea can be extended to two dimensions: in two dimensions the correspondence is between *pairs* of real numbers and points on a plane. One way to get this correspondence is by drawing two perpendicular lines, one horizontal and one vertical. These lines intersect at a point O called the **origin.** The horizontal line is the **x-axis,** and the vertical line is the **y-axis.**

Starting at the origin, the x-axis can be made into a number line by placing positive numbers to the right and negative numbers to the left. The y-axis can be made into a number line with positive numbers going up and negative numbers going down.

The *x*-axis and *y*-axis set up a **rectangular coordinate system,** or **Cartesian coordinate system** (named for one of its co-inventors, René Descartes; the other coinventor was Pierre de Fermat). The plane into which the coordinate system is introduced is the **coordinate plane,** or ***xy*-plane.** The *x*-axis and *y*-axis divide the plane into four regions, or **quadrants,** labeled as shown in Figure 7. The points on the *x*-axis and *y*-axis themselves belong to no quadrant.

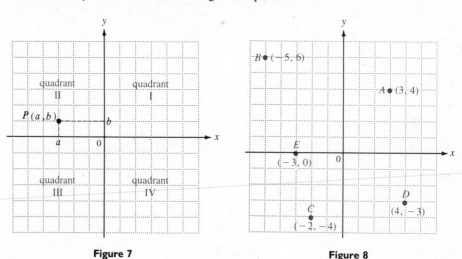

Figure 7   Figure 8

Find a pair of numbers corresponding to a given point *P* as follows. Start at *P* (see Figure 7), and draw a vertical line cutting the *x*-axis at *a*. Draw a horizontal line cutting the *y*-axis at *b*. Then point *P* has **coordinates** (a, b), where (a, b) is an *ordered pair* of numbers. An **ordered pair** of numbers consists of two numbers, written in parentheses, in which the sequence of the numbers is important. For example, $(4, 2)$ and $(2, 4)$ are not the same ordered pair since the sequence of the numbers is different.

As an example, Figure 8 shows the point *A* which corresponds to the ordered pair $(3, 4)$. Also in Figure 8, point *B* corresponds to the ordered pair $(-5, 6)$, *C* to $(-2, -4)$, *D* to $(4, -3)$, and *E* to $(-3, 0)$.

Example I Graph the set of all points (x, y) satisfying the inequality $|x| \leq 2$.

The inequality $|x| \leq 2$ means

$$-2 \leq x \leq 2.$$

The graph of the set of all ordered pairs (x, y) whose *x*-coordinate satisfies the inequalitly $-2 \leq x \leq 2$ is bounded by vertical lines through $(2, 0)$ and $(-2, 0)$. Graph the region satisfying $|x| \leq 2$ as in Figure 9. ∎

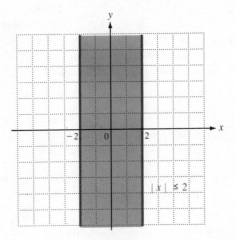

Figure 9

Distance Formula The Pythagorean theorem (in a right triangle with shorter sides a and b, and longest side c, $a^2 + b^2 = c^2$) may be used to obtain a formula for the distance between any two points in the plane. To get this formula, start with two points on a horizontal line, as in Figure 10(a). Use the symbol $P(x_1, y_1)$ to represent point P having coordinates (x_1, y_1). The distance between points $P(x_1, y_1)$ and $Q(x_2, y_1)$ can be found by subtracting the x-coordinates. (Absolute value is used to make sure that the distance is not negative—recall the work with distance in the previous section.) The distance between points P and Q is thus $|x_1 - x_2|$. If $d(P, Q)$ represents the distance between P and Q, then

$$d(P, Q) = |x_1 - x_2|.$$

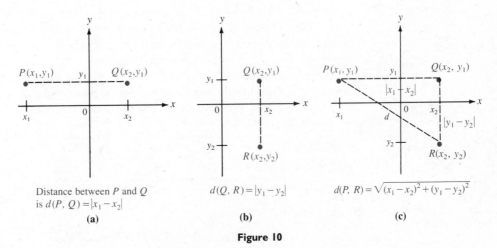

Figure 10

Figure 10(b) shows points $Q(x_2, y_1)$ and $R(x_2, y_2)$ on a vertical line. To find the distance between Q and R, subtract the y-coordinates, finding

$$d(Q, R) = |y_1 - y_2|.$$

Finally, Figure 10(c) shows two points, $P(x_1, y_1)$ and $R(x_2, y_2)$, which are *not* on a horizontal or vertical line. To find $d(P, R)$, construct the right triangle shown in the figure. One side of this triangle is horizontal and has length $|x_1 - x_2|$. The other side is vertical and has length $|y_1 - y_2|$. By the Pythagorean theorem,

$$[d(P, R)]^2 = |x_1 - x_2|^2 + |y_1 - y_2|^2.$$

Writing $|x_1 - x_2|^2$ as the equal expression $(x_1 - x_2)^2$ and $|y_1 - y_2|^2$ as $(y_1 - y_2)^2$ gives the following result, called the **distance formula.**

Distance Formula

> Suppose $P(x_1, y_1)$ and $R(x_2, y_2)$ are two points in a coordinate plane. Then the distance between P and R, written $d(P, R)$, is
>
> $$d(P, R) = \sqrt{(x_1 - x_2)^2 + (y_1 - y_2)^2}.$$

Example 2   Find the distance between $P(-8, 4)$ and $Q(3, -2)$.

According to the distance formula,

$$\begin{aligned}
d(P, Q) &= \sqrt{(-8 - 3)^2 + [4 - (-2)]^2} \\
&= \sqrt{(-11)^2 + 6^2} \\
&= \sqrt{121 + 36} \\
&= \sqrt{157}.
\end{aligned}$$

Using a calculator with a square root key gives 12.530 as an approximate value for $\sqrt{157}$. ∎

Example 3   Are the points $M(-2, 5)$, $N(12, 3)$, and $Q(10, -11)$ the vertices of a right triangle?

To decide whether or not the triangle determined by these three points is a right triangle, use the converse of the Pythagorean theorem: if the sides a, b, and c of a triangle satisfy $a^2 + b^2 = c^2$, then the triangle is a right triangle. A triangle with the three given points as vertices is shown in Figure 11. This triangle is a right triangle if the square of the length of the longest side equals the sum of the squares of the lengths of the other two sides. Use the distance formula to find the length of each side of the triangle.

$$d(M, N) = \sqrt{[12 - (-2)]^2 + (3 - 5)^2} = \sqrt{196 + 4} = \sqrt{200}$$
$$d(M, Q) = \sqrt{[10 - (-2)]^2 + (-11 - 5)^2} = \sqrt{144 + 256} = \sqrt{400} = 20$$
$$d(N, Q) = \sqrt{(10 - 12)^2 + (-11 - 3)^2} = \sqrt{4 + 196} = \sqrt{200}$$

By these results,

$$[d(M, Q)]^2 = [d(M, N)]^2 + [d(N, Q)]^2,$$

proving that the triangle is a right triangle with hypotenuse connecting M and Q. ∎

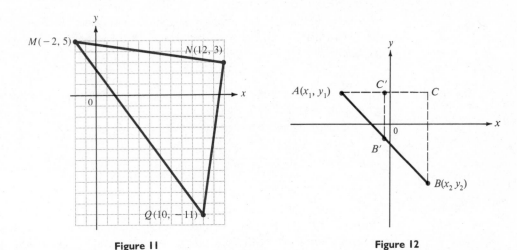

Figure 11   **Figure 12**

Midpoint Formula   The distance formula is used to find the distance between any two points in a plane, while the **midpoint formula** is used to find the coordinates of the midpoint of a line segment.

To develop this formula, let $A(x_1, y_1)$ and $B(x_2, y_2)$ be two different points in a plane (see Figure 12). Assume that A and B are not on a horizontal or vertical line. Let C be the intersection of the horizontal line through A and the vertical line through B. Let B' (read "B-prime") be the midpoint of segment AB. Draw a line through B' and parallel to segment BC. Let C' be the point where this line cuts segment AC. If the coordinates of B' are (x', y'), then C' has coordinates (x', y_1). Since B' is the midpoint of AB, point C' must be the midpoint of segment AC (why?), and

$$d(C, C') = d(C', A),$$

or

$$|x_2 - x'| = |x' - x_1|.$$

Because $d(C, C')$ and $d(C', A)$ must be positive, the only solutions for this equation are found if

$$x_2 - x' = x' - x_1,$$

or

$$x_2 + x_1 = 2x'.$$

Finally

$$x' = \frac{x_1 + x_2}{2},$$

so that the x-coordinate of the midpoint is the average of the x-coordinates of the endpoints of the segment. In a similar manner, the y-coordinate of the midpoint is $(y_1 + y_2)/2$, proving the following result.

Midpoint Formula

The coordinates of the midpoint of the line segment with endpoints having coordinates (x_1, y_1) and (x_2, y_2) is

$$\left(\frac{x_1 + x_2}{2}, \frac{y_1 + y_2}{2} \right).$$

In words: the coordinates of the midpoint of a segment are found by finding the *average* of the x-coordinates and the *average* of the y-coordinates of the endpoints of the segment.

Example 4   Find the coordinates of the midpoint M of the segment with endpoints having coordinates $(8, -4)$ and $(-9, 6)$.

Use the midpoint formula to find that the coordinates of M are

$$\left(\frac{8 + (-9)}{2}, \frac{-4 + 6}{2} \right) = \left(-\frac{1}{2}, 1 \right). \quad \blacksquare$$

Example 5   A line segment has an endpoint with coordinates $(2, -8)$ and a midpoint with coordinates $(-1, -3)$. Find the coordinates of the other endpoint of the segment.

The x-coordinate of the midpoint is found from $(x_1 + x_2)/2$. Here, the x-coordinate of the midpoint is -1. To find x_2, use this formula, with $x_1 = 2$, getting

$$-1 = \frac{2 + x_2}{2}$$

$$-2 = 2 + x_2$$

$$-4 = x_2.$$

In the same way, $y_2 = 2$; the endpoint has coordinates $(-4, 2)$. $\blacksquare$

The two formulas derived in this section can be used to prove various results from geometry.

Example 6   Prove that the diagonals of a parallelogram bisect each other.

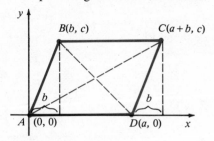

Figure 13

Figure 13 shows parallelogram $ABCD$ having diagonals AC and BD. The figure has been placed on a coordinate system with A at the origin and side AD along the x-axis. Assign coordinates (b, c) to B and $(a, 0)$ to D. Since DC is parallel to AB and is the same length as AB, use results from congruent triangles of geometry and write the coordinates of C as $(a + b, c)$. Show that the diagonals bisect each other by showing that they have the same midpoint. By the midpoint formula,

$$\text{midpoint of } AC = \left(\frac{a + b + 0}{2}, \frac{c + 0}{2}\right) = \left(\frac{a + b}{2}, \frac{c}{2}\right)$$

$$\text{midpoint of } BC = \left(\frac{a + b}{2}, \frac{c + 0}{2}\right) = \left(\frac{a + b}{2}, \frac{c}{2}\right).$$

The midpoints are the same, so AC and BD must bisect each other.   ■

1.2 Exercises

Plot the following points in the xy-plane. Identify the quadrant for each.

1. $A(6, -5)$ **2.** $B(8, 3)$ **3.** $C(-4, 7)$

4. $D(-9, -8)$ **5.** $E(0, -5)$ **6.** $F(-8, 0)$

Graph the set of all points satisfying the following conditions for ordered pairs (x, y).

7. $x = 0$ **8.** $x > 0$ **9.** $y \le 0$ **10.** $y = 0$

11. $xy < 0$ **12.** $\dfrac{x}{y} > 0$ **13.** $|x| = 4, y \ge 2$ **14.** $|y| = 3, x \ge 4$

15. $|y| < 2, x > 1$ **16.** $|x| < 3, y < -2$

17. $2 \le |x| \le 3, y \ge 2$ **18.** $1 \le |y| \le 4, x < 3$

Find the distance $d(P, Q)$ and the coordinates of the midpoint of segment PQ.

19. $P(5, 7), Q(13, -1)$ **20.** $P(-2, 5), Q(4, -3)$

21. $P(-8, -2), Q(-3, -5)$ **22.** $P(-6, -10), Q(6, 5)$

23. $P(\sqrt{2}, -\sqrt{5}), Q(3\sqrt{2}, 4\sqrt{5})$ **24.** $P(5\sqrt{7}, -\sqrt{3}), Q(-7, 8\sqrt{3})$

Give the distance between the following points rounded to the nearest thousandth.

25. $(5, 7), (2, 14)$ **26.** $(-4, 6), (8, -5)$

27. $(3, -7), (-5, 19)$ **28.** $(-9, -2), (-1, -15)$

Find the coordinates of the other endpoint of the segments with endpoints and midpoints having coordinates as given.

29. endpoint $(-3, 6)$, midpoint $(5, 8)$ **30.** endpoint $(2, -8)$, midpoint $(3, -5)$

31. endpoint $(6, -1)$, midpoint $(-2, 5)$ **32.** endpoint $(-5, 3)$, midpoint $(-7, 6)$

Decide whether or not the following points are the vertices of a right triangle.

33. $(-2, 5), (1, 5), (1, 9)$ **34.** $(-9, -2), (-1, -2), (-9, 11)$

35. $(-4, 0), (1, 3), (-6, -2)$ **36.** $(-8, 2), (5, -7), (3, -9)$

37. $(\sqrt{3}, 2\sqrt{3} + 3), (\sqrt{3} + 4, -\sqrt{3} + 3), (2\sqrt{3}, 2\sqrt{3} + 4)$

38. $(4 - \sqrt{3}, -2\sqrt{3}), (2 - \sqrt{3}, -\sqrt{3}), (3 - \sqrt{3}, -2\sqrt{3})$

Use the distance formula to decide whether or not the following points lie on a straight line.

39. $(0, 7)$, $(3, -5)$, $(-2, 15)$

40. $(1, -4)$, $(2, 1)$, $(-1, -14)$

41. $(0, -9)$, $(3, 7)$, $(-2, -19)$

42. $(1, 3)$, $(5, -12)$, $(-1, 11)$

Find all values of x or y such that the distance between the given points is as indicated.

43. $(x, 7)$ and $(2, 3)$ is 5

44. $(5, y)$ and $(8, -1)$ is 5

45. $(3, y)$ and $(-2, 9)$ is 12

46. $(x, 11)$ and $(5, -4)$ is 17

47. (x, x) and $(2x, 0)$ is 4

48. (y, y) and $(0, 4y)$ is 6

49. Show that the points $(-2, 2)$, $(13, 10)$, $(21, -5)$, and $(6, -13)$ are the vertices of a square.

50. Are the points $A(1, 1)$, $B(5, 2)$, $C(3, 4)$, $D(-1, 3)$ the vertices of a parallelogram? Of a rhombus (all sides equal in length)?

51. Use the distance formula and write an equation for all points that are 5 units from $(0, 0)$. Sketch a graph showing these points.

52. Write an equation for all points 3 units from $(-5, 6)$. Sketch a graph showing these points.

53. Find all points (x, y) with $x = y$ that are 4 units from $(1, 3)$.

54. Find all points satisfying $x + y = 0$ that are 8 units from $(-2, 3)$.

55. Write an equation for the points on the perpendicular bisector of the line segment with endpoints at $(0, 0)$ and $(-8, -10)$.

56. Let point A be $(-3, 0)$ and point B be $(3, 0)$. Write an expression for all points (x, y) such that the sum of the distances from A to (x, y) and from (x, y) to B is 8. Simplify the result so that no radicals are involved.

57. Let a be a positive number. Show that the distance between the points (ax_1, ay_1) and (ax_2, ay_2) is a times the distance between (x_1, y_1) and (x_2, y_2).

Use the midpoint formula and distance formula, as necessary, to prove each of the following.

58. The midpoint of the hypotenuse of a right triangle is equally distant from all three vertices.

59. The diagonals of a rectangle are equal in length.

60. The line segment connecting the midpoints of two adjacent sides of any quadrilateral is the same length as the line segment connecting the midpoints of the other two sides.

61. The diagonals of an isosceles trapezoid are equal in length.

62. If the diagonals of a parallelogram are equal in length, then the parallelogram is a rectangle.

63. Find the length of the hypotenuse in each right triangle of the figure.

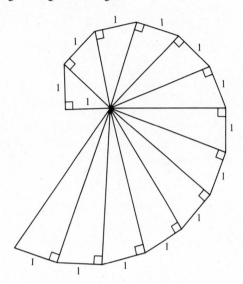

Find the value of x in each right triangle.

64.

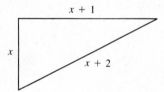

65.

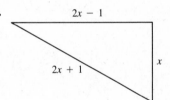

66.

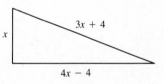

1.3 Graphs

For any set of ordered pairs of real numbers, there is a corresponding set of points in the coordinate plane. For each ordered pair (x, y) in the set, there is a corresponding point (x, y) in the coordinate plane. The set of all points in the plane corresponding to the set of ordered pairs is the **graph** of the set of ordered pairs. For now, we can find graphs only by identifying a reasonable number of ordered pairs. The points corresponding to these ordered pairs are then located, and used to try to decide on the shape of the entire graph. Later, more useful methods of identifying particular graphs are developed.

Example 1 Draw the graph of $S = \{(x, y)|y = -4x + 3\}$.

Use the equation $y = -4x + 3$ to obtain several sample ordered pairs belonging to S. Find these ordered pairs by selecting a number of values for x (or y) and then finding the corresponding values for the other variable. For example, if $x = -3$, then $y = -4(-3) + 3 = 15$, producing the ordered pair $(-3, 15)$. Additional ordered pairs found in this way are given in the following table.

x	-3	-2	-1	0	1	2	3
y	15	11	7	3	-1	-5	-9
ordered pair	$(-3, 15)$	$(-2, 11)$	$(-1, 7)$	$(0, 3)$	$(1, -1)$	$(2, -5)$	$(3, -9)$

The ordered pairs from this table lead to the points that have been plotted in Figure 14(a). These points suggest that the entire graph is a straight line, as drawn in Figure 14(b). ∎

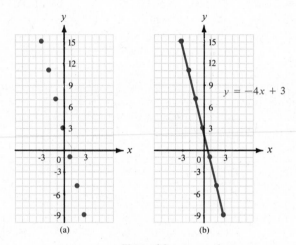

Figure I4

The set of ordered pairs in Example 1, $S = \{(x, y)|y = -4x + 3\}$ involves the equation $y = -4x + 3$. For each value of x that might be chosen, this equation can be used to find a corresponding value of y. By using all possible such values of x, the equation $y = -4x + 3$ leads to a set of ordered pairs (a, b) such that $b = -4a + 3$. The ordered pairs (a, b) are **solutions** of the equation $y = -4x + 3$. The set of all these solutions has a graph, the **graph of the equation.** Using this definition, the graph of the equation $y = -4x + 3$ is the same as the graph of the set $\{(x, y)|y = -4x + 3\}$.

Example 2 Graph the equation $y = x^2 + 2$.

As in the previous example, choose several values of x and find the corresponding values of y.

x	-4	-3	-2	-1	0	1	2	3	4
y	18	11	6	3	2	3	6	11	18

The ordered pairs obtained from this table were plotted in Figure 15(a), and a smooth curve was then drawn through the points as in Figure 15(b). This graph, called a **parabola,** will be studied in more detail later. The lowest point on this parabola, the point (0, 2), is called the **vertex** of the parabola. In $y = x^2 + 2$, any value at all may be chosen for x. However, any real value of x makes $x^2 \geq 0$, so that $x^2 + 2 \geq 2$. Since $y = x^2 + 2$, the value of y will always be at least equal to 2, or $y \geq 2$. ■

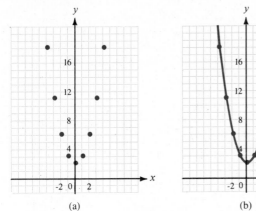

(a)   (b)

Figure 15

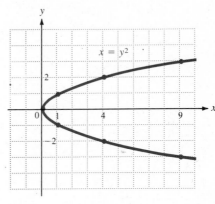

Figure 16

There is a danger in the method used in Examples 1 and 2—we might choose a few values for x, find the corresponding values of y, begin to sketch a graph through these few points, and then make a completely wrong guess as to the shape of the graph. For example, choosing only -1, 0, and 1 as values of x in Example 2 above would produce only the three points $(-1, 3)$, $(0, 2)$, and $(1, 3)$. These three points would not be nearly enough to give sufficient information to determine the proper graph for $y = x^2 + 2$. However, this section involves only elementary graphs, and when more complicated graphs are presented later, more accurate methods of working with them will be developed.

Example 3   Graph $x = y^2$.

Since y is squared, it is probably easier to choose values of y and then to find the corresponding values of x. For example, choosing the value 2 for y gives $x = 2^2 = 4$. Choosing -2 for y gives $x = (-2)^2 = 4$, the same result. The following table shows the values of x corresponding to various values of y.

y	0	1	-1	2	-2	3	-3
x	0	1	1	4	4	9	9

The ordered pairs from this table were used to get the points plotted in Figure 16. (Don't forget that x always goes first in the ordered pair.) A smooth curve was then drawn through the resulting points. This curve is a parabola with vertex (0, 0), opening to the right. Here, y can take on any value. Since $x = y^2$, then $x \geq 0$. ■

Example 4   Graph $y = |x|$.

Start with a table.

x	-4	-3	-2	-1	0	1	2	3	4
y	4	3	2	1	0	1	2	3	4

Use this table to get the points in Figure 17. The graph drawn through these points is made up of portions of two straight lines. Here x may represent any real number, while $y \geq 0$.   ■

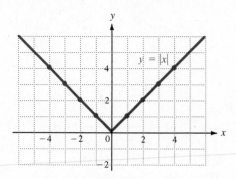

Figure 17

Example 5   Graph $xy = -4$.

Make a table of values. Since this graph is more complicated than the ones above, it is a good idea to use more points. Also, neither x nor y can be equal to 0 so it is a good idea to make several choices for x that are close to 0.

x	-8	-4	-2	-1	$-\dfrac{1}{2}$	$-\dfrac{1}{4}$	$-\dfrac{1}{16}$	$-\dfrac{1}{64}$
y	$\dfrac{1}{2}$	1	2	4	8	16	64	256

x	$\dfrac{1}{64}$	$\dfrac{1}{16}$	$\dfrac{1}{4}$	$\dfrac{1}{2}$	1	2	4	8
y	-256	-64	-16	-8	-4	-2	-1	$-\dfrac{1}{2}$

As x approaches 0 from the left, y gets larger and larger. As x approaches 0 from the right, y gets more and more negative. If $x = 0$, there is no value of y; therefore the graph cannot cross the y-axis. Also, if $y = 0$, there is no value of x, so the graph cannot cross the x-axis either. The variables x and y can take on any value except 0; here $x \neq 0$ and $y \neq 0$. The table was used to get enough points to draw the curve in Figure 18.   ■

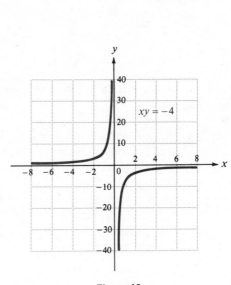

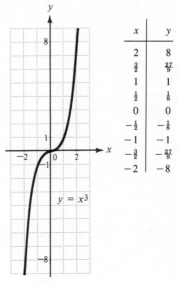

x	y
2	8
$\frac{3}{2}$	$\frac{27}{8}$
1	1
$\frac{1}{2}$	$\frac{1}{8}$
0	0
$-\frac{1}{2}$	$-\frac{1}{8}$
-1	-1
$-\frac{3}{2}$	$-\frac{27}{8}$
-2	-8

Figure 18   **Figure 19**

Example 6 Graph $y = x^3$.

A table of values and the graph are shown in Figure 19. Both the variables x and y can take on any values at all. ■

Circles   A **circle** is the set of all points in a plane which lie a given distance from a given point. The given distance is the **radius** of the circle and the given point is the **center.** The equation of a circle can be found by using the distance formula discussed in Section 2 of this chapter.

For example, Figure 20 shows a circle of radius 3 with center at the origin. To find the equation of this circle, let (x, y) be any point on the circle. The distance between (x, y) and the center of the circle, $(0, 0)$, is given by

$$\sqrt{(x - 0)^2 + (y - 0)^2}.$$

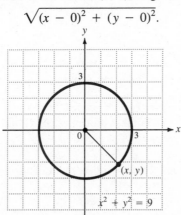

Figure 20

Since this distance equals the radius, 3,

$$\sqrt{(x - 0)^2 + (y - 0)^2} = 3$$
$$\sqrt{x^2 + y^2} = 3$$
$$x^2 + y^2 = 9.$$

As suggested by the graph, the possible values of x are $-3 \leq x \leq 3$, while the possible values of y are $-3 \leq y \leq 3$.

Example 7   Find an equation for the circle having radius 6 and center at $(-3, 4)$.

This circle is shown in Figure 21. Its equation can be found by using the distance formula. Start by letting (x, y) be any point on the circle. The distance from (x, y) to $(-3, 4)$ is given by

$$\sqrt{[x - (-3)]^2 + (y - 4)^2} = \sqrt{(x + 3)^2 + (y - 4)^2}.$$

This same distance is given by the radius, 6. Therefore,

$$\sqrt{(x + 3)^2 + (y - 4)^2} = 6$$

or

$$(x + 3)^2 + (y - 4)^2 = 36.$$

The graph in Figure 21 suggests that the possible values of x are $-9 \leq x \leq 3$ while the possible values of y are $-2 \leq y \leq 10$.   ■

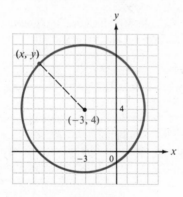

Figure 21

Generalizing the work of Example 7 to a circle of radius r and center (h, k) would give the following result.

Center-Radius Form of the Equation of a Circle

The circle with center (h, k) and radius r has equation

$$(x - h)^2 + (y - k)^2 = r^2,$$

the **center-radius form** of the equation of a circle. As a special case,

$$x^2 + y^2 = r^2$$

is the equation of a circle with radius r and center at the origin.

Starting with the center-radius form of the equation of a circle, $(x - h)^2 + (y - k)^2 = r^2$, and squaring $x - h$ and $y - k$ gives a result of the form

$$x^2 + y^2 + cx + dy + e = 0, \qquad (*)$$

where c, d, and e are real numbers, the **general form** of the equation of a circle. Also, starting with an equation similar to (*), the process of *completing the square* can be used to get an equation of the form

$$(x - h)^2 + (y - k)^2 = m$$

for some number m. If $m > 0$, then $r^2 = m$, and the graph is that of a circle with radius $\sqrt{m}$. If $m = 0$, the graph is the single point (h, k), while there is no graph if $m < 0$.

Example 8 Is $x^2 - 6x + y^2 + 10y + 25 = 0$ the equation of a circle?

Since this equation has the form of equation (*) above, it either represents a circle, a single point, or no points at all. To decide which, complete the square on x and y separately. Start with

$$(x^2 - 6x \qquad) + (y^2 + 10y \qquad) = -25.$$

Half of -6 is -3, and $(-3)^2 = 9$. Also, half of 10 is 5, and $5^2 = 25$. Add 9 and 25 on the left, and to compensate, add 9 and 25 on the right:

$$(x^2 - 6x + 9) + (y^2 + 10y + 25) = -25 + 9 + 25$$
$$(x - 3)^2 + (y + 5)^2 = 9.$$

Since $9 > 0$, the equation represents a circle with center at $(3, -5)$ and radius 3. ∎

Example 9 Is $x^2 + 10x + y^2 - 4y + 33 = 0$ the equation of a circle?

Completing the square as above gives

$$(x^2 + 10x + 25) + (y^2 - 4y + 4) = -33 + 25 + 4$$
$$(x + 5)^2 + (y - 2)^2 = -4.$$

Since $-4 < 0$, there are no ordered pairs (x, y), with x and y real numbers, satisfying the equation. The graph of the given equation would contain no points. ∎

1.3 Exercises

Graph each of the following.

1. $y = 8x - 3$

2. $y = 2x + 7$

3. $y = 3x$

4. $y = -2x$

5. $3y + 4x = 12$

6. $5y - 3x = 15$

7. $y = 3x^2$

8. $y = 5x^2$

9. $y = -x^2$

10. $y = -2x^2$

11. $y = x^2 - 8$

12. $y = x^2 + 6$

13. $y = 4 - x^2$

14. $y = -2 - x^2$

15. $xy = -9$

16. $xy = 25$

17. $4x = y^2$

18. $9x = y^2$

19. $16y^2 = -x$

20. $4y^2 = -x$

21. $y^2 = x + 2$

22. $y^2 = -5 + x$

23. $y = x^3 - 3$

24. $y = x^3 + 4$

25. $y = 1 - x^3$

26. $y = -5 - x^3$

27. $2y = x^4$

28. $y = -x^4$

29. $y = |x| + 4$ **30.** $y = |x| - 3$ **31.** $y = |x| - 2$ **32.** $y = -|x| + 1$

33. $y = 3 - |x|$ **34.** $y = -4 + |x|$ **35.** $y = |x + 3|$ **36.** $y = -|x - 2|$

37. $x^2 + y^2 = 36$ **38.** $x^2 + y^2 = 81$ **39.** $(x - 2)^2 + y^2 = 36$

40. $x^2 + (y + 3)^2 = 49$ **41.** $(x - 4)^2 + (y + 3)^2 = 4$ **42.** $(x + 3)^2 + (y - 2)^2 = 16$

43. $(x + 2)^2 + (y - 5)^2 = 12$ **44.** $(x - 4)^2 + (y - 3)^2 = 8$

Find equations for each of the following circles.

45. center $(1, 4)$, radius 3 **46.** center $(-2, 5)$, radius 4

47. center $(-8, 6)$, radius 5 **48.** center $(3, -2)$, radius 2

49. center $(-1, 2)$, passing through $(2, 6)$ **50.** center $(2, -7)$, passing through $(-2, -4)$

51. center $(-3, -2)$, tangent to the x-axis **52.** center $(5, -1)$, tangent to the y-axis

For each of the following that are equations of circles, give the center and radius of the circle.

53. $x^2 + 6x + y^2 + 8y = -9$ **54.** $x^2 - 4x + y^2 + 12y + 4 = 0$

55. $x^2 - 12x + y^2 + 10y + 25 = 0$ **56.** $x^2 + 8x + y^2 - 6y = -16$

57. $x^2 + 8x + y^2 - 14y + 65 = 0$ **58.** $x^2 - 2x + y^2 + 1 = 0$

59. $x^2 + y^2 = 2y + 48$ **60.** $x^2 + 4x + y^2 = 21$

61. $x^2 - 2.84x + y^2 + 1.4y + 1.8664 = 0$ **62.** $x^2 + 7.4x + y^2 - 3.8y + 16.09 = 0$

63. Find an equation of the circle having the points $(3, -5)$ and $(-7, 2)$ as endpoints of a diameter.

64. Suppose a circle is tangent to both axes, has its center in the third quadrant, and has a radius of $\sqrt{2}$. Find an equation for the circle.

65. One circle has center at $(3, 4)$ and radius 5. A second circle has center at $(-1, -3)$ and radius 4. Do the circles cross?

66. Does the circle with radius 6 and center at $(0, 5)$ cross the circle with center at $(-5, -4)$ with radius 4?

Use the appropriate formula to find the circumference and area of each of the circles having equations as follows.

67. $(x - 2)^2 + (y + 4)^2 = 25$ **68.** $(x + 3)^2 + (y - 1)^2 = 9$

69. $x^2 + 2x + y^2 - 4y + 1 = 0$ **70.** $x^2 - 6x + y^2 + 8y = -16$

The **unit circle** is a circle centered at the origin and having radius 1. Show that each of the following points lies on the unit circle. Only an approximation is possible in Exercises 75 and 76.

71. $(-1/2, \sqrt{3}/2)$ **72.** $(-\sqrt{2}/2, -\sqrt{2}/2)$ **73.** $(-\sqrt{39}/8, 5/8)$

74. $(11/13, 4\sqrt{3}/13)$ **75.** $(-.47691285, .87895059)$ **76.** $(.21514257, -.97658265)$

77. Decide if each of the following points is *inside, on,* or *outside* the circle with center $(1, -4)$ and radius 6:
    (a) $(3, -2)$ **(b)** $(9, 1)$ **(c)** $(7, -4)$ **(d)** $(0, 9)$.

78. Find any points of intersection of the circles having equations $x^2 - 4x + y^2 - 10y = -4$ and $x^2 + 8x + y^2 - 10y = -40$, given that $x = -3$ for the point of intersection.

1.4 Functions

Suppose X is the set of all students studying this book every Monday evening at the local pizza parlor. Let Y be the set of integers between 0 and 100. To each student in set X can be associated a number from Y which represents the score the student received on the last test. Typical associations between students in set X and scores in set Y are shown in Figure 22.

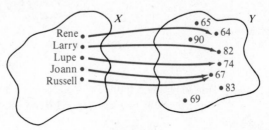

Figure 22

A correspondence such as the one shown in Figure 22 is called a **function** or a **mapping,** defined as follows.

Function

> A **function** from a set X to a set Y is a correspondence that assigns to each element of X exactly one element of Y.

The set X in this definition is called the **domain** of the function.

Three things should be noticed about the function shown in Figure 22 above.

First, there is a single score associated with each student. That is, each element in X corresponds to exactly one element in Y.

Second, the same score may correspond to more than one student. In the example above, two students (Joann and Russell) have a score of 67.

Third, not every element of Y need be used; for example, none of the students had a score of 83 or 65.

While the sets X and Y above were different, in many cases X and Y have many elements in common. In fact, it is not unusual for X and Y to be equal.

Example 1   Decide whether or not the following diagrams represent functions from X to Y.

(a)

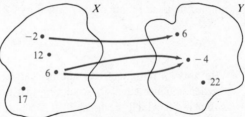

Figure 23

The diagram in Figure 23 does not represent a function from X to Y: there is no arrow leading from 17 in the domain, so that there is no element in Y corresponding to the element 17 in X.

(b)

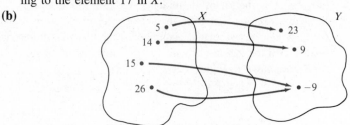

Figure 24

This diagram (Figure 24) does represent a function from X to Y since there is a single element in Y corresponding to each element in X.

(c)

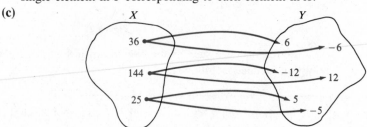

Figure 25

Since each element of set X corresponds to *two* elements in set Y, the diagram in Figure 25 does not represent a function from X to Y.   ■

It is common to use the letters f, g, and h to name functions. If f is a function and x is an element in the domain X, then $f(x)$ represents the element in Y that corresponds to x in X. For example, if f is used to name the function in Figure 24 above, then

$$f(5) = 23, \quad f(14) = 9, \quad f(15) = -9, \quad \text{and} \quad f(26) = -9.$$

For a given element x in set X, the corresponding element $f(x)$ in set Y is called the **value** or **image** of f at x. The set of all possible values of $f(x)$ makes up the **range** of the function. Throughout this book, if the domain for a function specified by a formula is not given, it will be assumed to be the largest possible set of real numbers for which the formula is meaningful, unless otherwise specified. For example, suppose function f is defined by

$$f(x) = \frac{-4x}{2x - 3}.$$

With this formula, any real number can be used for x except $x = 3/2$, which makes the denominator equal 0. Assuming that the domain is the largest possible set of real numbers for which the formula is meaningful makes the domain $\{x \mid x \neq 3/2\}$.

Example 2   Find the domain and range for the functions defined by the following rules.

(a) $f(x) = x^2$

Any number may be squared, so the domain is the set of all real numbers. Since $x^2 \geq 0$ for every value of x, the range, written in interval notation, is $[0, +\infty)$.

(b) $f(x) = \sqrt{6 - x}$

Since $6 - x$ must be greater than or equal to 0 for the square root to exist, x can take on any value less than or equal to 6, so that the domain is $(-\infty, 6]$. The radical indicates the nonnegative square root, so the interval $[0, +\infty)$ is the range.

(c) $f(x) = \sqrt{x^2 + x - 6}$

The domain includes those values of x that make

$$x^2 + x - 6 \geq 0.$$

Factor to get

$$(x + 3)(x - 2) \geq 0.$$

By the methods of algebra, solve this quadratic inequality to get the domain $(-\infty, -3] \cup [2, +\infty)$. Since the radical exists only when the radicand, $x^2 + x - 6$, takes on nonnegative values, the range is $[0, +\infty)$. ■

For most of the functions in this book, the domain can be found with algebraic methods already discussed. The range, however, must often be found with graphing (see Figure 27 below), complicated algebra, or calculus.

Based on the definition of a function, functions f and g are **equal** if and only if f and g have exactly the same domains and $f(x) = g(x)$ for every value of x in the domain.

Suppose a function is defined by $f(x) = -3x + 2$. To emphasize that this statement is used to find values in the range of f, it is common to write

$$y = -3x + 2.$$

When a function is written in the form $y = f(x)$, x is called the **independent variable,** and y the **dependent variable.** There is no reason to restrict the variables to x or y—different areas of study use different variables. For example, it is common to use t for time in physics, or p for price in management.

By the definition, a function f is a rule that assigns to each element of one set exactly one element of a second set. For a particular value of x in the first set, the corresponding element in the second set is written $f(x)$. There is a distinction between f and $f(x)$: f is the function or rule, while $f(x)$ is the value obtained by applying the rule to an element x. However, it is very common to abbreviate

"the function defined by the rule $y = f(x)$"

as simply

"the function $y = f(x)$."

Example 3 Let $g(x) = 3\sqrt{x}$ and $h(x) = 1 + 4x$. Find each of the following.

(a) $g(16)$

To find $g(16)$, replace x in $g(x) = 3\sqrt{x}$ with 16, getting
$g(16) = 3\sqrt{16} = 3 \cdot 4 = 12$.

(b) $h(-3) = 1 + 4(-3) = -11$

(c) $g(-4)$ does not exist; -4 is not in the domain of g since $\sqrt{-4}$ is not a real number

(d) $h(\pi) = 1 + 4\pi$

(e) $g(m) = 3\sqrt{m}$, if m represents a nonnegative real number

(f) $g[h(3)]$

First find $h(3)$, as follows.

$$h(3) = 1 + 4 \cdot 3 = 1 + 12 = 13$$

Now, $g[h(3)] = g(13) = 3\sqrt{13}$. ∎

Example 4 Let $f(x) = 2x^2 - 3x$, and find the quotient

$$\frac{f(x + h) - f(x)}{h}, \quad h \neq 0,$$

which is important in calculus.

To find $f(x + h)$, replace x with $x + h$, to get

$$\frac{f(x + h) - f(x)}{h} = \frac{2(x + h)^2 - 3(x + h) - (2x^2 - 3x)}{h}$$

$$= \frac{2(x^2 + 2xh + h^2) - 3x - 3h - 2x^2 + 3x}{h}$$

$$= \frac{2x^2 + 4xh + 2h^2 - 3x - 3h - 2x^2 + 3x}{h}$$

$$= \frac{4xh + 2h^2 - 3h}{h}$$

$$= 4x + 2h - 3. \quad ∎$$

A function can be thought of as a set of ordered pairs; in fact, a function f produces the set of ordered pairs $\{(x, f(x)) | x$ is in the domain of $f\}$. By the definition of function, for each x which appears in the first position of an ordered pair, there will be exactly one value of $f(x)$ in the second position.

In addition, a set of ordered pairs in which any two ordered pairs that have the same first entry x also have equal values for the second entry must be a function. The following *alternate definition of a function* is based on this.

Alternate Definition of a Function

A **function** is a set of ordered pairs in which two ordered pairs cannot have the same first entries and different second entries.

The idea of a function as a set of ordered pairs can be used to define the **graph of a function:** the graph of a function f is the set of all points in the plane of the form $(x, f(x))$, where x is in the domain of f. The graph of a function f is the same as the graph of the equation $y = f(x)$.

There is a quick way to tell if a given graph is the graph of a function or not. Figure 26 shows two graphs. In the graph of part (a), for any x that might be chosen, exactly one value of $f(x)$ or y could be found, showing that the graph is the graph of a function. On the other hand, the graph in part (b) is not the graph of a function. For example, the vertical line through x_1 leads to two different values of y, namely y_1 and y_2. This example suggests the **vertical line test** for a function.

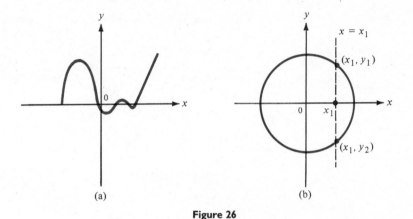

(a)   (b)

Figure 26

Vertical Line Test

If each vertical line cuts a graph in no more than one point, the graph is the graph of a function.

The graph of a function can be used to find the domain and range of the function, as shown in Figure 27.

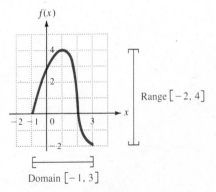

Figure 27

Odd and Even Functions The idea of symmetry is very useful when drawing graphs. A function whose graph is symmetric with respect to the y-axis is called an *even function,* while a function having a graph symmetric to the origin is an *odd function.*

Odd and Even Functions

Suppose x and $-x$ are both in the domain of a function f. Then

f is an **odd function** if $f(-x) = -f(x)$ for every x in the domain of f,

f is an **even function** if $f(-x) = f(x)$ for every x in the domain of f.

Example 5 Decide if the functions defined as follows are odd, even, or neither.

(a) $f(x) = 8x^4 - 3x^2$

Replacing x with $-x$ gives

$$f(-x) = 8(-x)^4 - 3(-x)^2 = 8x^4 - 3x^2 = f(x).$$

Since $f(x) = f(-x)$ for each x in the domain of the function, f is an even function.

(b) $f(x) = 6x^3 - 9x$

Here

$$f(-x) = 6(-x)^3 - 9(-x) = -6x^3 + 9x = -f(x).$$

This function is odd.

(c) $f(x) = \dfrac{1}{x - 3}$

Replacing x with $-x$ produces

$$f(-x) = \frac{1}{-x - 3}$$

which equals neither $f(-x)$ nor $-f(x)$. This function is neither odd nor even. ∎

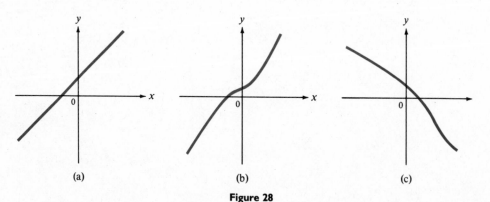

(a)   (b)   (c)

Figure 28

The exponents on the terms in parts (a) and (b) of Example 5 show the origin of the names *odd* and *even*.

Increasing and Decreasing Functions Intuitively, a function is *increasing* if its graph moves up as it goes to the right. The functions graphed in Figure 28(a) and (b) are increasing functions. On the other hand, a function is *decreasing* if its graph goes down as it goes to the right. The function graphed in Figure 28(c) is a decreasing function.

The function of Figure 29 is neither an increasing nor a decreasing function. However, this function is increasing on both the interval $(-\infty, -2]$ and the interval $[4, +\infty)$; also, f is decreasing on the interval $[-2, 4]$.

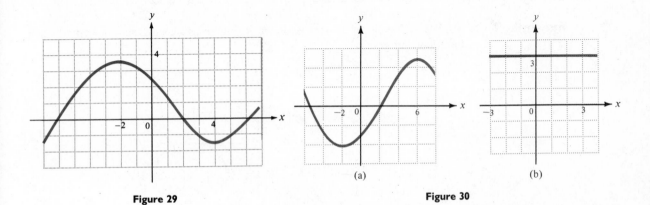

Figure 29

(a)

(b)

Figure 30

The idea of a function increasing and decreasing on an interval is defined below, where I represents any interval of real numbers.

Increasing and Decreasing Functions

Let f be a function, with the interval I a subset of the domain of f. Let x_1 and x_2 be in I. Then

f is **increasing** on I if $f(x_1) < f(x_2)$ whenever $x_1 < x_2$,   and

f is **decreasing** on I if $f(x_1) > f(x_2)$ whenever $x_1 < x_2$.

Example 6 Find where the following functions are increasing or decreasing.

(a) The function graphed in Figure 30(a) is increasing on the interval $[-2, 6]$. It is decreasing on $(-\infty, -2]$ and $[6, +\infty)$.

(b) The function in Figure 30(b) is never increasing or decreasing. This function is an example of a **constant function**, a function defined by $f(x) = k$, for k a real number. ■

Recall from algebra that graphs of functions can be translated, as summarized below.

Translations of the Graph of a Function

Let f be a function, and let c be a positive number.

To graph:	shift the graph of $y = f(x)$ by c units:
$y = f(x) + c$	upward
$y = f(x) - c$	downward
$y = f(x + c)$	left
$y = f(x - c)$	right

Example 7 A graph of a function $y = f(x)$ is shown in Figure 31. Using this graph, find each of the following graphs.

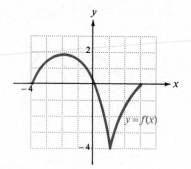

Figure 31

(a) $y = f(x) + 3$

This graph is the same as the graph in Figure 31, translated 3 units upward. See Figure 32(a).

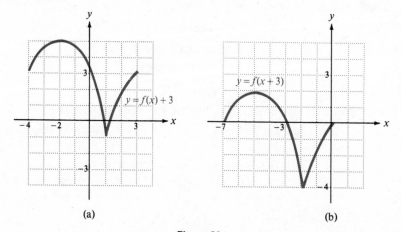

(a)   (b)

Figure 32

(b) $y = f(x + 3)$

To get the graph of $y = f(x + 3)$, translate the graph of $y = f(x)$ to the left 3 units. See Figure 32(b). ∎

1.4 Exercises

For each of the following, find the indicated function values: **(a)** $f(-2)$,   **(b)** $f(0)$,
(c) $f(1)$,   **(d)** $f(4)$.

1.

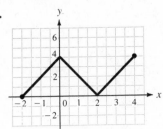

2.

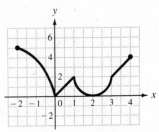

3.

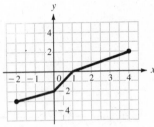

4.

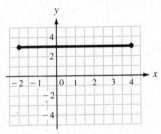

Let $f(x) = 3x - 1$ and $g(x) = |x^2 - 8|$. Find each of the following.

5. $f(0)$	**6.** $f(-1)$	**7.** $f(-3) + 2$	**8.** $f(4) - 5$
9. $g(2)$	**10.** $g(0)$	**11.** $f(a^2)$	**12.** $g(b^3)$
13. $f(1) + g(1)$	**14.** $f(-2) - g(-2)$	**15.** $[g(5)]^2$	**16.** $[f(-1)]^2$
17. $f(-2m)$	**18.** $f(-11p)$	**19.** $f(5a - 2)$	**20.** $f(3 + 2k)$
21. $g(5p - 2)$	**22.** $g(-6k + 1)$	**23.** $f(p) + g(2p)$	**24.** $g(1 - r) - f(1 - r)$
25. $g(m) \cdot f(m)$	**26.** $\dfrac{f(5r)}{g(r)}$	**27.** $f\left(\dfrac{1}{m + p}\right)$	**28.** $f\left(\dfrac{-4 + r}{4 + r}\right)$

Let $f(x) = -4.6x^2 - 8.9x + 1.3$. Find each of the following function values.

29. $f(3)$ **30.** $f(-5)$ **31.** $f(-4.2)$ **32.** $f(-1.8)$

Give the domain and range of the functions defined as follows. Give only the domain in Exercises 45 and 46.

33. $f(x) = 2x - 1$	**34.** $g(x) = 3x + 5$	**35.** $g(x) = x^4$
36. $h(x) = (x - 2)^2$	**37.** $f(x) = \sqrt{8 + x}$	**38.** $f(x) = -\sqrt{x + 6}$
39. $h(x) = \sqrt{16 - x^2}$	**40.** $m(x) = \sqrt{x^2 - 25}$	**41.** $k(x) = (x - 4)^{1/2}$

42. $f(x) = (3x + 2)^{1/2}$

43. $g(x) = \sqrt{\dfrac{1}{x^2 + 25}}$

44. $z(x) = -\sqrt{\dfrac{4}{x^2 + 1}}$

45. $g(x) = \dfrac{2}{x^2 - 3x + 2}$

46. $h(x) = \dfrac{-4}{x^2 + 5x + 4}$

47. $r(x) = \sqrt{x^2 - 4x - 5}$

48. $r(x) = \sqrt{x^2 + 7x + 10}$

49. $f(x) = |x - 4|$

50. $k(x) = -|2x - 7|$

51. $g(x) = -\sqrt{2x + 5}$

52. $h(x) = \sqrt{1 - x}$

53. $f(x) = \sqrt{36 - x^2}$

54. $g(x) = -\sqrt{25 - x^2}$

55. $k(x) = -\sqrt{x^2 - 9}$

56. $z(x) = \sqrt{x^2 - 100}$

57.

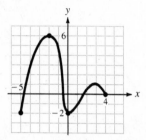

58.

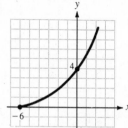

59.

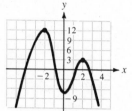

60.

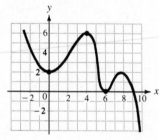

61.

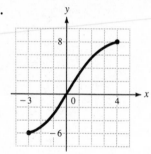

62.

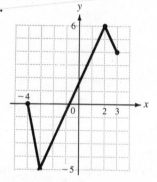

For the functions defined as follows, find **(a)** $f(x + h)$, **(b)** $f(x + h) - f(x)$, and
(c) $\dfrac{f(x + h) - f(x)}{h}$. (Assume $h \ne 0$.)

63. $f(x) = x^2 - 4$

64. $f(x) = 8 - 3x^2$

65. $f(x) = 6x + 2$

66. $f(x) = 4x + 11$

67. $f(x) = 2x^3 + x^2$

68. $f(x) = -4x^3 - 8x$

Find where the functions graphed or defined as follows are increasing or decreasing. In
Exercises 75–82, first graph the functions.

69.

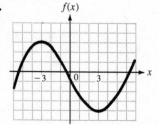

70.

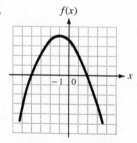

71.

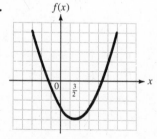

72.

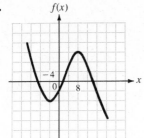

73.

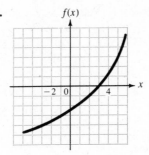

74.

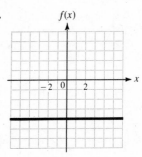

75. $f(x) = -4x + 2$ **76.** $f(x) = 5x - 1$ **77.** $f(x) = x^2 + 4$ **78.** $f(x) = -x^2 - 3$

79. $f(x) = -|x + 2|$ **80.** $f(x) = |3 - x|$ **81.** $f(x) = x + |x|$ **82.** $f(x) = |x| - x$

Decide whether the functions defined as follows are even, odd, or neither.

83. $f(x) = x^2$

84. $f(x) = x^3$

85. $f(x) = x^4 + x^2 + 5$

86. $f(x) = x^3 - x + 1$

87. $f(x) = 2x + 3$

88. $f(x) = |x|$

89. $f(x) = \dfrac{2}{x - 6}$

90. $f(x) = \dfrac{8}{x}$

Graph on the same coordinate axes the graphs of $y = f(x)$ for the given values of c.

91. $f(x) = x^2 + c$; $c = -1$, $c = 2$

92. $f(x) = (x + c)^2$; $c = -1$, $c = 2$

93. $f(x) = |x - c|$; $c = -2$, $c = 1$

94. $f(x) = c - |x|$; $c = 2$, $c = -3$

Let the graph of a function $y = f(x)$ be as shown. Sketch the graph of each of the following.

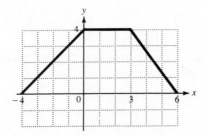

95. $y = f(x) + 4$

96. $y = 3 - f(x)$

97. $y = f(x - 1)$

98. $y = f(x + 2)$

99. $y = f(x + 3) - 2$

100. $y = f(x - 1) - 4$

101. A box is made from a piece of metal 12 by 16 inches by cutting squares of side x from each corner. (See the sketch.) Give the volume of the box as a function of x.

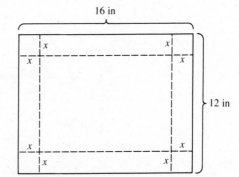

102. A cone has a radius of 6 inches and a height of 9 inches. (See the figure.) The cone is filled with water having a depth of h inches. The radius of the surface of the water is r inches. Use similar triangles to give r as a function of h.

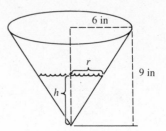

Suppose $f(x) = x^2 + 5x$. Find all values of t so that

103. $f(t + 1) = f(2t)$ **104.** $f(2t + 3) = f(t + 2)$

Let $f(x) = x^2 - 3x$. Decide which of the folllowing results hold for all values of x, as long as no denominator is 0.

105. $f\left(\dfrac{1}{x}\right) = \dfrac{1}{f(x)}$ **106.** $f(x) \cdot \dfrac{1}{f(x)} = 1$

107. $f(2x) = 2 \cdot f(x)$ **108.** $f(x + 5) = f(x) + f(5)$

109. Give the area of a circle as a function of its radius; also, give the circumference as a function of the radius.

110. Write the area of a circle as a function of the diameter of the circle; then write the circumference as a function of the diameter.

111. A rectangle is inscribed in a circle of radius r. Let x represent the length of one side of the rectangle. Give the area of the rectangle as a function of r.

112. The height of a cone is half the radius of the base. Give the volume of the cone as a function of the radius of the base.

I.5 **Composite and Inverse Functions**

The sketch in Figure 33 shows a function f which assigns to each element x of set X some element y of set Y. Suppose also that a function g takes each element of set Y and assigns a value z of set Z. Using both f and g, then, an element x in X is assigned to an element z in Z. The result of this process is a new function h, which takes an element x in X and assigns an element z in Z.

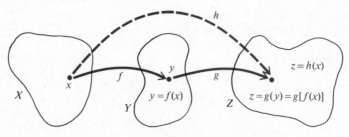

Figure 33

This function h is called the *composition* of functions g and f, written $g \circ f$, and defined as follows.

Composition of Functions

> Let f and g be functions. The **composite function,** or **composition,** of g and f, written $g \circ f$, is defined by
>
> $$(g \circ f)(x) = g[f(x)],$$
>
> for all x in the domain of f such that $f(x)$ is in the domain of g.

Example 1   Let $f(x) = 4x + 1$ and $g(x) = 2x^2 + 5x$. Find each of the following.

(a) $(g \circ f)(x)$

By definition, $(g \circ f)(x) = g[f(x)]$. Using the given functions,

$$
\begin{aligned}
(g \circ f)(x) &= g[f(x)] \\
&= g(4x + 1) \\
&= 2(4x + 1)^2 + 5(4x + 1) \\
&= 2(16x^2 + 8x + 1) + 20x + 5 \\
&= 32x^2 + 16x + 2 + 20x + 5 \\
&= 32x^2 + 36x + 7.
\end{aligned}
$$

(b) $(f \circ g)(x)$

Use the definition above with f and g interchanged, so that $(f \circ g)(x)$ becomes $f[g(x)]$. Then

$$
\begin{aligned}
(f \circ g)(x) &= f[g(x)] \\
&= f(2x^2 + 5x) \\
&= 4(2x^2 + 5x) + 1 \\
&= 8x^2 + 20x + 1. \quad \blacksquare
\end{aligned}
$$

As this example shows, it is not always true that $f \circ g = g \circ f$. In fact, $f \circ g$ is very rarely equal to $g \circ f$. In Example 1, the domain of both composite functions is the set of all real numbers.

Example 2   Let $f(x) = 1/x$ and $g(x) = \sqrt{3 - x}$. Find $f \circ g$ and $g \circ f$. Give the domain of each.

First find $f \circ g$.

$$
\begin{aligned}
(f \circ g)(x) &= f[g(x)] \\
&= f(\sqrt{3 - x}) \\
&= \frac{1}{\sqrt{3 - x}}
\end{aligned}
$$

The radicand $\sqrt{3 - x}$ represents a nonzero real number only when $3 - x > 0$, or $x < 3$, so that the domain of $f \circ g$ is the interval $(-\infty, 3)$.

Use the same functions to find $g \circ f$, as follows.

$$(g \circ f)(x) = g[f(x)]$$

$$= g\left(\frac{1}{x}\right) = \sqrt{3 - \frac{1}{x}} = \sqrt{\frac{3x - 1}{x}}$$

The domain of $g \circ f$ is the set of all real numbers x such that $(3x - 1)/x \geq 0$. By the methods of solving quadratic inequalities, the domain of $g \circ f$ is the set $(-\infty, 0) \cup [1/3, +\infty)$.   ∎

Addition and subtraction are inverse operations: starting with a number x, adding 5, and then subtracting 5 gives x back as a result. Also, some functions are inverses of each other. For example, it turns out that the functions

$$f(x) = 8x \qquad \text{and} \qquad g(x) = \frac{1}{8}x$$

are inverses of each other. To see why, choose a value of x such as $x = 12$ and find $f(12)$.

$$f(12) = 8 \cdot 12 = 96$$

Calculating $g(96)$ gives

$$g(96) = \frac{1}{8} \cdot 96 = 12,$$

so that $(g \circ f)(12) = g[f(12)] = 12$. Also, finding $g(12)$ and then $f[g(12)]$ would show that $(f \circ g)(12) = 12$. For these two functions f and g, it turns out that for *any* value of x,

$$(f \circ g)(x) = 8\left(\frac{1}{8}x\right) = x \qquad \text{and} \qquad (g \circ f)(x) = \frac{1}{8}(8x) = x.$$

As we shall see, this condition makes f and g inverses of each other.

This section shows how to start with a function such as $f(x) = 8x$ and use it to obtain the inverse function $g(x) = (1/8)x$, if an inverse exists. Not all functions have inverses: the only functions with an inverse function are *one-to-one functions*.

One-to-one Functions   Given the function defined by $y = x^2$, it is possible for two different values of x to lead to the same value of y. For example, the value $y = 4$ is obtained from either of two values of x: both $2^2 = 4$ and $(-2)^2 = 4$. On the other hand, for the function defined by $y = 6x$, a given value of y can be found from exactly one value of x. For this function, if $y = 30$, then $x = 5$; there is no other value of x that will produce a value of 30 for y.

This second function, defined by $y = 6x$, is an example of a **one-to-one function**: a function is one-to-one if each element in the range is obtained from exactly one element in the domain, or

One-to-one Function	a function f is **one-to-one** if $f(a) = f(b)$ implies that $a = b$.

Example 3   Decide whether the functions defined as follows are one-to-one.

(a) $f(x) = -4x + 12$

Suppose $f(a) = f(b)$, or

$$-4a + 12 = -4b + 12.$$

Then $-4a = -4b,$

or $a = b.$

The fact that $f(a) = f(b)$ implies $a = b$, which makes the function one-to-one.

(b) $g(x) = \sqrt{25 - x^2}$

Start with $g(a) = g(b)$, or

$$\sqrt{25 - a^2} = \sqrt{25 - b^2}.$$

Squaring both sides gives

$$25 - a^2 = 25 - b^2,$$

or $a^2 = b^2.$

From $a^2 = b^2$, it is *not* possible to conclude that $a = b$, since a might equal $-b$ instead. This means the function g is not one-to-one. As a numerical example, $g(3) = g(-3)$, but $3 \neq -3$.   ■

There is a useful graphical test for deciding whether or not a function is one-to-one. Figure 34(a) shows the graph of a function defined by $y = f(x)$ cut by a horizontal line. As the graph suggests, $f(x_1) = f(x_2) = f(x_3)$, even though x_1, x_2, and x_3 are all distinct values. Since one value of y can be obtained from more than one value of x, the function is not one-to-one. On the other hand, drawing horizontal lines on the graph of Figure 34(b) shows that a given value of y can be obtained from only one value of x, making the function one-to-one.

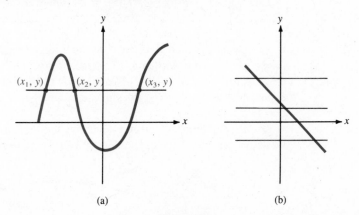

(a)   (b)

Figure 34

The examples suggest the **horizontal line test** for one-to-one functions.

**Horizontal Line
Test**

If no horizontal line cuts the graph of a function in more than one point,
then the function is **one-to-one.**

Inverse Functions We saw above that the functions defined as $f(x) = 8x$ and
$g(x) = (1/8)x$ have the property that

$$(f \circ g)(x) = x \quad \text{and} \quad (g \circ f)(x) = x.$$

In other words, if $y = f(x)$, then $g(y) = x$. For example, $f(5) = 40$ and $g(40) =$
5. See the sketch in Figure 35.

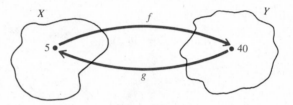

Figure 35

Functions f and g have the following property: starting with a value of x in the
domain of f and finding $f(x)$, then evaluating g at $f(x)$, the result is x. Because of
this property, f and g are called *inverse functions* of each other.

Inverse Functions

Let f be a one-to-one function with domain X and range Y. Let g be a
function with domain Y and range X. Then g is the **inverse function** of f if

$$(f \circ g)(x) = x \quad \text{for every } x \text{ in } Y,$$

and $\qquad (g \circ f)(x) = x \quad \text{for every } x \text{ in } X.$

A special notation is often used for inverse functions: if g is the inverse of a
function f, then g is written as f^{-1} (read "f-inverse.") In the example above, $f(x) =$
$8x$, and $g(x) = f^{-1}(x) = (1/8)x$. Do not confuse the -1 in f^{-1} with a negative
exponent. The symbol f^{-1} does not represent $1/f$; it is used for the inverse of function f.

By the definition of inverse function, the domain of f equals the range of f^{-1},
while the range of f equals the domain of f^{-1}. See Figure 36.

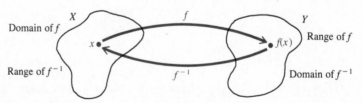

Figure 36

Example 4   Let functions f and g be defined by $f(x) = x^3 - 1$ and $g(x) = \sqrt[3]{x + 1}$, respectively. Is g the inverse function of f?

First check that f is one-to-one. Since it is, now find $(f \circ g)(x)$ and $(g \circ f)(x)$.

$$(f \circ g)(x) = f[g(x)] = (\sqrt[3]{x + 1})^3 - 1 = x + 1 - 1 = x$$

$$(g \circ f)(x) = g[f(x)] = \sqrt[3]{(x^3 - 1) + 1} = \sqrt[3]{x^3} = x.$$

Since both $(f \circ g)(x) = x$ and $(g \circ f)(x) = x$, and since the domain of f equals the range of g, and the range of f equals the domain of g, function g is indeed the inverse of function f, so that f^{-1} is given by

$$f^{-1}(x) = \sqrt[3]{x + 1}. \quad \blacksquare$$

Keep in mind that a function f can have an inverse function f^{-1} if and only if f is one-to-one. Since increasing and decreasing functions are one-to-one, they must have inverse functions.

Given a one-to-one function f and an x in the domain of f, the corresponding value in the range of f is found by means of the equation $y = f(x)$. With the inverse function f^{-1}, the value of y can be used to produce x, since $x = f^{-1}(y)$. Therefore, the equation for f^{-1} can be found by solving $y = f(x)$ for x.

For example, let $f(x) = 7x - 2$. Then $y = 7x - 2$. The function f is one-to-one, so that f^{-1} exists. Solve $y = 7x - 2$ for x, as follows:

$$y = 7x - 2$$

$$7x = y + 2$$

$$x = \frac{y + 2}{7}$$

or, since $x = f^{-1}(y)$,

$$f^{-1}(y) = \frac{y + 2}{7}.$$

For a given x in X (the domain of f), the equation $y = 7x - 2$ produces a value of $f(x)$ in Y (the range of f). The function $f^{-1}(y) = (y + 2)/7$ takes a value of y in Y and produces a value x in X. Since it is customary to use x for the domain element of a function, replace y with x in f^{-1} to get

$$f^{-1}(x) = \frac{x + 2}{7}.$$

Check that $(f \circ f^{-1})(x) = x$ and $(f^{-1} \circ f)(x) = x$, so that f^{-1} is indeed the inverse of f.

In summary, find the equation of an inverse function with the following steps.

**Finding an
Equation for f^{-1}**

1. Check that the function f defined by $y = f(x)$ is a one-to-one function.
2. Solve for x. Let $x = f^{-1}(y)$.
3. Exchange x and y to get $y = f^{-1}(x)$.
4. Check the domains and ranges: the domain of f and the range of f^{-1} should be equal, as should the domain of f^{-1} and the range of f.

Example 5   For each of the functions defined as follows, find any inverse functions.

(a) $f(x) = \dfrac{4x + 6}{5}$

This function is one-to-one and so has an inverse. Let $y = f(x)$, and solve for x.

$$y = \frac{4x + 6}{5}$$
$$5y = 4x + 6$$
$$5y - 6 = 4x$$
$$\frac{5y - 6}{4} = x$$

Finally, exchange x and y, and let $y = f^{-1}(x)$, to get

$$\frac{5x - 6}{4} = y,$$

or
$$f^{-1}(x) = \frac{5x - 6}{4}.$$

Verify that the domain of f is the range of f^{-1} and the range of f is the domain of f^{-1}.

(b) $f(x) = x^3 - 1$

Two different values of $f(x)$ come from two different values of x, so the function is one-to-one and has an inverse. To find the inverse, first solve $y = x^3 - 1$ for x, as follows.

$$y = x^3 - 1$$
$$y + 1 = x^3$$

$$\sqrt[3]{y + 1} = x$$

Finally, exchange x and y, and let $y = f^{-1}(x)$, to get
$$\sqrt[3]{x + 1} = y,$$

or
$$f^{-1}(x) = \sqrt[3]{x + 1}.$$

Check that the domain of f^{-1} is the range of f and the range of f^{-1} is the domain of f.

(c) $f(x) = x^2$

The two different x-values 4 and -4 give the same value of y, namely 16, showing that the function is not one-to-one and has no inverse function.   ■

Suppose f and f^{-1} are inverse functions of each other. Suppose that $f(a) = b$, for real numbers a and b. Then by the definition of inverse, $f^{-1}(b) = a$. This shows that if a point (a, b) is on the graph of f, then (b, a) will belong to the graph of f^{-1}. As shown in Figure 37, the points (a, b) and (b, a) are **symmetric with respect to the line $y = x$.** Thus, the graph of f^{-1} can be obtained from the graph of f by reflecting the graph of f about the line $y = x$.

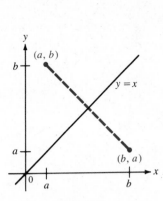

Figure 37

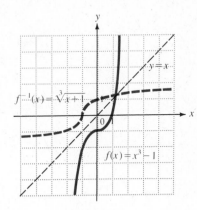

Figure 38

For example, Figure 38 shows the graph of $f(x) = x^3 - 1$ as a solid line and the graph of $f^{-1}(x) = \sqrt[3]{x + 1}$ as a dashed line. These graphs are symmetric with respect to the line $y = x$.

Example 6 Let $f(x) = \sqrt{x + 5}$ with domain $[-5, +\infty)$. Find $f^{-1}(x)$.

The function f is one-to-one and has an inverse function. To find this inverse function, start with

$$y = \sqrt{x + 5}$$

and solve for x, to get

$$y = \sqrt{x + 5}$$
$$y^2 = x + 5$$
$$y^2 - 5 = x.$$

Exchanging x and y gives

$$x^2 - 5 = y.$$

We cannot just give $x^2 - 5$ as $f^{-1}(x)$. In the definition of f above, the domain was given as $[-5, +\infty)$. The range of f is $[0, +\infty)$. Since the range of f equals the domain of f^{-1}, the function f^{-1} must be given as

$$f^{-1}(x) = x^2 - 5, \quad \text{domain } [0, +\infty).$$

As a check, the range of f^{-1}, $[-5, +\infty)$ equals the domain of f. Graphs of f and f^{-1} are shown in Figure 39. The line $y = x$ is included on the graph to show that the graphs of f and f^{-1} are mirror images with respect to this line. ■

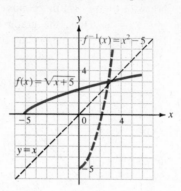

Figure 39

1.5 Exercises

Let $f(x) = 4x^2 - 2x$ and let $g(x) = 8x + 1$. Find each of the following.

1. $(f \circ g)(2)$ **2.** $(f \circ g)(-5)$ **3.** $(g \circ f)(2)$

4. $(g \circ f)(-5)$ **5.** $(f \circ g)(k)$ **6.** $(g \circ f)(5z)$

Find $f \circ g$ and $g \circ f$ for each of the pairs of functions defined as follows.

7. $f(x) = 8x + 12,\quad g(x) = 3x - 1$

8. $f(x) = -6x + 9,\quad g(x) = 5x + 7$

9. $f(x) = 5x + 3,\quad g(x) = -x^2 + 4x + 3$

10. $f(x) = 4x^2 + 2x + 8,\quad g(x) = x + 5$

11. $f(x) = -x^3 + 2,\quad g(x) = 4x$ **12.** $f(x) = 2x,\quad g(x) = 6x^2 - x^3$

13. $f(x) = \dfrac{1}{x},\quad g(x) = x^2$ **14.** $f(x) = \dfrac{2}{x^4},\quad g(x) = 2 - x$

15. $f(x) = \sqrt{x + 2},\quad g(x) = 8x^2 - 6$ **16.** $f(x) = 9x^2 - 11x,\quad g(x) = 2\sqrt{x + 2}$

17. $f(x) = \dfrac{1}{x - 5},\quad g(x) = \dfrac{2}{x}$ **18.** $f(x) = \dfrac{8}{x - 6},\quad g(x) = \dfrac{4}{3x}$

19. $f(x) = \sqrt{x + 1},\quad g(x) = \dfrac{-1}{x}$ **20.** $f(x) = \dfrac{8}{x},\quad g(x) = \sqrt{3 - x}$

The graphs of functions f and g are shown on the following page. Use these graphs to find the values in Exercises 21–24.

21. $(f \circ g)(2)$ **22.** $(g \circ f)(2)$ **23.** $(g \circ f)(-4)$ **24.** $(f \circ g)(-2)$

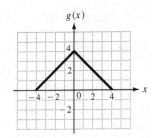

For each of the pairs of functions defined as follows, show that $(f \circ g)(x) = x$ and $(g \circ f)(x) = x$.

25. $f(x) = 3x, \quad g(x) = \dfrac{1}{3}x$

26. $f(x) = \dfrac{3}{4}x, \quad g(x) = \dfrac{4}{3}x$

27. $f(x) = 8x - 11, \quad g(x) = \dfrac{x + 11}{8}$

28. $f(x) = \dfrac{x - 3}{4}, \quad g(x) = 4x + 3$

29. $f(x) = x^3 + 6, \quad g(x) = \sqrt[3]{x - 6}$

30. $f(x) = \sqrt[5]{x - 9}, \quad g(x) = x^5 + 9$

31. Suppose the population P of a certain species of fish depends on the number x (in hundreds) of a smaller kind of fish which serves as its food supply, so that

$$P(x) = 2x^2 + 1.$$

Suppose, also, that the number x (in hundreds) of the smaller species of fish depends upon the amount a (in appropriate units) of its food supply, a kind of plankton. Suppose

$$x = f(a) = 3a + 2.$$

Find $(P \circ f)(a)$, the relationship between the population P of the large fish and the amount a of plankton available.

32. Suppose the demand for a certain brand of vacuum cleaner is given by

$$D(p) = \dfrac{-p^2}{100} + 500,$$

where p is the price in dollars. If the price, in terms of the cost, c, is expressed as

$$p(c) = 2c - 10,$$

find the demand in terms of the cost.

33. An oil well off the Gulf Coast is leaking, with the leak spreading oil over the surface as a circle. At any time t, in minutes, after the beginning of the leak, the radius of the circular oil slick on the surface is $r(t) = 4t$ ft. Let $A(r) = \pi r^2$ represent the area of a circle of radius r. Find and interpret $(A \circ r)(t)$.

34. When a thermal inversion layer is over a city (such as happens often in Los Angeles), pollutants cannot rise vertically but are trapped below the layer and must disperse horizontally. Assume that a factory smokestack begins emitting a pollutant at 8 A.M. Assume that the pollutant disperses horizontally, forming a circle. If t represents the time, in hours, since the factory began emitting pollutants ($t = 0$ represents 8 A.M.), assume that the radius of the circle of pollution is $r(t) = 2t$ mi. Let $A(r) = \pi r^2$ represent the area of a circle of radius r. Find and interpret $(A \circ r)(t)$.

Which of the functions graphed or defined as follows are one-to-one?

35.

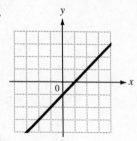

36.

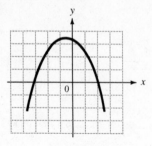

37.

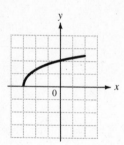

38. $y = 4x - 5$

39. $y = 6 - x$

40. $y = -x^2$

41. $y = (x - 2)^2$

42. $y = \sqrt{36 - x^2}$

43. $y = -\sqrt{100 - x^2}$

44. $y = x^3 - 1$

45. $y = -\sqrt[3]{x + 5}$

46. $y = (\sqrt{x} + 1)^2$

47. $y = (3 - 2\sqrt{x})^2$

48. $y = \dfrac{1}{x + 2}$

49. $y = \dfrac{-4}{x - 8}$

Which of the pairs of functions graphed or defined as follows are inverses of each other?

50.

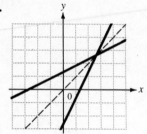

51.

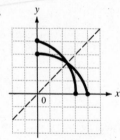

52.

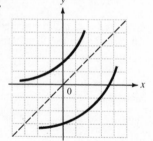

53. $f(x) = -\dfrac{3}{11}x, \quad g(x) = -\dfrac{11}{3}x$

54. $f(x) = 2x + 4, \quad g(x) = \dfrac{1}{2}x - 2$

55. $f(x) = 5x - 5, \quad g(x) = \dfrac{1}{5}x + 1$

56. $f(x) = 8x - 7, \quad g(x) = \dfrac{x + 8}{7}$

57. $f(x) = \dfrac{1}{x + 1}, \quad g(x) = \dfrac{x - 9}{12}$

58. $f(x) = \dfrac{1}{x + 1}, \quad g(x) = \dfrac{1 - x}{x}$

59. $f(x) = x^2 + 3$, domain $[0, +\infty)$, and $g(x) = \sqrt{x - 3}$, domain $[3, +\infty)$

60. $f(x) = \sqrt{x + 8}$, domain $[-8, +\infty)$, and $g(x) = x^2 - 8$, domain $[0, +\infty)$

Graph the inverse of each one-to-one function.

61.

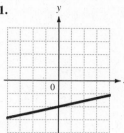

62.

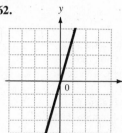

63.

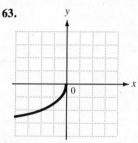

64.

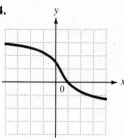

For each of the functions defined as follows that is one-to-one, write an equation for the inverse function in the form of $y = f^{-1}(x)$, and then graph f and f^{-1}.

65. $y = 4x - 5$

66. $y = 3x - 4$

67. $y = -\dfrac{2}{5}x$

68. $y = \dfrac{1}{3}x$

69. $y = -x^3 - 2$

70. $y = x^3 + 1$

71. $3x + y = 9$

72. $x + 4y = 12$

73. $y = -x^2 + 2$

74. $y = x^2$

75. $xy = 4$

76. $y = \dfrac{1}{x}$

77. $y = \dfrac{-6x + 5}{3x - 1}$

78. $y = \dfrac{8x + 3}{4x - 1}$

79. $f(x) = \sqrt{6 + x}$, domain $[-6, +\infty)$

80. $f(x) = 4 - x^2$, domain $(-\infty, 0]$

The graph of a function f is shown in the figure. Use the graph to find each of the following values.

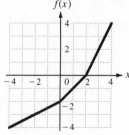

81. $f^{-1}(4)$

82. $f^{-1}(2)$

83. $f^{-1}(0)$

84. $f^{-1}(-2)$

85. $f^{-1}(-3)$

86. $f^{-1}(-4)$

Let $f(x) = x^2 + 5x$, for $x \ge -5/2$. Find each of the following, rounding to the nearest hundredth.

87. $f^{-1}(7)$

88. $f^{-1}(-3)$

Let $f(x) = -4x + 3$, while $g(x) = 2x^3 - 4$. Find each of the following.

89. $(f \circ g)^{-1}(x)$

90. $(f^{-1} \circ g^{-1})(x)$

91. $(g^{-1} \circ f^{-1})(x)$

92. Show that a one-to-one function has exactly one inverse function.

93. Let f be an odd one-to-one function. What can you say about f^{-1}?

Chapter I Summary

Key Words

number line
coordinate
inequalities
absolute value
distance formula
midpoint formula
parabola
vertex
function
mapping
domain

range
independent variable
dependent variable
vertical line test for a
    function
increasing function
decreasing function
composition of functions
one-to-one function
horizontal line test
inverse function

Review Exercises

Write the following numbers in numerical order, from smallest to largest.

1. $|6 - 4|$, $-|-2|$, $|8 + 1|$, $-|3 - (-2)|$

2. $\dfrac{5}{6}$, $\dfrac{1}{2}$, $-\dfrac{2}{3}$, $-\dfrac{5}{4}$, $-\dfrac{3}{8}$

3. $\sqrt{15}$, $-\sqrt{2}$, $\sqrt{169}$, $-|-8 + \sqrt{13}|$

4. $\sqrt{7}$, $-\sqrt{8}$, $-|\sqrt{16}|$, $|-\sqrt{12}|$

Write without absolute value bars.

5. $|3 - \sqrt{7}|$

6. $|\sqrt{8} - 3|$

7. $|m - 3|$, if $m > 3$

8. $|-6 - x^2|$

In each of the following exercises, the coordinates of three points are given.
Find   **(a)** $d(A, B)$ **(b)** $d(A, B) + d(B, C)$.

9. A, -2; B, -1; C, 10

10. A, -8; B, -12; C, -15

Under what conditions are the following statements true?

11. $|x| = x$

12. $|a + b| = -a - b$

13. $|x + y| = -|x| - |y|$

14. $|x| \leq 0$

15. $d(A, B) = 0$

16. $d(A, B) + d(B, C) = d(A, C)$

Graph the set of points satisfying each condition below.

17. $x < 0$

18. $y \geq 0$

19. $xy > 0$

20. $x = 2$

Find the distance $d(P, Q)$ and the midpoint of segment PQ.

21. $P(3, -1)$ and $Q(-4, 5)$

22. $P(-8, 2)$ and $Q(3, -7)$

23. Find the other endpoint of a line segment having one end at $(-5, 7)$ and having midpoint at $(1, -3)$.

24. Are the points $(5, 7)$, $(3, 9)$, $(6, 8)$ the vertices of a right triangle?

25. Find all possible values of k so that $(-1, 2)$, $(-10, 5)$, and $(-4, k)$ are the vertices of a right triangle.

26. Find all possible values of x so that the distance between $(x, -9)$ and $(3, -5)$ is 6.

27. Find all points (x, y) with $x = 6$ so that (x, y) is 4 units from $(1, 3)$.

28. Find all points (x, y) with $x + y = 0$ so that (x, y) is 6 units from $(-2, 3)$.

Prove each of the following.

29. The lines joining each vertex of a triangle with the midpoint of the opposite side meet in a point.

30. The line segment connecting midpoints of two sides of a triangle is half as long as the third side.

Graph each of the following. Give the domain for those graphs that are graphs of functions.

31. $x + y = 4$ **32.** $3x - 5y = 20$ **33.** $y = \frac{1}{2}x^2$ **34.** $y = 3 - x^2$

35. $y = \dfrac{-8}{x}$ **36.** $y = -2x^3$ **37.** $y = \sqrt{x - 7}$ **38.** $(x - 3)^2 + y^2 = 16$

Find equations for each of the circles below.

39. Center $(-2, 3)$, radius 5 **40.** Center $(\sqrt{5}, -\sqrt{7})$, radius $\sqrt{3}$

41. Center $(-8, 1)$, passing through $(0, 16)$ **42.** Center $(3, -6)$, tangent to the x-axis

Find the center and radius of each of the following that are circles.

43. $x^2 - 4x + y^2 + 6y + 12 = 0$ **44.** $x^2 - 6x + y^2 - 10y + 30 = 0$

45. $x^2 + 7x + y^2 + 3y + 1 = 0$ **46.** $x^2 + 11x + y^2 - 5y + 46 = 0$

Give the domain of the functions defined as follows.

47. $y = -4 + |x|$ **48.** $y = 3x^2 - 1$ **49.** $y = (x - 4)^2$ **50.** $y = 8 - x$

51. $y = \dfrac{8 + x}{8 - x}$ **52.** $y = -\sqrt{\dfrac{5}{x^2 + 9}}$ **53.** $y = \sqrt{49 - x^2}$ **54.** $y = -\sqrt{x^2 - 4}$

The graph of a function f is shown in the figure below. Sketch the graph of each of the functions defined as follows.

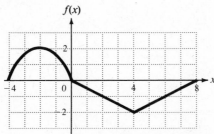

55. $y = f(x) + 3$ **56.** $y = f(x) - 4$ **57.** $y = f(x - 2)$ **58.** $y = f(x + 4)$

59. $y = f(x + 3) - 2$ **60.** $y = f(x - 1) + 4$ **61.** $y = |f(x)|$

For each of the following, find $\dfrac{f(x + h) - f(x)}{h}$, if $h \neq 0$.

62. $f(x) = -2x^2 + 4x - 3$ **63.** $f(x) = -x^3 + 2x^2$ **64.** $f(x) = \sqrt{x}$

Let $f(x) = \sqrt{x} - 2$ and $g(x) = x^2$. Find each of the following.

65. $(f \circ g)(x)$ **66.** $(g \circ f)(x)$ **67.** $(f \circ g)(-6)$

68. $(f \circ g)(2)$ **69.** $(g \circ f)(3)$ **70.** $(g \circ f)(24)$

Find all intervals where the functions defined as follows are increasing or decreasing.

71. $y = -(x + 1)^2$ **72.** $y = -3 + |x|$ **73.** $y = \dfrac{|x|}{x}$ **74.** $y = |x| + x^2$

Which of the functions graphed or defined as follows are one-to-one?

75.

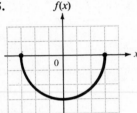

76.

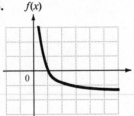

77.

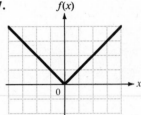

78. $y = \dfrac{8x - 9}{5}$ **79.** $y = -x^2 + 11$ **80.** $y = \sqrt{5 - x}$

81. $y = \sqrt{100 - x^2}$ **82.** $y = -\sqrt{1 - \dfrac{x^2}{100}}$, $x \geq 0$

For each of the functions defined as follows that is one-to-one, write an equation for the inverse function in the form $y = f^{-1}(x)$ and then graph f and f^{-1}.

83. $f(x) = 12x + 3$ **84.** $f(x) = \dfrac{2}{x - 9}$ **85.** $f(x) = x^3 - 3$

86. $f(x) = \sqrt{25 - x^2}$, domain $[0, 5]$ **87.** $f(x) = -\sqrt{x - 3}$

88. If the point $P(x, y)$ is on the line through $P_1(x_1, y_1)$ and $P_2(x_2, y_2)$ such that $d(P_1, P)/d(P_1, P_2) = k$, prove that the coordinates of P are given by $x = x_1 + k(x_2 - x_1)$ and $y = y_1 + k(y_2 - y_1)$, where

$0 < k < 1$ if P is between P_1 and P_2

$k > 1$ if P is not between P_1 and P_2 and is closer to P_2

$k < 0$ if P is not between P_1 and P_2 and is closer to P_1.*

*Exercises 88–93 from *The Calculus with Analytic Geometry*, 4th edition by Louis Leithold.
Copyright © 1981 by Louis Leithold. Reprinted by permission of Harper and Row, Publishers, Inc.

89. Find formulas for $(f \circ g)(x)$ if

$$f(x) = \begin{cases} 0 & \text{if } x < 0 \\ 2x & \text{if } 0 \le x \le 1 \\ 0 & \text{if } x > 1 \end{cases} \quad \text{and} \quad g(x) = \begin{cases} 1 & \text{if } x < 0 \\ x/2 & \text{if } 0 \le x \le 1 \\ 1 & \text{if } x > 1. \end{cases}$$

90. Find formulas for $(g \circ f)(x)$ for the functions in Exercise 89.

91. Find formulas for $(f \circ g)(x)$ if

$$f(x) = \begin{cases} 0 & \text{if } x < 0 \\ x^2 & \text{if } 0 \le x \le 1 \\ 0 & \text{if } x > 1 \end{cases} \quad \text{and} \quad g(x) = \begin{cases} 1 & \text{if } x < 0 \\ 2x & \text{if } 0 \le x \le 1 \\ 1 & \text{if } x > 1. \end{cases}$$

92. Given $f(x) = \begin{cases} x & \text{if } x < 1 \\ x^2 & \text{if } 1 \le x \le 9 \\ 27\sqrt{x} & \text{if } 9 < x, \end{cases}$

prove that f has an inverse function and find $f^{-1}(x)$.

93. Determine the value of the constant k so that the function defined by

$$f(x) = \frac{x + 5}{x + k}$$

will be its own inverse.

2

Trigonometric Functions

Many different types of functions are discussed throughout this book, including linear, quadratic, polynomial, exponential, and logarithmic. This chapter introduces the trigonometric functions, which differ in a fundamental way from those of algebra: the trigonometric functions describe a *periodic* or *repetitive* relationship.

An example of a periodic relationship is shown by this electrocardiogram, a graph of the human heartbeat. The EKG shows electrical impulses from the heart. Each small square represents .04 seconds, and each large square represents .2 seconds. How often does this (abnormal) heart beat?

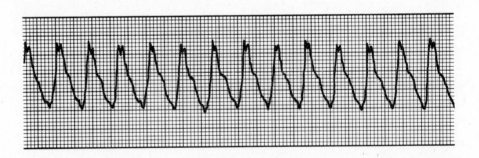

Trigonometric functions describe many natural phenomena, making them important in the study of electronics, optics, heat, X-rays, acoustics, and seismology, for example. Trigonometric functions occur again and again in calculus and are key to the study of navigation and surveying.

2.1 The Sine and Cosine Functions

A circle with center at the origin and radius one unit is called a **unit circle.** Consider starting at the point $(1, 0)$ on the unit circle and letting (x, y) be any other point on the circle. Measuring in a counterclockwise direction along the arc of the circle from $(1, 0)$ to (x, y) gives a positive real number s as the length of this arc. There is also a negative number that gives the length of the arc from $(1, 0)$ to (x, y), measured in a clockwise direction. An arc with $s > 0$ is shown in Figure 1.

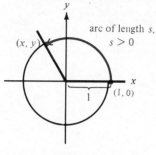

Figure 1

The circumference of a circle of radius r is given by $C = 2\pi r$. Since $r = 1$, the circumference of a unit circle is 2π, so that s may take values from 0 to 2π or from -2π to 0. The possible values of s can be extended to include all real numbers by allowing arcs which wrap around the circle more than once. For example, $s = 3\pi$ would correspond to the same point on the circle as $s = \pi$ or $s = -\pi$. Figure 2 shows how s can take any real number as a value by wrapping a real number line around the unit circle.

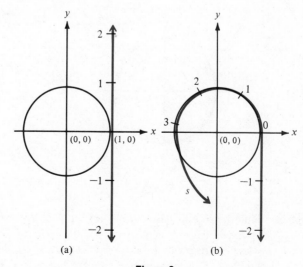

(a)   (b)

Figure 2

Example 1   Find the coordinates of the points on the unit circle corresponding to the following arc lengths.

(a) π

Since π is one-half the circumference of the unit circle, an arc of length π which starts at $(1, 0)$ would end at the point $(-1, 0)$, as shown in Figure 3.

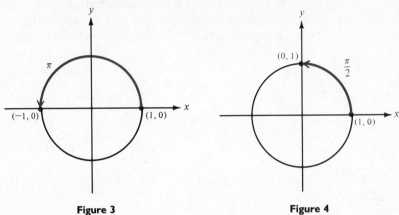

Figure 3   Figure 4

(b) $\pi/2$

An arc of length $\pi/2$ corresponds to the point $(0, 1)$. See Figure 4.

(c) $\dfrac{3\pi}{2}$

As Figure 5 shows, the arc of length $\dfrac{3\pi}{2}$ corresponds to $(0, -1)$.   ■

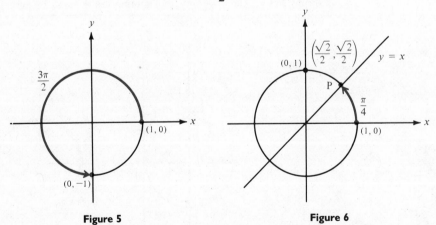

Figure 5   Figure 6

Example 2   Find the coordinates of the point P on the unit circle which corresponds to $s = \pi/4$.

Since the arc length from $(1, 0)$ to $(0, 1)$ is $\pi/2$, the point P on the unit circle which corresponds to $s = \pi/4$ is halfway between $(1, 0)$ and $(0, 1)$, as shown in Figure 6. This point also lies on the line $y = x$. Since the equation of the unit circle is $x^2 + y^2 = 1$, replacing y with x gives

$$x^2 + x^2 = 1,$$

or
$$2x^2 = 1$$

$$x^2 = \frac{1}{2}.$$

Since P is in the first quadrant, use the positive square root. This gives

$$x = \frac{1}{\sqrt{2}} = \frac{\sqrt{2}}{2}.$$

Also, since $y = x$,
$$y = \frac{\sqrt{2}}{2}.$$

The point P on the unit circle which corresponds to $s = \pi/4$ is $(\sqrt{2}/2, \sqrt{2}/2)$. ■

Example 3 Find the coordinates of the point Q on the unit circle which corresponds to each of the following.

(a) $s = 3\pi/4$

Figure 7 shows the point Q and the point P from Example 2 above. Since $3\pi/4 = \pi - \pi/4$, by symmetry Q has the same y-coordinate as P, but an x-coordinate with opposite sign. Thus, the coordinates of Q are $(-\sqrt{2}/2, \sqrt{2}/2)$.

(b) $s = -\pi/4$

As Figure 8 shows, this time the coordinates of Q are $(\sqrt{2}/2, -\sqrt{2}/2)$.

(c) $s = 5\pi/4$

Since $5\pi/4 = \pi + \pi/4$, use symmetry again to find that the coordinates are $(-\sqrt{2}/2, -\sqrt{2}/2)$. See Figure 9. ■

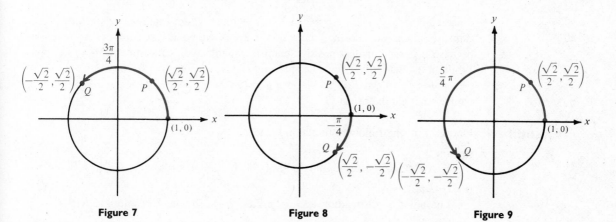

Figure 7   **Figure 8**   **Figure 9**

The correspondence between an arc of length s and a point (x, y) on the unit circle can be used to define two functions. Each value of s leads to a unique value of x and a unique value of y. The function which associates each arc length s with the y-value of the corresponding point on the unit circle is called the **sine function,** abbreviated $\sin s = y$. Also, the function which associates each arc length with the x-value of the corresponding point is called the **cosine function,** which is abbreviated $\cos s = x$. In summary,

Sine and Cosine Functions

> let s be any real number and let the point (x, y) on the unit circle correspond to s. Then
> $$\sin s = y$$
> and
> $$\cos s = x.$$

By these definitions, the coordinates of the point associated with an arc of length s can be written as $(\cos s, \sin s)$.

Both the sine and cosine functions have the set of real numbers as domain. The values of $\cos s$ (or x) range from 1 when $s = 0$ to -1 when $s = \pi$. $\cos s$ is 0 when s is $\pi/2$ or $3\pi/2$. Also, $\sin s$ ranges from -1 when $s = 3\pi/2$ to 1 when $s = \pi/2$, with $\sin s = 0$ when s is 0 or π.

Because the circumference of the unit circle is 2π, adding 2π to any value of s leaves the corresponding values of $\sin s$ and $\cos s$ unchanged. That is,

> for every real number s,
> $$\sin s = \sin(s + 2\pi)$$
> and
> $$\cos s = \cos(s + 2\pi).$$

Sine and cosine, which describe a cyclic relationship, are examples of **periodic functions.** For sine and cosine, the number 2π is called the **period** of the function, because it is the smallest positive number which satisfies the statements in the box. While it is also true, for example, that $\sin s = \sin(s + 4\pi)$ and $\sin s = \sin(s + 6\pi)$, the period is 2π since 2π can be shown to be the *smallest* positive number that can be used.

Example 4   Find each of the following function values.

(a) $\sin 0$

If $s = 0$, the corresponding point is $(1, 0)$. Since $\sin s = y$, $\sin 0 = 0$.

(b) $\cos \pi/2$

The point corresponding to $s = \pi/2$ is $(0, 1)$. Since, by definition, $\cos s = x$, $\cos \pi/2 = 0$.

(c) $\cos \pi/4$

Example 2 showed that $(\sqrt{2}/2, \sqrt{2}/2)$ corresponds to $\pi/4$. From this, the value of $\cos \pi/4 = \sqrt{2}/2$ (also, $\sin \pi/4 = \sqrt{2}/2$).   ■

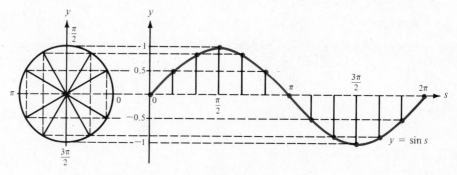

Figure 10

While sine and cosine functions will not be graphed in detail until later in this chapter, it is beneficial to sketch the basic graphs now. This is done with the definitions of sine and cosine in terms of the unit circle. A unit circle with various arc lengths s is shown in Figure 10. The vertical dotted lines give the corresponding values of y, which is sin s. Projecting horizontally gives points on the graph of $y = $ sin s. Figure 10 shows only *one period* of the graph; the complete graph would extend indefinitely to the right and to the left. (The process used to get the graph in Figure 10 could have come from measuring successive positions of bicycle pedals, just one application of sine.) A similar process can be used to obtain the graph of $y = $ cos s, shown in Figure 11. Again, the portion shown is only one period of the graph.

The following chart summarizes the intervals where sine and cosine are increasing and decreasing. The results in the chart can be found by inspecting the graphs of Figure 10 and Figure 11.

Interval	$[0, \pi/2]$	$[\pi/2, \pi]$	$[\pi, 3\pi/2]$	$[3\pi/2, 2\pi]$
Quadrant	I	II	III	IV
sin s	increases from 0 to 1	decreases from 1 to 0	decreases from 0 to -1	increases from -1 to 0
cos s	decreases from 1 to 0	decreases from 0 to -1	increases from -1 to 0	increases from 0 to 1

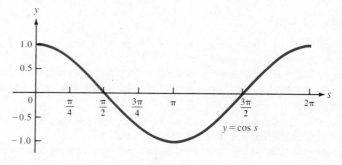

Figure 11

2.1 Exercises

For each of the following arc lengths, find the coordinates of the corresponding point on the unit circle.

1. $13\pi/4$

2. $15\pi/4$

3. π

4. -3π

5. 4π

6. 17π

7. -2π

8. $-\dfrac{\pi}{2}$

9. $\dfrac{5\pi}{2}$

10. $\dfrac{11\pi}{2}$

11. $\dfrac{-11\pi}{4}$

12. $\dfrac{-7\pi}{4}$

13. 2.25π

14. 1.5π

15. $.75\pi$

16. -2.25π

In each of the following exercises, the point which corresponds to arc length s is given. Use symmetry to find the coordinates of the point that corresponds to (a) $-s$; (b) $s + 2\pi$; (c) $s + \pi$; (d) $\pi - s$.

17. $\left(\dfrac{2}{3}, \dfrac{\sqrt{5}}{3}\right)$

18. $\left(\dfrac{5}{13}, \dfrac{12}{13}\right)$

19. $\left(\dfrac{4}{5}, \dfrac{3}{5}\right)$

20. $\left(\dfrac{3}{4}, \dfrac{\sqrt{7}}{4}\right)$

21. $\left(-\dfrac{1}{2}, \dfrac{\sqrt{3}}{2}\right)$

22. $\left(\dfrac{\sqrt{3}}{2}, -\dfrac{1}{2}\right)$

23. $\left(-\dfrac{2}{5}, \dfrac{-\sqrt{21}}{5}\right)$

24. $\left(\dfrac{-3}{5}, \dfrac{-4}{5}\right)$

For each of the following arc lengths, find $\sin s$ and $\cos s$.

25. 0

26. $\dfrac{\pi}{2}$

27. $\dfrac{\pi}{4}$

28. $\dfrac{3\pi}{4}$

29. $\dfrac{3\pi}{2}$

30. 987π

31. -1423π

32. -2π

33. $-\dfrac{\pi}{4}$

34. $\dfrac{-\pi}{2}$

35. $\dfrac{-3\pi}{4}$

36. $\dfrac{-7\pi}{4}$

Identify the quadrant in which arcs having the following lengths would terminate.

37. $s = 10$

38. $s = 18$

39. $s = 36$

40. $s = 92$

41. $s = -896.1$

42. $s = -1046.001$

Show that the following points lie (approximately) on the unit circle.

43. $(.39852144, .91715902)$

44. $(-.81745602, -.57599102)$

Solve the following problems.

45. (a) Use a 30°–60° triangle to show that the arc length $\pi/3$ corresponds to a point on the unit circle which is also on the line $y = \sqrt{3}\,x$.
   (b) Find the coordinates of the point on the unit circle which corresponds to $\pi/3$.
   (c) Use symmetry to find the coordinates of the points on the unit circle which correspond to $-\pi/3$, $2\pi/3$, and $4\pi/3$.

46. (a) Use a 30°–60° triangle to show that the arc length $\pi/6$ corresponds to a point on the unit circle which is also on the line $\sqrt{3}\,y = x$.
   (b) Find the coordinates of the point on the unit circle which corresponds to $\pi/6$.
   (c) Use symmetry to find the coordinates of the points on the unit circle which correspond to $-\pi/6$, $5\pi/6$, and $7\pi/6$.

Use the results of Exercises 45 and 46 to find each of the following.

47. $\sin \pi/3$ **48.** $\cos \pi/3$ **49.** $\cos 2\pi/3$ **50.** $\sin 2\pi/3$

51. $\sin 4\pi/3$ **52.** $\cos - \pi/3$ **53.** $\sin 5\pi/6$ **54.** $\cos 5\pi/6$

Find $\sin s$ and $\cos s$ if s corresponds to the following points on the unit circle.

55. (m, p^2) **56.** (r^3, t^2) **57.** $(3/8, z)$ **58.** $(k, -2/5)$

59. $(5a, 3b)$ **60.** $(2p, 7q)$

Find the value of x or y as appropriate for each of the following points on the unit circle. Then, assuming that the real number s corresponds to the given point, find $\sin s$ and $\cos s$.

61. $\left(\dfrac{3}{5}, y\right), \ y > 0$ **62.** $\left(x, \dfrac{7}{25}\right), \ x > 0$ **63.** $\left(x, \dfrac{5}{8}\right), \ x < 0$ **64.** $\left(\dfrac{2}{3}, y\right), \ y < 0$

65. $\left(-\dfrac{1}{\sqrt{13}}, y\right), \ y > 0$ **66.** $\left(x, -\dfrac{2}{\sqrt{15}}\right), \ x < 0$ **67.** $\left(\dfrac{3}{\sqrt{11}}, y\right), \ y < 0$ **68.** $\left(x, \dfrac{3}{\sqrt{19}}\right), \ x > 0$

69. $\left(x, \dfrac{b}{\sqrt{a^2 + b^2}}\right), \ x < 0, \ a < 0$ **70.** $\left(\dfrac{2p}{\sqrt{4p^2 + t^2}}, y\right), \ y > 0, \ t < 0$

2.2   Further Trigonometric Functions

In the previous section, the sine and cosine functions were defined as $\sin s = y$ and $\cos s = x$, where (x, y) is the point on the unit circle corresponding to an arc of length s. Four additional functions can now be derived from these two basic functions: the **tangent, cotangent, cosecant,** and **secant** functions, abbreviated as **tan, cot, csc,** and **sec,** respectively. (The definitions of sine and cosine are included for reference.)

Trigonometric or Circular Functions

Let s be any real number and let the point (x, y) on the unit circle correspond to s. Then

$$\sin s = y \qquad\qquad \csc s = \frac{1}{y} \quad (y \neq 0)$$

$$\cos s = x \qquad\qquad \sec s = \frac{1}{x} \quad (x \neq 0)$$

$$\tan s = \frac{y}{x} \quad (x \neq 0) \qquad \cot s = \frac{x}{y} \quad (y \neq 0).$$

These six functions are called the **trigonometric functions** or the **circular functions.**

Example 1 Find $\tan \pi/4$, $\cot \pi/4$, $\csc \pi/4$, and $\sec \pi/4$.

Example 2 of the last section showed that the point $(\sqrt{2}/2, \sqrt{2}/2)$ corresponds to an arc of length $s = \pi/4$, so that $x = \sqrt{2}/2$ and $y = \sqrt{2}/2$. By their definitions,

$$\tan \frac{\pi}{4} = \frac{y}{x} = \frac{\sqrt{2}/2}{\sqrt{2}/2} = 1$$

$$\cot \frac{\pi}{4} = \frac{x}{y} = \frac{\sqrt{2}/2}{\sqrt{2}/2} = 1$$

$$\csc \frac{\pi}{4} = \frac{1}{y} = \frac{1}{\sqrt{2}/2} = \frac{2}{\sqrt{2}} = \sqrt{2}$$

$$\sec \frac{\pi}{4} = \frac{1}{x} = \frac{1}{\sqrt{2}/2} = \frac{2}{\sqrt{2}} = \sqrt{2}. \quad \blacksquare$$

Like the sine and cosine functions, the four new trigonometric functions are periodic. The cosecant and secant functions have the same period as sine and cosine, 2π. The tangent and cotangent functions, however, have a period of π, as shown in the next chapter.

Example 2 Find the values of the trigonometric functions for an arc of length $\pi/2$.

The point which corresponds to an arc of length $\pi/2$ is $(0, 1)$ (see Figure 12), so that $x = 0$ and $y = 1$. Use the definitions of the various functions.

$$\sin \frac{\pi}{2} = y = 1 \qquad\qquad \cot \frac{\pi}{2} = \frac{x}{y} = \frac{0}{1} = 0$$

$$\cos \frac{\pi}{2} = x = 0 \qquad\qquad \csc \frac{\pi}{2} = \frac{1}{y} = \frac{1}{1} = 1$$

$$\tan \frac{\pi}{2} = \frac{y}{x} = \frac{1}{0} \quad \text{(undefined)} \qquad \sec \frac{\pi}{2} = \frac{1}{x} = \frac{1}{0} \quad \text{(undefined)} \quad \blacksquare$$

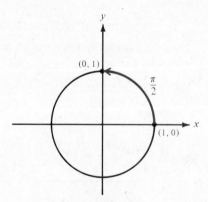

Figure 12

Several important properties of the trigonometric functions can be obtained from the definitions of the functions. First, recall that the *reciprocal* of a nonzero number a is $1/a$. (There is no reciprocal for 0.) The numbers $1/x$ and $x/1$ are reciprocals. From the definitions of trigonometric functions, $1/x$ is sec s and $x/1$ (or x) is cos s, so that cos s and sec s are reciprocals. In the same way, sine and cosecant are reciprocals as are tangent and cotangent. In summary, we have the following relationships, called the **reciprocal identities.**

Reciprocal Identities

$$\sin s = \frac{1}{\csc s} \quad \text{and} \quad \csc s = \frac{1}{\sin s},$$

$$\cos s = \frac{1}{\sec s} \quad \text{and} \quad \sec s = \frac{1}{\cos s},$$

$$\tan s = \frac{1}{\cot s} \quad \text{and} \quad \cot s = \frac{1}{\tan s}.$$

These identities hold whenever the denominators are not zero.

Example 3 Suppose $\tan s = 3/4$. Find cot s.
Since $\cot s = 1/\tan s,$

$$\cot s = \frac{1}{3/4} = \frac{4}{3}. \quad \blacksquare$$

The definitions can be used to determine the signs of the values of the trigonometric functions in each of the four quadrants. For example, if s terminates in quadrant I, then both x and y are positive, and all the trigonometric functions have positive values. In quadrant II, x is negative, and y is positive, so cos $s = x$ is negative, sin $s = y$ is positive, tan $s = y/x$ is negative, and so on. The signs of the values of the trigonometric functions in the various quadrants are summarized as follows.

Signs of Values of Trigonometric Functions

II sine and cosecant have positive values	I all functions have positive values
III tangent and cotangent have positive values	IV cosine and secant have positive values

Example 4 Suppose s terminates in quadrant II and $\sin s = 2/3$. Find the values of the other trigonometric functions.

Since $\sin s = y$, then $y = 2/3$. To find x, use the fact that on the unit circle $x^2 + y^2 = 1$.

$$x^2 + \left(\frac{2}{3}\right)^2 = 1$$

$$x^2 + \frac{4}{9} = 1$$

$$x^2 = \frac{5}{9}$$

$$x = \frac{\pm\sqrt{5}}{3}$$

Since s terminates in quadrant II, x must be negative, so $x = -\sqrt{5}/3$.

The values of the remaining functions can now be found from their definitions.

$$\cos s = -\frac{\sqrt{5}}{3}$$

$$\tan s = \frac{2/3}{-\sqrt{5}/3} = \frac{2}{-\sqrt{5}} = \frac{-2\sqrt{5}}{5}$$

$$\cot s = \frac{-\sqrt{5}/3}{2/3} = \frac{-\sqrt{5}}{2}$$

$$\sec s = \frac{1}{-\sqrt{5}/3} = \frac{-3}{\sqrt{5}} = \frac{-3\sqrt{5}}{5}$$

$$\csc s = \frac{1}{2/3} = \frac{3}{2} \quad \blacksquare$$

Several other relationships among the trigonometric functions can be derived from the definitions of these functions. The equation of the unit circle is $x^2 + y^2 = 1$; since $\cos s = x$ and $\sin s = y$,

$$(\cos s)^2 + (\sin s)^2 = 1.$$

It is customary to write $(\sin s)^2$ as $\sin^2 s$, giving $\sin^2 s + \cos^2 s = 1$. Starting with $x^2 + y^2 = 1$, divide both sides by x^2, and then by y^2, to get two additional identities: $1 + \tan^2 s = \sec^2 s$, and $\cot^2 s + 1 = \csc^2 s$. Finally, two more identities, for $\tan s$ and $\cot s$ are derived from the equalities $\cos s = x$ and $\sin s = y$. These last few identities (whose proofs are included as Exercises 73–76 below), make up the **fundamental identities**.

Fundamental Identities

$$\sin^2 s + \cos^2 s = 1 \qquad\qquad \tan s = \frac{\sin s}{\cos s} \quad (\cos s \neq 0)$$

$$1 + \tan^2 s = \sec^2 s$$

$$\cot^2 s + 1 = \csc^2 s \qquad\qquad \cot s = \frac{\cos s}{\sin s} \quad (\sin s \neq 0)$$

These relationships can be used to find all values of the trigonometric functions for a particular value of s, given the value of one function and the quadrant in which the arc of length s terminates.

Example 5 Suppose that $\tan t = 1/4$, and t terminates in quadrant III. Find the other five trigonometric function values for t.

Since $\cot t = 1/\tan t$,

$$\cot t = \frac{1}{1/4} = 4.$$

Use the identity $1 + \tan^2 t = \sec^2 t$ and substitute $1/4$ for $\tan t$.

$$1 + \left(\frac{1}{4}\right)^2 = \sec^2 t$$

$$1 + \frac{1}{16} = \sec^2 t$$

$$\frac{17}{16} = \sec^2 t$$

The arc of length t terminates in quadrant III, where $\sec t$ takes on negative values, so taking square roots of both sides of $17/16 = \sec^2 t$ gives

$$-\frac{\sqrt{17}}{4} = \sec t.$$

Find $\cos t$ from the identity $\cos t = 1/\sec t$:

$$\cos t = -\frac{4\sqrt{17}}{17}.$$

Finally, use $\qquad\qquad \tan t = \frac{\sin t}{\cos t},$

or $\sin t = (\tan t)(\cos t)$ to get

$$\sin t = \left(\frac{1}{4}\right)\left(-\frac{4\sqrt{17}}{17}\right) = -\frac{\sqrt{17}}{17}.$$

The reciprocal of $\sin t$ gives $\csc t$:

$$\csc t = 1/\sin t = -\sqrt{17}. \quad \blacksquare$$

2.2 Exercises

For each of the following, find tan s, cot s, sec s, and csc s. (Do not use tables or a calculator.)

1. sin s = 1/2, cos s = $\sqrt{3}/2$ **2.** sin s = 3/4, cos s = $\sqrt{7}/4$ **3.** sin s = 4/5, cos s = $-3/5$

4. sin s = $-1/2$, cos s = $-\sqrt{3}/2$ **5.** sin s = $-\sqrt{3}/2$, cos s = 1/2   **6.** sin s = 12/13, cos s = 5/13

For each of the following, find the values of the six trigonometric functions. For Exercises 13–18, use the results of Exercises 45 and 46 of Section 6.1. (Do not use tables or calculator.)

7. π **8.** $3\pi/2$ **9.** $3\pi/4$ **10.** $5\pi/4$

11. $-\pi/4$ **12.** $-\pi/2$ **13.** $\pi/6$ **14.** $\pi/3$

15. $2\pi/3$ **16.** $5\pi/6$ **17.** $7\pi/6$ **18.** $4\pi/3$

Complete the following table of signs of the values of the trigonometric functions. (Do not use tables or calculator.)

	Quadrant	sin	cos	tan	cot	sec	csc
19.	I	+	+			+	+
20.	II	+		−	−		
21.	III						
22.	IV			−			−

Find the value of x or y, as appropriate, for each of the following points on the unit circle. Then, assuming that the real number s corresponds to the given point, find the six trigonometric function values for s.

23. $\left(x, \dfrac{1}{4}\right)$, $x > 0$ **24.** $\left(\dfrac{3}{8}, y\right)$, $y < 0$ **25.** $\left(-\dfrac{3}{7}, y\right)$, $y > 0$

26. $\left(x, -\dfrac{4}{9}\right)$, $x > 0$ **27.** $\left(x, -\dfrac{\sqrt{3}}{2}\right)$, $x > 0$ **28.** $\left(x, \dfrac{\sqrt{2}}{2}\right)$, $x < 0$

29. $\left(-\dfrac{1}{\sqrt{7}}, y\right)$, $y > 0$ **30.** $\left(x, -\dfrac{3}{\sqrt{11}}\right)$, $x < 0$ **31.** $\left(x, \dfrac{a}{\sqrt{a^2 + b^2}}\right)$, $x > 0, b > 0$

32. $\left(\dfrac{t}{\sqrt{4p^2 + t^2}}, y\right)$, $y < 0, p > 0$ **33.** (x, q), $x < 0, 0 < q < 1$ **34.** (p, y), $y > 0, -1 < p < 0$

Decide whether each of the following statements is *possible* or *impossible*. (Use the range for sine and cosine given in the previous section, and the reciprocal identities.)

35. sin t = 2   **36.** cos s = -1.001 **37.** csc t = 2

38. sec s = -1.001 **39.** cos s + 1 = .6   **40.** tan t − 1 = 4.2

41. sin t = 1/2 and csc t = 2   **42.** cos s = 3/4 and sec s = 4/3   **43.** tan s = 2 and cot s = -2

44. sec t = 1/5 and cos t = 2   **45.** sin s = 3.251924 and csc s = .3075103   **46.** tan t = 4.67129 and cot t = .214074

Decide what quadrant(s) s must terminate in to satisfy the following conditions for $0 \le s < 2\pi$.

47. $\sin s > 0$, $\cos s < 0$ **48.** $\cos s > 0$, $\tan s > 0$ **49.** $\sec s < 0$, $\csc s < 0$

50. $\tan s > 0$, $\cot s > 0$ **51.** $\cos s < 0$ **52.** $\tan s > 0$

53. $\csc s < 0$ **54.** $\sin s > 0$

For each of the following find the values of the other trigonometric functions.

55. $\cos s = \dfrac{-4}{9}$, s terminates in quadrant II   **56.** $\csc s = 2$, s terminates in quadrant II

57. $\tan s = \dfrac{3}{2}$, $\csc s = \dfrac{\sqrt{13}}{3}$ **58.** $\sec s = -2$, $\cot s = \dfrac{\sqrt{3}}{3}$

59. $\sin t = \dfrac{\sqrt{5}}{5}$, $\cos t < 0$ **60.** $\tan t = -\dfrac{3}{5}$, $\sec t > 0$

61. $\sec s = -\sqrt{7}/2$, $\tan s = \sqrt{3}/2$ **62.** $\csc t = -\sqrt{17}/3$, with $\cot t = 2\sqrt{2}/3$

63. $\sin t = .164215$, with t in quadrant II   **64.** $\cot s = -1.49586$, with s in quadrant IV

65. $\tan t = .642193$, with t in quadrant III   **66.** $\cos s = -.425847$, with s in quadrant III

67. $\sin s = a$, s is in quadrant I   **68.** $\tan t = m$, t is in quadrant III

69. $\cos s = -.428193$, s terminates in quadrant II   **70.** $\sec t = 28.4096$, t terminates in quadrant IV

71. $\csc t = -10.4349$, $\sec t > 0$ **72.** $\cot t = -.139725$, $\cos t > 0$

Use the definitions of the trigonometric functions to prove the following statements.

73. $1 + \tan^2 s = \sec^2 s$ **74.** $\cot^2 s + 1 = \csc^2 s$

75. $\tan s = \dfrac{\sin s}{\cos s}$, $\cos s \ne 0$ **76.** $\cot s = \dfrac{\cos s}{\sin s}$, $\sin s \ne 0$

The figure shows a quarter circle of radius 1 with FB tangent to the circle at B and EC tangent at C. Give each of the following lengths in terms of s.

77. OA **78.** AG

79. OD **80.** OF

81. BF **82.** CD

83. arc BG **84.** arc CG

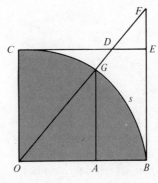

Prove that each of the following is true for any positive integer n.

85. $\cos n\pi = (-1)^n$ **86.** $\sin (2n + 1)\pi/2 = (-1)^n$

2.3   Angles and the Unit Circle

A very basic idea in trigonometry is that of *angle*. To define an angle, start with Figure 13, showing a line through the two points A and B. This line is named **line AB.** The portion of the line between A and B, including points A and B themselves, is called **line segment AB.** The portion of line AB that starts at A and continues through B, and on past B, is called **ray AB.** Point A is the endpoint of the ray.

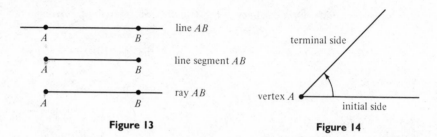

line AB

line segment AB

ray AB

Figure 13

terminal side

vertex A

initial side

Figure 14

An **angle** is formed by rotating a ray around its endpoint. The initial position of the ray is the **initial side** of the angle, while the location of the ray at the end of its rotation is the **terminal side** of the angle. The endpoint of the ray is the **vertex** of the angle. Figure 14 shows the initial and terminal sides of an angle with vertex A.

If the rotation of an angle is counterclockwise, the angle is **positive.** If the rotation is clockwise, the angle is **negative.** Figure 15 shows angles of both types.

An angle can be named by using the name of its vertex. For example, the angle in Figure 15(b) can be called angle C. Also, an angle can be named by using three letters. For example, the angle in Figure 15(b) could be named angle ACB. (Put the vertex in the middle.)

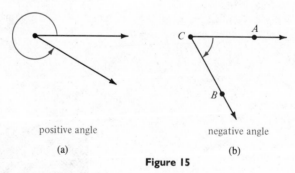

positive angle

(a)

negative angle

(b)

Figure 15

An angle is in **standard position** if its vertex is at the origin of a coordinate system and its initial side is along the positive x-axis. The two angles in Figure 16 are in standard position. An angle in standard position is said to lie in the quadrant where its terminal side lies.

Example 1   Find the quadrants for the angles in Figure 16.

The angle in Figure 16(a) is a quadrant I angle. The angle in Figure 16(b) is a quadrant II angle.   ■

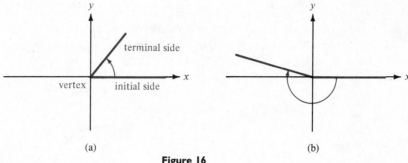

Figure 16

An angle in standard position whose terminal side coincides with the *x*-axis or *y*-axis is called a **quadrantal angle.** Two angles with the same initial side and the same terminal side, but different amounts of rotation, are called **coterminal angles.** Figure 17 shows two examples of coterminal angles.

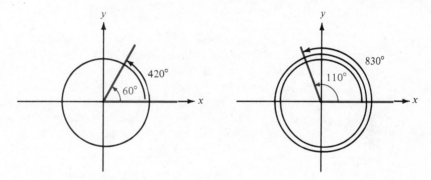

Figure 17

There are two systems commonly used to measure angles. In most work in applied trigonometry, angles are measured in degrees. Degree measure has remained unchanged since the Babylonians developed it over 4000 years ago. To use degree measure, assign 360 degrees to the rotation of a ray through a complete circle. As Figure 18 shows, the terminal side corresponds with its initial side when it makes a complete rotation. **One degree,** 1°, represents 1/360 of a rotation. One sixtieth of a degree is called a *minute,* and one sixtieth of a minute is a *second*. The measure 12° 42′ 38″ represents 12 degrees, 42 minutes, 38 seconds.

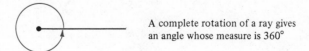

A complete rotation of a ray gives
an angle whose measure is 360°

Figure 18

An angle having a measure between 0° and 90° is called an **acute angle.** An angle whose measure is exactly 90° is a **right angle.** An angle measuring more than 90° but less than 180° is an **obtuse angle,** and an angle of exactly 180° is a **straight angle.**

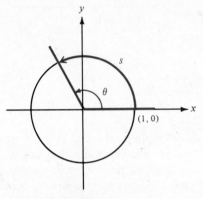

Figure 19

Degree measure is based on an arbitrary assignment of 360° to one complete rotation of a ray. In advanced work in trigonometry, and in calculus, angles are measured in a more natural system of measurement: radians. Radian measure simplifies many formulas. To see how to measure an angle in radians, start with Figure 19, which shows an angle θ (θ is the Greek letter theta) in standard position on a unit circle. As shown in the figure, angle θ determines an arc of positive length s on the unit circle. This real number s can be used as a measure of θ:

$$\theta = s \text{ radians}$$

or simply $\theta = s$.

Several angles and their radian measures are shown in Figure 20.

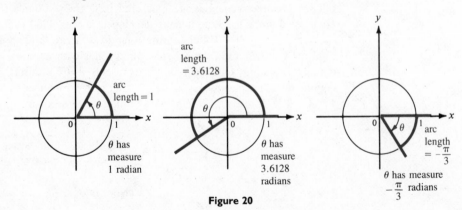

Figure 20

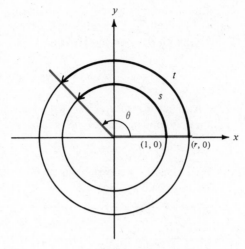

Figure 21

While the radian measure of an angle was defined in terms of a unit circle, radian measures can be found for angles from a circle of any positive radius r. To see how, start with the **central angle** (an angle whose vertex is the center of a circle) θ in Figure 21. Let θ cut an arc of length s on the unit circle and an arc of length t on the circle of radius r, where $r > 0$. By the definition of radian measure above, $\theta = s$. From geometry, the arc lengths s and t have the same ratios as the radii of the respective circles, or

$$\frac{t}{s} = \frac{r}{1},$$

from which

$$s = \frac{t}{r}.$$

Since $\theta = s$,

$$\theta = \frac{t}{r}.$$

In summary,

Radian Measure

suppose a circle has a radius $r > 0$. Let θ be a central angle of the circle. If θ cuts off an arc of length t on the circle, then the radian measure of θ is

$$\theta = \frac{t}{r}.$$

The radian measure of an angle is the ratio of the arc length cut by the angle to the radius of the circle. In this ratio, the units of measure "divide out," leaving only a number. For this reason, radian measure is just a real number—there are no units associated with a radian measure.

Example 2 Find the radian measure of a central angle which cuts off an arc of length 8 inches on a circle with a radius of 5 inches.

Start with $\theta = t/r$ and replace t with 8 and r with 5 to get

$$\theta = \frac{8}{5} = 1.6 \text{ radians.} \quad \blacksquare$$

An angle of measure 360° would correspond to an arc that went entirely around the unit circle. This makes the radian measure of a 360° angle 2π, or

$$360° = 2\pi \text{ radians,}$$

giving a basis for comparing degree measure and radian measure. Dividing both sides of this last result by 2 gives

Degree-Radian Correspondence

$$180° = \pi \text{ radians.}$$

Since π radians = 180°, divide both sides by π to find that

$$1 \text{ radian} = \frac{180°}{\pi},$$

or, approximately, 1 radian = 57° 17′ 45″.

On the other hand, dividing the equation 180° = π radians on both sides by 180 gives

$$1° = \frac{\pi}{180} \text{ radians,}$$

or, approximately, 1° = .0174533 radians.

Example 3 Convert each of the following degree measures to radians.

(a) 45°

Since 1° = $\pi/180$ radians,

$$45° = 45\left(\frac{\pi}{180}\right) \text{ radians} = \frac{45\pi}{180} \text{ radians} = \frac{\pi}{4} \text{ radians.}$$

The word *radian* is often omitted, so the result could be written as just 45° = $\pi/4$.

(b) 240°

$$240° = 240\left(\frac{\pi}{180}\right) = \frac{4\pi}{3}. \quad \blacksquare$$

Example 4 Convert each of the following radian measures to degrees.

(a) $\dfrac{9\pi}{4}$

Since 1 radian = 180°/π,

$$\frac{9\pi}{4} \text{ radians} = \frac{9\pi}{4}\left(\frac{180°}{\pi}\right) = 405°.$$

(b) $\dfrac{11\pi}{3}$ radians $= \dfrac{11\pi}{3}\left(\dfrac{180°}{\pi}\right) = 660°.$ ■

Many calculators will convert back and forth from radian measure to degree measure. A difficulty that often comes up is that most calculators work with decimal degrees, rather than degrees, minutes, and seconds. The next example shows how to handle this.

Example 5   **(a)** Convert 146° 18′ 34″ to radians.
Since $1' = 1/60°$ and $1'' = 1/60' = 1/3600°$,

$$146° \ 18' \ 34'' = 146° + \frac{18°}{60} + \frac{34}{3600}°$$
$$= 146° + .3° + .00944°$$
$$= 146.30944°.$$

Now, activate the calculator keys that convert from degrees to radians to get

$$146° \ 18' \ 34'' \approx 2.55358 \text{ radians.}$$

(b) Convert .97682 radians to degrees.
Enter .97682 in your calculator and use the keys that convert from radians to degrees to get

$$.97682 \text{ radians} = 55.967663°.$$

This result may be converted to degree-minute-second measure if desired.

$$55.967663° = 55° + (.967663)(60')$$
$$= 55° + 58.05978'$$
$$= 55° + 58' + (.05978)(60'')$$
$$= 55° + 58' + 4''$$
$$= 55° \ 58' \ 4''$$ ■

A calculator could be used to show that 30 radians is about the same as 1719°. Figure 22 shows angles of measure 30 radians and 30 degrees; the figure shows that these angles are not at all equal in size.

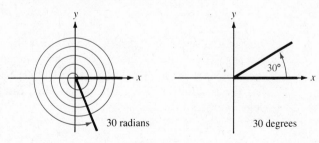

30 radians   30 degrees

Figure 22

The relationship $\theta = t/r$, found above, gives a way to find an arc length on a circle when the central angle and radius are known. This is useful in applications as the next example shows. (In this example, $\theta = t/r$ is rewritten as $t = r\theta$.)

Example 6 Reno, Nevada, is approximately due north of Los Angeles. The latitude of Reno is 40° N, while that of Los Angeles is 34° N. (The N means that the location is north of the equator.) If the radius of the earth is 6400 km, find the north-south distance between the two cities.

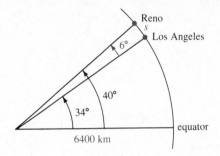

Figure 23

Latitude gives the measure of a central angle with vertex at the earth's center whose initial side goes through the earth's equator and whose terminal side goes through the location in question. As shown in Figure 23, the central angle between Reno and Los Angeles is 6°. The distance between the two cities can thus be found by the formula $s = r\theta$, after 6° is first converted to radians.

$$6° = 6\left(\frac{\pi}{180}\right) = \frac{\pi}{30} \text{ radians}$$

The distance between the two cities is

$$s = r\theta$$

$$s = 6400\left(\frac{\pi}{30}\right) \text{ km}$$

$$\approx 670 \text{ km.} \quad \blacksquare$$

Linear and Angular Velocity Radian measure is very useful for discussing linear and angular velocity of a point. To see how, suppose that point P moves at a constant speed along a circle of radius r and center O. See Figure 24. The measure of how fast the position of P is changing is called **linear velocity.** If v represents linear velocity, then

$$v = \frac{s}{t},$$

where s is the length of the arc cut by point P at time t. (This formula is just a restatement of the familiar result $d = rt$.)

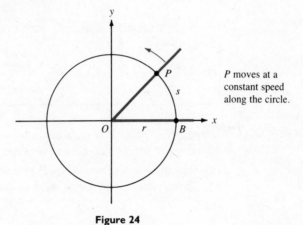

Figure 24

Look at Figure 24 again. As point P moves along the circle, ray OP rotates around the origin. Since the ray OP is the terminal side of angle POB, the measure of the angle changes as P moves along the circle. The measure of how fast angle POB is changing is called **angular velocity.** Angular velocity, written ω, can be given as

$$\omega = \frac{\theta}{t}, \qquad \theta \text{ in radians,}$$

where θ is the measure of angle POB at time t. The angle measure θ must be in radians, with ω expressed as radians per unit of time. Angular velocity is used in physics and engineering, among other applications.

As shown above, the length s of the arc cut on a circle of radius r by a central angle of measure θ radians is $s = r\theta$. Using this formula, the formula for linear velocity, $v = s/t$, becomes

$$v = \frac{r\theta}{t}$$

or
$$v = r\omega.$$

This last formula relates linear and angular velocity.

It was mentioned above that a radian is a "pure number," with no units associated with it. This is why the product of the length r, measured in units such as centimeters, and ω, measured in units such as radians per second, is velocity, v, measured in units such as centimeters per second.

The formulas for velocity are summarized on the following page.

Angular and Linear Velocity		
Angular velocity	$\omega = \dfrac{\theta}{t}$	(ω in radians per unit time)
Linear velocity	$v = \dfrac{s}{t}$	
	$v = \dfrac{r\theta}{t}$	(θ in radians)
	$v = r\omega$	

Example 7 Suppose that point P is on a circle with a radius of 10 cm, and ray OP is rotating with angular velocity of $\pi/18$ radians per sec.

(a) Find the angle generated by P in 6 sec.

The velocity of ray OP is $\pi/18$ radians per sec. Since $\omega = \theta/t$, then in 6 sec

$$\frac{\pi}{18} = \frac{\theta}{6},$$

or $\theta = 6(\pi/18) = \pi/3$ radians.

(b) Find the distance traveled by P along the circle in 6 sec.

In 6 sec P generates an angle of $\pi/3$ radians. Since $s = r\theta$,

$$s = 10\left(\frac{\pi}{3}\right) = \frac{10\pi}{3} \text{ cm.}$$

(c) Find the linear velocity of P.

Since $v = s/t$, then in 6 sec

$$v = \frac{10\pi/3}{6} = \frac{5\pi}{9} \text{ cm per sec.} \quad \blacksquare$$

2.3 Exercises

Find the angles of smallest positive measure coterminal with the following angles.

1. $-40°$
2. $-98°$
3. $-125°$
4. $-203°$
5. $450°$
6. $489°$
7. $539°$
8. $699°$

Convert each of the following degree measures to radians. Leave answers as multiples of π.

9. $60°$
10. $30°$
11. $90°$
12. $120°$
13. $135°$
14. $270°$
15. $300°$
16. $390°$
17. $405°$
18. $20°$
19. $140°$
20. $320°$

Convert each of the following radian measures to degrees.

21. $\pi/3$
22. $8\pi/3$
23. $7\pi/4$
24. $2\pi/3$
25. $11\pi/6$
26. $15\pi/4$
27. $-\pi/6$
28. $-\pi/4$
29. 5π
30. 7π
31. $7\pi/20$
32. $17\pi/20$

Convert each of the following degree measures to radians. Round to four decimal places.

33. 139° 10′

34. 174° 50′

35. 64.29°

36. 85.04°

37. −29° 42′ 36″

38. −157° 11′ 9″

39. −209° 46′ 15″

40. −387° 05′ 09″

Convert each of the following radian measures to degrees. Write answers with four decimal places and also to the nearest minute.

41. 2

42. 5

43. 1.74

44. 3.06

45. .0912

46. .3417

47. 9.84763

48. 5.01095

Find the measure of the central angle in radians for each of the following.

49. $r = 8$ inches, $t = 12$ inches

50. $r = 18$ mm, $t = 6$ mm

51. $r = 16.4$ m, $t = 20.1$ m

52. $r = 5.80$ cm, $t = 12.3$ cm

53. $r = 1.93470$ cm, $t = 5.98421$ cm

54. $r = 294.893$ m, $t = 122.097$ m

Find the distance in miles between the following pairs of cities whose latitudes are given. Assume the cities are on a north-south line and that the radius of the earth is 4.0×10^3 miles. Give answers to two significant digits.

55. Grand Portage, Minnesota, 44° N, and New Orleans, Louisiana, 30° N

56. Farmersville, California, 36° N and Penticton, British Columbia 49° N

57. New York City, 41° N, and Lima, Peru, 12° S

58. Halifax, Nova Scotia, 45° N, and Buenos Aires, Argentina, 34° S

Find the arc length cut by each of the following angles.

59. $r = 8.00$ in, $\theta = \pi$ radians

60. $r = 72.0$ ft, $\theta = \pi/8$ radians

61. $r = 12.3$ cm, $\theta = 2\pi/3$ radians

62. $r = .892$ cm, $\theta = 11\pi/10$ radians

63. $r = 4.82$ m, $\theta = 60°$

64. $r = 71.9$ cm, $\theta = 135°$

65. $r = 58.402$ m, $\theta = 52.417°$

66. $r = 39.4$ cm, $\theta = 68.059°$

Work the following exercises.

67. **(a)** How many inches will the weight on the left rise if the pulley is rotated through an angle of 71° 50′ ?

(b) Through what angle, to the nearest minute, must the pulley be rotated to raise the weight 6 in?

68. Find the radius of the pulley on the right if a rotation of 51.6° raises the weight 11.4 cm.

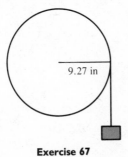

9.27 in

Exercise 67

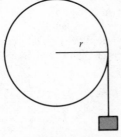

r

Exercise 68

Find the value of the indicated variable in each of the following.

69. θ, if $\omega = \pi/4$ radians per min, $t = 5$ min

70. θ, if $\omega = 2\pi/3$ radians per sec, $t = 3$ sec

71. ω, if $\theta = 2\pi/5$ radians, $t = 10$ sec

72. ω, if $\theta = 3\pi/4$ radians, $t = 8$ sec

73. t, if $\theta = 3\pi/8$ radians, $\omega = \pi/24$ radians per min

74. t, if $\theta = 2\pi/9$ radians, $\omega = 5\pi/27$ radians per min

75. v, if $r = 8$ cm, $\omega = 9\pi/5$ radians per sec

76. v, if $r = 12$ m, $\omega = 2\pi/3$ radians per sec

77. ω, if $v = 18$ ft per sec, $r = 3$ ft

78. ω, if $v = 9$ m per sec, $r = 5$ m

79. v, if $r = 24.93215$ cm, $\omega = .372914$ radians per sec

80. ω, if $v = 107.692$ m per sec, $r = 58.7413$ m

The formula $\omega = \theta/t$ can be rewritten as $\theta = \omega t$. Using ωt for θ changes $s = r\theta$ to $s = r\omega t$. Use this formula to find the values of the missing variables in each of the following.

81. $r = 6$ cm, $\omega = \pi/3$ radians per sec, $t = 9$ sec

82. $r = 9$ yd, $\omega = 2\pi/5$ radians per sec, $t = 12$ sec

83. $s = 6\pi$ cm, $r = 2$ cm, $\omega = \pi/4$ radians per sec

84. $s = 12\pi/5$ m, $r = 3/2$ m, $\omega = 2\pi/5$ radians per sec

85. $s = 3\pi/4$ km, $r = 2$ km, $t = 4$ sec

86. $s = 8\pi/9$ m, $r = 4/3$ m, $t = 12$ sec

87. $r = 37.6584$ cm, $\omega = .714213$ radians per sec, $t = .924473$ sec

88. $s = 5.70201$ m, $r = 8.92399$ m, $\omega = .614277$ radians per sec

89. Find ω for the hour hand of a clock.

90. Find ω for the second hand of a clock.

91. Find ω for a line from the center to the edge of a phonograph record revolving 33 1/3 times per minute.

92. Find the distance traveled in 9 seconds by a point on the edge of a circle of radius 6 cm, which is rotating through $\pi/3$ radians per second.

93. A point on the edge of a circle travels $8\pi/9$ m in 12 seconds. The radius of the circle is 4/3 m. Find ω.

94. Find the number of seconds it would take for a point on the edge of a circle to move $12\pi/5$ m, if the radius of the circle is 1.5 m and the circle is rotating through $2\pi/5$ radians per second.

95. The earth revolves on its axis once every 24 hr. Assuming that the earth's radius is 6400 km, find the following:
(a) angular velocity of the earth in radians per day and radians per hr
(b) linear velocity at the North Pole or South Pole
(c) linear velocity at Quito, Equador, a city on the equator
(d) linear velocity at Salem, Oregon (halfway from the equator to the North Pole).

96. Eratosthenes (*ca.* 230 B.C.)* made a famous measurement of the earth. He observed at
Syene [the modern Aswan], at noon and at the summer solstice, that a vertical stick had
no shadow, while at Alexandria (on the same meridian as Syene) the sun's rays were
inclined 1/50 of a complete circle to the vertical. See the figure. He then calculated the
circumference of the earth from the known distance of 5000 stades between Alexandria
and Syene. Obtain Eratosthenes' result of 250,000 stades for the circumference of the
earth. There is reason to suppose that a stade is about equal to 516.7 feet. Assuming this,
calculate from the above result the polar diameter of the earth in miles. (The actual polar
diameter of the earth, to the nearest mile, is 7900 miles.)

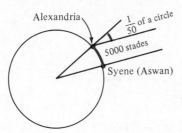

2.4 **Trigonometric Functions of Angles**

The trigonometric functions defined so far have sets of real numbers for their domains. These domains can be extended to include angle measures by using the radian
measure of the angle. For example, Figure 25 shows an angle with a measure θ in
degrees or s in radians, and by the various definitions given so far, $\sin \theta = \sin s$,
$\cos \theta = \cos s$, and so on. Thus, the trigonometric functions lead to the same function
values whether the domain represents arc lengths or angle measures.

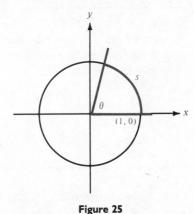

Figure 25

*From *A Survey of Geometry*, Vol. I by Howard Eves. Copyright © 1963 by Allyn & Bacon, Inc.

Example 1   Find the following trigonometric function values.

(a) $\sin 90°$

As shown in the previous section, $90° = \pi/2$ radians. Then

$$\sin 90° = \sin \frac{\pi}{2} = 1.$$

(b) $\tan 45°$

Since $45° = \pi/4$ radians, and using results from Section 2.2,

$$\tan 45° = \tan \frac{\pi}{4} = 1. \quad \blacksquare$$

The trigonometric functions were defined earlier in terms of the coordinates of a point P on the unit circle. These definitions can now be extended so that P need not be on the unit circle. To do this, start with Figure 26 showing an angle θ in standard position, a point P' on the unit circle, and a point P which lies r units from the origin. Triangles OPQ and $OP'Q'$ are both right triangles and OP has length r. From the figure,

$$\frac{y'}{1} = \frac{y}{r} \quad \text{and} \quad \frac{x'}{1} = \frac{x}{r}.$$

Since $\sin \theta = y'$ and $\cos \theta = x'$, this last result gives

$$\sin \theta = \frac{y}{r} \quad \text{and} \quad \cos \theta = \frac{x}{r}.$$

With this result, the definitions of the trigonometric functions can be generalized as follows.

Trigonometric Functions of an Angle

Let (x, y) be a point other than the origin on the terminal side of an angle θ in standard position. Let r be the distance from the origin to (x, y). Then the **trigonometric functions** of θ are defined as follows. (Assume no denominators are 0.)

$$\sin \theta = \frac{y}{r} \qquad \cos \theta = \frac{x}{r} \qquad \tan \theta = \frac{y}{x}$$

$$\csc \theta = \frac{r}{y} \qquad \sec \theta = \frac{r}{x} \qquad \cot \theta = \frac{x}{y}.$$

With these new definitions of the trigonometric functions, the values of the trigonometric functions for first quadrant angles can be thought of as ratios of the sides of a right triangle. (A right triangle has a 90° angle.)

Example 2   The terminal side of an angle α goes through the point $(8, 15)$. Find the values of the trigonometric functions of α.

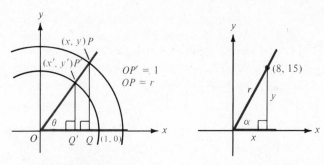

Figure 26   **Figure 27**

Figure 27 shows angle α and the triangle formed by dropping a perpendicular from the point $(8, 15)$. Since $(8, 15)$ is on the terminal side of the angle, let $x = 8$ and $y = 15$. Find r with the Pythagorean theorem.

$$r^2 = x^2 + y^2 \qquad \text{or} \qquad r = \sqrt{x^2 + y^2}$$

(Recall that $\sqrt{a}$ represents the *nonnegative* square root of a.) Letting $x = 8$ and $y = 15$,

$$r = \sqrt{8^2 + 15^2}$$
$$= \sqrt{64 + 225}$$
$$r = \sqrt{289}.$$

A calculator gives $r = 17$. The values of the trigonometric functions of angle α are now found by the definitions given above.

$$\sin \alpha = \frac{y}{r} = \frac{15}{17} \qquad \cos \alpha = \frac{x}{r} = \frac{8}{17} \qquad \tan \alpha = \frac{y}{x} = \frac{15}{8}$$

$$\csc \alpha = \frac{r}{y} = \frac{17}{15} \qquad \sec \alpha = \frac{r}{x} = \frac{17}{8} \qquad \cot \alpha = \frac{x}{y} = \frac{8}{15} \quad \blacksquare$$

The definitions of the trigonometric functions of angles, together with some results from geometry, can be used to find the values of the trigonometric functions for 30°, 45°, and 60°. (The methods of Exercises 45 and 46 of Section 6.1 could also be used.) The values of the trigonometric functions of 30° and 60° are found from a 30°–60° right triangle. Such a triangle can be obtained from an **equilateral triangle,** a triangle with all sides equal in length. Each angle of an equilateral triangle has a measure of 60°. See Figure 28(a).

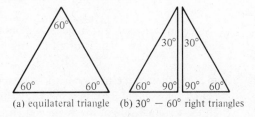

(a) equilateral triangle   (b) 30° − 60° right triangles

Figure 28

Bisecting one angle of an equilateral triangle gives two right triangles, each of which has angles of 30°, 60°, and 90°, as shown in Figure 28(b). If the hypotenuse of one of these right triangles has a length of 2, then the shortest side will have a length of 1. (Why?) If x represents the length of the medium side, then, by the Pythagorean theorem,

$$2^2 = 1^2 + x^2$$
$$4 = 1 + x^2$$
$$3 = x^2$$
$$\sqrt{3} = x.$$

The length of the medium side is $\sqrt{3}$. In summary,

30°–60° Right Triangles

> in a 30°–60° right triangle, the hypotenuse is always twice as long as the shortest side (the side opposite the 30° angle), and the medium side is $\sqrt{3}$ times as long as the shortest side.

Example 3   Find the trigonometric function values for 30°.

Start with a 30° angle in standard position, as shown in Figure 29. Choose a point P on the terminal side of the angle so that $r = 2$. By the work above, P will have coordinates $(\sqrt{3}, 1)$, with $x = \sqrt{3}$, $y = 1$, and $r = 2$. By the definitions of the trigonometric functions,

$$\sin 30° = \frac{1}{2} \qquad \tan 30° = \frac{\sqrt{3}}{3} \qquad \sec 30° = \frac{2\sqrt{3}}{3}$$

$$\cos 30° = \frac{\sqrt{3}}{2} \qquad \cot 30° = \sqrt{3} \qquad \csc 30° = 2. \quad \blacksquare$$

If you have a calculator which finds trigonometric function values at the touch of a key, you may wonder why we spend so much time in finding values for special angles. We do this because a calculator gives only *approximate* values in most cases, while we need *exact* values. For example, a calculator might give the tangent of 30° as

$$\tan 30° \approx 0.5773502692$$

($\approx$ means "is approximately equal to"); however, we found the *exact* value:

$$\tan 30° = \frac{\sqrt{3}}{3}.$$

Since an exact value is frequently more useful than an approximation, you should be able to give exact values of all the trigonometric functions for the special angles.

Example 4   Find the values of the trigonometric functions for 210°.

Draw an angle of 210° in standard position, as shown in Figure 30. Choose point P on the terminal side of the angle so that $r = 2$. From the 30°–60° right triangle in

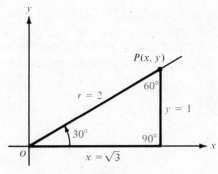

Figure 29

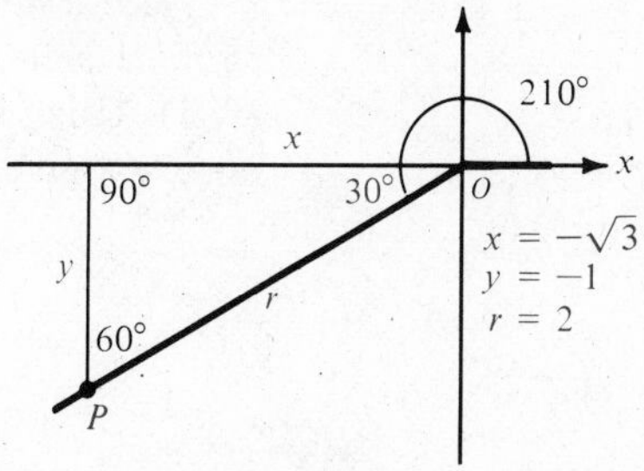

Figure 30

Figure 30, the coordinates of point P are $(-\sqrt{3}, -1)$. This makes $x = -\sqrt{3}$, $y = -1$, and $r = 2$, with

$$\sin 210° = -\frac{1}{2} \qquad \tan 210° = \frac{\sqrt{3}}{3} \qquad \sec 210° = -\frac{2\sqrt{3}}{3}$$

$$\cos 210° = -\frac{\sqrt{3}}{2} \qquad \cot 210° = \sqrt{3} \qquad \csc 210° = -2. \quad \blacksquare$$

In Section 2.1 the trigonometric function values for $\pi/4$, $3\pi/4$, $-\pi/4$, and so on, were found using a unit circle. Since $\pi/4$ corresponds to an angle of 45°, these function values can also be found using a 45°–45° right triangle, such as the one of Figure 31. This triangle has two sides of equal length (since it has two angles of equal measure) and so is an **isosceles** triangle.

If the shorter sides each have length 1 and if r represents the length of the hypotenuse, then

$$1^2 + 1^2 = r^2$$
$$2 = r^2$$
$$\sqrt{2} = r.$$

Generalizing from this example,

45°–45° Right Triangles | in a 45°–45° right triangle, the hypotenuse is $\sqrt{2}$ times as long as either of the shorter sides.

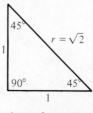

45° — 45° right triangle

Figure 31

Example 5 Find the values of the trigonometric functions for 45°.

Place a 45° angle in standard position, as in Figure 32. Choose point P on the terminal side of the angle so that $r = \sqrt{2}$. Then the coordinates of P become $(1, 1)$. Using the definitions of the trigonometric functions with $x = 1$, $y = 1$, and $r = \sqrt{2}$ gives

$$\sin 45° = \frac{\sqrt{2}}{2} \qquad \tan 45° = 1 \qquad \sec 45° = \sqrt{2}$$

$$\cos 45° = \frac{\sqrt{2}}{2} \qquad \cot 45° = 1 \qquad \csc 45° = \sqrt{2}.$$

The denominators were rationalized for the values of $\sin 45°$ and $\cos 45°$. ■

A summary of the trigonometric function values for certain special angles is given inside the back covers of this book.

Figure 33 shows an acute angle θ in standard position. A right triangle has been drawn. By the work above, $\sin \theta = y/r$. It is convenient to call y the length of the *side opposite* angle θ, with r the length of the *hypotenuse*. Also, x is the length of the *side adjacent* to θ. Using these terms,

Trigonometric Functions of an Acute Angle

if θ is an acute angle of a right triangle, then

$$\sin \theta = \frac{\text{side opposite}}{\text{hypotenuse}} \qquad\qquad \csc \theta = \frac{\text{hypotenuse}}{\text{side opposite}}$$

$$\cos \theta = \frac{\text{side adjacent}}{\text{hypotenuse}} \qquad\qquad \sec \theta = \frac{\text{hypotenuse}}{\text{side adjacent}}$$

$$\tan \theta = \frac{\text{side opposite}}{\text{side adjacent}} \qquad\qquad \cot \theta = \frac{\text{side adjacent}}{\text{side opposite}}.$$

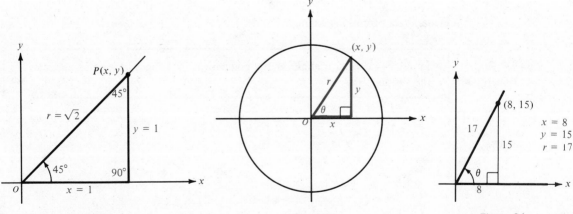

Figure 32   Figure 33   Figure 34

Example 6 Find the values of sin θ, cos θ, and tan θ for angle θ of Figure 34.

As shown in the figure, the length of the hypotenuse is 17, the length of the side opposite θ is 15, and the length of the side adjacent to θ is 8, with

$$\sin \theta = \frac{15}{17}, \quad \cos \theta = \frac{8}{17}, \quad \text{and} \quad \tan \theta = \frac{15}{8}. \quad \blacksquare$$

2.4 Exercises

Find the values of the six trigonometric functions for the following angles. Do not use tables or a calculator.

1. 120°	**2.** 135°	**3.** 150°	**4.** 225°
5. $4\pi/3$	**6.** $5\pi/3$	**7.** 330°	**8.** 390°
9. 420°	**10.** 495°	**11.** 510°	**12.** 570°
13. π	**14.** $3\pi/2$	**15.** $-\pi/2$	**16.** $-\pi$

Complete the following table. Do not use tables or a calculator.

	θ in degrees	θ in radians	sin θ	cos θ	tan θ	cot θ	sec θ	csc θ
17.	30°	$\pi/6$	1/2	$\sqrt{3}/2$			$2\sqrt{3}/3$	2
18.	45°	$\pi/4$			1	1		
19.	60°	$\pi/3$		1/2	$\sqrt{3}$		2	
20.	120°	$2\pi/3$	$\sqrt{3}/2$		$-\sqrt{3}$			$2\sqrt{3}/3$
21.	135°	$3\pi/4$	$\sqrt{2}/2$	$-\sqrt{2}/2$			$-\sqrt{2}$	$\sqrt{2}$
22.	150°	$5\pi/6$		$-\sqrt{3}/2$	$-\sqrt{3}/3$			2
23.	210°	$7\pi/6$	$-1/2$		$\sqrt{3}/3$	$\sqrt{3}$		-2
24.	240°	$4\pi/3$	$-\sqrt{3}/2$	$-1/2$			-2	$-2\sqrt{3}/3$

Find the values of the six trigonometric functions for θ where the point given below is on the terminal side of angle θ in standard position.

25. $(-3, 4)$	**26.** $(-4, -3)$	**27.** $(24, 7)$	**28.** $(-7, 24)$
29. $(-12, -5)$	**30.** $(6, 8)$	**31.** $(-9, -12)$	**32.** $(5, -12)$
33. $(2\sqrt{2}, -2\sqrt{2})$	**34.** $(-2\sqrt{2}, 2\sqrt{2})$	**35.** $(\sqrt{5}, -2)$	**36.** $(-\sqrt{7}, \sqrt{2})$
37. $(-\sqrt{13}, \sqrt{3})$	**38.** $(-\sqrt{11}, -\sqrt{5})$	**39.** $(\sqrt{15}, -\sqrt{10})$	**40.** $(-\sqrt{12}, \sqrt{13})$

41. $(8.7691, -3.2473)$ **42.** $(-5.1021, 7.6132)$ **43.** $(-.04716, -.03219)$

44. $(126.89, 104.21)$ **45.** $(9.713\sqrt{12.4}, -8.765\sqrt{10.2})$ **46.** $(-5.114\sqrt{286}, 2.1094\sqrt{395})$

47. The terminal side lies on the line $y = 5x$, with $x < 0$.

48. The terminal side lies on the line $y + 3x = 0$, with $x > 0$.

Find the values of the six trigonometric functions of θ in the right triangles in Exercises 49–52.

49.
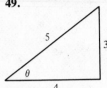

50.
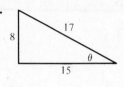

51.
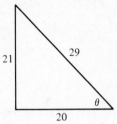

52.
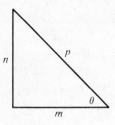

Find sin A, cos A, and tan A for the following right triangles.

53.
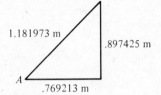

54.
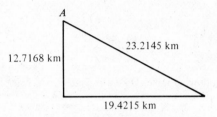

Evaluate each of the following. (Do not use tables or calculator.)

55. $\sin^2 120° + \cos^2 120°$

56. $\sin^2 225° + \cos^2 225°$

57. $2 \tan^2 120° + 3 \sin^2 150° - \cos^2 180°$

58. $\cot^2 135° - \sin 30° + 4 \tan 45°$

59. $\sin^2 225° - \cos^2 270° + \tan 60°$

60. $\cot^2 90° - \sec^2 180° + \csc^2 135°$

61. $\cos^2 60° + \sec^2 150° - \csc^2 210°$

62. $\cot^2 135° + \tan^4 60° - \sin^4 180°$

63. $\sec 30° - \sin 60° + \cos 210°$

64. $\cot 30° + \tan 60° - \sin 240°$

Answer *true* or *false* for each of the following.

65. $\sin 30° + \sin 60° = \sin(30° + 60°)$

66. $\sin(30° + 60°) = \sin 30° \cdot \cos 60° + \sin 60° \cdot \cos 30°$

67. $\cos 60° = 2 \cos^2 30° - 1$

68. $\cos 60° = 2 \cos 30°$

69. $\sin 120° = \sin 150° - \sin 30°$

70. $\sin 210° = \sin 180° + \sin 30°$

71. $\sin 120° = \sin 180° \cdot \cos 60° - \sin 60° \cdot \cos 180°$

72. $\cos 300° = \cos 240° \cdot \cos 60° - \sin 240° \cdot \sin 60°$

73. $\cos 150° = \cos 120° \cdot \cos 30° - \sin 120° \cdot \sin 30°$

74. $\sin 120° = 2 \sin 60° \cdot \cos 60°$

Find all values of the angle θ, when $0° \le \theta < 360°$, for which the following are true.

75. $\sin \theta = \dfrac{1}{2}$

76. $\cos \theta = \dfrac{\sqrt{3}}{2}$

77. $\tan \theta = \sqrt{3}$

78. $\sec \theta = \sqrt{2}$

79. $\cos \theta = -\dfrac{1}{2}$ **80.** $\cot \theta = -\dfrac{\sqrt{3}}{3}$ **81.** $\sin \theta = -\dfrac{\sqrt{3}}{2}$ **82.** $\cos \theta = -\dfrac{\sqrt{2}}{2}$

83. $\cos \theta = 0$ **84.** $\sin \theta = 1$ **85.** $\cot \theta$ is undefined   **86.** $\csc \theta$ is undefined

Use a calculator with sine and tangent keys and find each of the following. (Be sure to set the machine for degree measure.) Then explain why these answers are not really ''correct'' if the exact value has been requested.

87. $\sin 45°$ **88.** $\tan 60°$

Find the exact value of each labeled part in each of the following figures.

89.
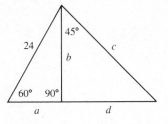

90.
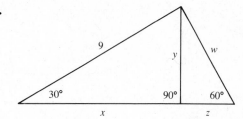

91.
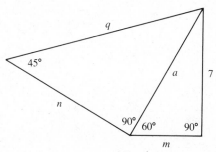

92.
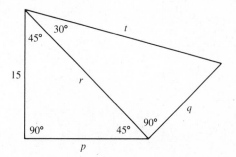

2.5 Values of Trigonometric Functions

Up to now, values of the trigonometric functions have been found only for certain special values of θ. The methods used to find these special values were primarily based on geometry. For angles whose function values cannot be easily found by geometry, approximate values can be found with a calculator or by using special tables, such as Table 3 in the back of the book. To use a calculator, first recall that

$$1 \text{ minute} = 1' = 60 \text{ seconds} = \frac{1}{60}^{\circ}$$

$$1 \text{ second} = 1'' = \frac{1}{60}' = \frac{1}{3600}^{\circ}.$$

For example, an angle of 12° 25′ 56″ is

$$12 + \frac{25}{60} + \frac{56}{3600} \text{ degrees.}$$

Most calculators work in **decimal degrees.** For example, 56.832° is

$$56\frac{832}{1000} \text{ of a degree.}$$

When using a calculator to find the values for an angle given in degrees, minutes, and seconds, it is often necessary to convert to decimal degrees.

Example 1 Use a calculator with sine, cosine, and tangent keys to find each of the following. Round to five decimal places. Make sure the calculator is set for degree measure.

(a) sin 49° 12′

Convert 49° 12′ to decimal degrees.

$$49° \ 12′ = 49\frac{12°}{60} = 49.2°$$

Then press the sine key.

$$\sin 49.2° = 0.75700$$

(b) $\tan 132° \ 41′ = \tan 132\frac{41°}{60}$

$$= \tan \left(132 + \frac{41}{60}\right)°$$

$$= \tan 132.68333°$$

$$= -1.08432$$

(c) $\sec 97° \ 58′ \ 37″ = \sec\left(97 + \frac{58}{60} + \frac{37}{3600}\right)° = \sec 97.97694°$

Calculators do not have secant keys. However,

$$\sec \theta = \frac{1}{\cos \theta}$$

for all angles θ when $\cos \theta$ is not 0. Use this identity and find sec 97.97694° by pressing the cosine key, and then pressing the $1/x$ key to get the reciprocal.

$$\sec 97.97694° = \frac{1}{\cos 97.97694°} = -7.20593$$

(d) cot 51.4283°

This angle is already in decimal degrees. Find cot θ from the identity cot $\theta = 1/\tan \theta$.

$$\cot 51.4283° = \frac{1}{\tan 51.4283°} = 0.79748. \quad ■$$

Using Tables The next few examples show how to use tables to find the values of the trigonometric functions. These values can be found from Table 3 in the back of the book. Table 3 gives the values of the trigonometric functions for values of θ where $0° \le \theta \le 90°$ or $0 \le \theta \le \pi/2$. Most of the values in Table 3 are four decimal place approximations. However, for convenience, the equality symbol is used with the understanding that the values are actually approximations. The table is designed to give the value of θ in the first two columns (in degrees, then radians) for $0° \le \theta \le 45°$ (or $0 \le \theta \le \pi/4$) and in the last two columns for $45° \le \theta \le 90°$ (or $\pi/4 \le \theta \le \pi/2$). When locating values for $45° \le \theta \le 90°$ (or $\pi/4 \le \theta \le \pi/2$), read *up* the table and refer to the names of the trigonometric functions at the bottom of the page. Function values of real numbers can be found by using the "radian" column of Table 3. A small portion of Table 3 is reproduced here.

θ (degrees)	θ (radians)	$\sin \theta$	$\cos \theta$	$\tan \theta$	$\cot \theta$	$\sec \theta$	$\csc \theta$	
36°00′	.6283	.5878	.8090	.7265	1.376	1.236	1.701	.9425	**54°00′**
10	.6312	.5901	.8073	.7310	1.368	1.239	1.695	.9396	50
20	.6341	.5925	.8056	.7355	1.360	1.241	1.688	.9367	40
30	.6370	.5948	.8039	.7400	1.351	1.244	1.681	.9338	30
40	.6400	.5972	.8021	.7445	1.343	1.247	1.675	.9308	20
50	.6429	.5995	.8004	.7490	1.335	1.249	1.668	.9279	10
37°00′	.6458	.6018	.7986	.7536	1.327	1.252	1.662	.9250	**53°00′**
		$\cos \theta$	$\sin \theta$	$\cot \theta$	$\tan \theta$	$\csc \theta$	$\sec \theta$	θ (radians)	θ (degrees)

Example 2 Use the portion of Table 3 above or the complete Table 3 in the back of the book to find the following function values.

(a) $\sin 36° \, 40'$

For angles between $0°$ and $45°$, read down the *left* of the table and use the function names at the *top* of the table. Doing this here gives

$$\sin 36° \, 40' = .5972.$$

(b) $\csc 53° \, 40'$

Use the right "degree" column of the table for angles between $45°$ and $90°$. Use the function names at the bottom. Notice that $53° \, 40'$ is above the entry for $53°$. From the table,

$$\csc 53° \, 40' = 1.241.$$

(c) $\tan 82° \, 00' = 7.115$

(d) $\sin .2676 = 0.2644$ (Use the "radian" column of the table.)

(e) $\cot 1.2043 = 0.3839$ ∎

When the required value is not in the table, linear interpolation may be used, as shown in the next example. (The basic ideas of linear interpolation are explained in the appendix at the end of Chapter 7.

Example 3 Find each of the following.

(a) $\tan 40° \, 52'$

The value of tan 40° 52′ lies 2/10 or .2 of the way from the value of tan 40° 50′ to the value of tan 41° 00′.

$$10\left\{2\left\{\begin{matrix}\tan 40° \ 50′ = .8642\\ \tan 40° \ 52′ = \ ? \\ \tan 41° \ 00′ = .8693\end{matrix}\right\}.0051\right.$$

$$\tan 40° \ 52′ = .2(.8693 - .8642) + .8642$$
$$= .2(.0051) + .8642$$
$$= .8652$$

(b) cos 63° 34′
Work as follows.

$$10\left\{4\left\{\begin{matrix}\cos 63° \ 30′ = .4462\\ \cos 63° \ 34′ = \ ? \\ \cos 63° \ 40′ = .4436\end{matrix}\right\}.0026\right.$$

$$\cos 63° \ 34′ = .4462 - (.4)(.4462 - .4436)$$
$$= .4462 - .4(.0026)$$
$$= .4452$$

Here it was necessary to *subtract* from .4462 rather than add because the cosine function *decreases* as θ increases for $0 < \theta < \pi/2$. ■

Example 4 Find a value of θ satisfying each of the following.

(a) sin θ = .5807; θ in degrees
Use Table 3 and read columns having sine at either the top or the bottom. Since .5807 is found in a column having sine at the top, use angles at the left to give the value of θ.

$$\theta = 35° \ 30′$$

(b) tan θ = 2.699; θ in degrees
The number 2.699 is in a column having tangent at the bottom, with angles at the right used to find the value of θ.

$$\theta = 69° \ 40′ ■$$

Table 3 gives values only for angles between 0° and 90°, inclusive, or for real numbers between 0 and $\pi/2$, inclusive. For values outside this range, some of the earlier work with trigonometric functions must be used. As will be shown, the trigonometric function values for $90° < \theta < 360°$ can be found by referring to the appropriate value of θ in the interval $0° < \theta < 90°$ and affixing the correct sign. For any value θ in the interval $90° < \theta < 360°$, the positive acute angle made by the terminal side of angle θ and the *x*-axis is called the **reference** or **related angle** for θ; this new angle is written $\theta′$. See Figure 35. For example, if $\theta = 135°$, the reference angle $\theta′$ is 45°. If $\theta = 200°$, then $\theta′ = 20°$.

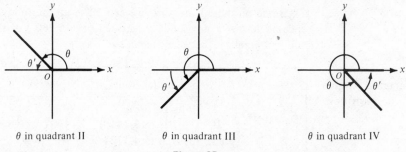

θ in quadrant II θ in quadrant III θ in quadrant IV

Figure 35

For radians, a similar definition could be given for **reference numbers.** For example, if $s = -\pi/6$, then the reference number s' is $s' = \pi/6$. Also, if $s = 5\pi/4$, then $s' = \pi/4$.

Example 5   Find the reference angle or reference number for each of the following.

(a) 218°

As shown in Figure 36, the positive acute angle made by the terminal side of this angle and the x-axis is $218° - 180° = 38°$.

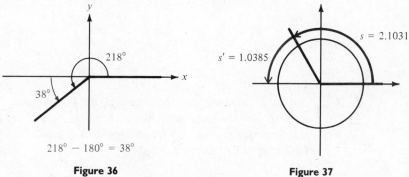

Figure 36   **Figure 37**

(b) 321° 10′

The positive acute angle made by the terminal side of this angle and the x-axis is $360° - 321° 10'$. Write 360° as 359° 60′ so that the reference angle is

$$359° \ 60' - 321° \ 10' = 38° \ 50'.$$

By the way, an angle of $-38° \ 50'$, which has the same terminal ray as 321° 10′, also has a reference angle of 38° 50′.

(c) $s = 2.1031$

The reference number is found by subtracting from π. See Figure 37. Using the four-decimal-place approximation 3.1416 for π gives

$$s' = 3.1416 - 2.1031 = 1.0385. \quad \blacksquare$$

Based on these examples, we can make up the following table to find the reference angle θ' for any angle θ between $0°$ and $360°$, or the reference number s' for any real number s between 0 and 2π.

Reference Angles or Numbers

θ or s in quadrant	θ' is	s' is
I (0 to 1.5708)	θ	s
II (1.5708 to 3.1416)	$180° - \theta$	$\pi - s$
III (3.1416 to 4.7124)	$\theta - 180°$	$s - \pi$
IV (4.7124 to 6.2832)	$360° - \theta$	$2\pi - s$

Figure 38 shows an angle θ and reference angle θ' drawn so that θ' is in standard position. Point P, with coordinates (x_1, y_1), has been located on the terminal side of angle θ. Let r be the distance from O to P. Choose point P' on the terminal side of angle θ', so that the distance from O to P' is also r. Let P' have the coordinates (x_2, y_2). By congruent triangles, verify that

$$x_1 = -x_2 \quad \text{and} \quad y_1 = y_2.$$

Thus,

$$\sin \theta = \frac{y_1}{r} = \frac{y_2}{r} = \sin \theta'$$

$$\cos \theta = \frac{x_1}{r} = \frac{-x_2}{r} = -\cos \theta'$$

$$\tan \theta = \frac{y_1}{x_1} = \frac{y_2}{-x_2} = -\tan \theta',$$

and so on, proving that the values of the trigonometric functions of the reference angle θ' are the same as those of angle θ, except perhaps for signs. While this statement was proven only for an angle θ in quadrant II, similar results can be proven for angles in the other quadrants.

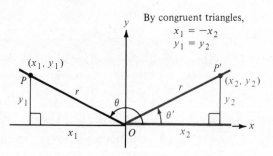

Figure 38

Based on this work, the values of the trigonometric functions for any angle θ can be found from a table by finding the function value for an angle between $0°$ and $90°$. Do this with the following steps. (Similar steps can be used to find trigonometric function values for any real number s.)

**Finding
Trigonometric
Function Values
from a Table**

1. If $\theta \geq 360°$, or if $\theta < 0°$, add or subtract 360° as many times as needed to get an angle at least 0° but less than 360°.
2. Find the reference angle θ'. (Use the table given above.)
3. Find the necessary values of the trigonometric functions for the reference angle θ'.
4. Find the correct signs for the values found in Step 3. This gives the trigonometric values for angle θ.

To find function values with most calculators, just enter the angle in decimal degrees and press the appropriate function key, along with the reciprocal key if needed.

Example 6   Find each of the following values.

(a) tan 315°

Begin by finding the reference angle for 315°, as in Figure 39. Since 315° is in quadrant IV, subtract 315° from 360°.

$$360° - 315° = 45°$$

The value of tangent for quadrant IV angles is negative, and tan 45° = 1, so

$$\tan 315° = -\tan 45° = -1.$$

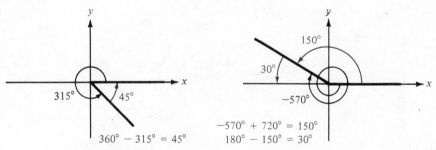

Figure 39   **Figure 40**

(b) cos(−570°)

To begin, find the smallest possible positive angle coterminal to −570°. See Figure 40. Add 2 · 360°, or 720°, to −570°.

$$-570° + 720° = 150°$$

Since 150° is in quadrant II, find its reference angle by subtracting 150° from 180°.

$$180° - 150° = 30°$$

As found earlier, cos 30° = $\sqrt{3}/2$ and the values of cosine for quadrant II angles are negative, so

$$\cos(-570°) = \cos 150° = -\cos 30° = -\sqrt{3}/2.$$

(c) cot 600°

$$600° - 360° = 240°$$

Verify that $\theta' = 240° - 180° = 60°$, and that

$$\cot 600° = \cot 60° = \sqrt{3}/3.$$

(d) sin 3.7845

Since $s = 3.7845$ is in quadrant III, the reference number s' is found by subtracting π, or 3.1416, from s.

$$s' = 3.7845 - 3.1416 = 0.6429$$

The values of sine are negative in quadrant III, so

$$\sin 3.7845 = -\sin 0.6429 = -0.5995. \quad\blacksquare$$

As shown earlier, the values of the trigonometric functions can be obtained from the lengths of the sides of a right triangle. The next example shows an application of this.

Example 7 A surveyor gathered the data shown on the triangle of Figure 41. Find the lengths of sides a and b of the triangle.

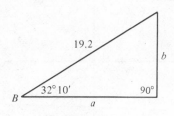

Figure 41

Let us first find b. The sine of angle B (the angle opposite b) is given by

$$\sin B = \frac{\text{side opposite}}{\text{hypotenuse}}.$$

Substitute in the known values.

$$\sin 32° 10' = \frac{b}{19.2}.$$

Use Table 3 or a calculator to find the value of sin 32° 10′, then solve for b.

$$.5324 = \frac{b}{19.2}$$
$$19.2(.5324) = b$$
$$10.2 = b$$

Use the cosine of angle B to find that $a = 16.3$.   ∎

Right triangle applications are studied in more detail in Chapter 5.

2.5 Exercises

Find the reference angle or reference number for each of the following. Use 3.1416 as an approximation for π.

1. 215°

2. 550°

3. −143°

4. −12°

5. −110° 10′

6. −429° 30′

7. 3.21

8. 2.4

9. 5.9690

10. −1.7861

11. −4.0230

12. −2.4580

Use Table 3 or a calculator to get a value for each of the following. Use 3.1416 as an approximation for π.

13. sin 39° 20′

14. cos 58° 40′

15. sin (−38° 40′)

16. csc (−168° 30′)

17. cos (−124° 50′)

18. sec 274° 30′

19. sec 1.9024

20. cot 3.1998

21. sin 7.5835

22. tan 6.4752

23. cos (−4.0230)

24. cot (−3.8426)

Use interpolation to find each of the following values.

25. tan 29° 42′

26. sin 56° 38′

27. tan 49° 17′

28. sin 78° 32′

29. sin .1635

30. cos (−1.3870)

31. tan (−.3034)

32. csc .2518

Use Table 3 to find a value of θ in the interval $0° \leq \theta \leq 90°$ satisfying each of the following. Give θ to the nearest ten minutes.

33. sin θ = .8480

34. tan θ = 1.473

35. cos θ = .8616

36. cot θ = 1.257

37. sin θ = .7214

38. sec θ = 2.749

39. tan θ = 6.435

40. sin θ = .2784

Find a value of s, where $0 \leq s \leq 1.5708$, that makes each of the following true.

41. tan s = .2126

42. cos s = .7826

43. sin s = .9918

44. cot s = .2994

45. cot s = .0963

46. csc s = 1.021

47. tan s = 1.621

48. cos s = .9272

49. sin s = .4067

50. cos s = .6626

51. tan s = .3607

52. sec s = 1.758

53. The values of the trigonometric functions repeat every 2π. For this reason, trigonometric functions are used to describe things that repeat periodically. For example, the maximum afternoon temperature in a given city might be approximated by

$$t = 60 - 30 \cos x\pi/6,$$

where t represents the maximum afternoon temperature in month x, with $x = 0$ representing January, $x = 1$ representing February, and so on. Find the maximum afternoon temperature for each of the following months: **(a)** January **(b)** April **(c)** May **(d)** June **(e)** August **(f)** October.

54. The temperature in Fairbanks is approximated by

$$T(x) = 37 \sin\left[\frac{2\pi}{365}(x - 101)\right] + 25,$$

where $T(x)$ is the temperature in degrees Fahrenheit on day x, with $x = 1$ corresponding to January 1 and $x = 365$ corresponding to December 31.* Use a calculator to estimate the temperature on the following days: **(a)** March 1 (day 60)   **(b)** April 1 (day 91)   **(c)** Day 150   **(d)** June 15   **(e)** September 1   **(f)** October 31.

In calculus courses $\sin x$ and $\cos x$ are expressed as infinite sums, where x is measured in radians. For all real numbers x and all natural numbers n, the following hold.

$$\cos x = 1 - \frac{x^2}{2!} + \frac{x^4}{4!} - \cdots + (-1)^{n-1}\frac{x^{2n-2}}{(2n-2)!} + \cdots$$

and

$$\sin x = x - \frac{x^3}{3!} + \frac{x^5}{5!} - \cdots + (-1)^{n-1}\frac{x^{2n-1}}{(2n-1)!} + \cdots$$

where $2! = 2 \cdot 1;$ $3! = 3 \cdot 2 \cdot 1;$ $4! = 4 \cdot 3 \cdot 2 \cdot 1;$

and $n! = n(n - 1)(n - 2) \ldots (2)(1).$

Using these formulas, $\sin x$ and $\cos x$ can be evaluated to any desired degree of accuracy. For example, let us find $\cos .5$ using the first three terms of this result and compare with the result using the first four terms. Using the first three terms,

$$\cos .5 = 1 - \frac{(.5)^2}{2!} + \frac{(.5)^4}{4!}$$

$$= 1 - \frac{.25}{2} + \frac{.0625}{24}$$

$$= 1 - .125 + .00260$$

$$= .87760.$$

If we use four terms, the result is

$$\cos .5 = 1 - \frac{(.5)^2}{2!} + \frac{(.5)^4}{4!} - \frac{(.5)^6}{6!}$$

$$= 1 - .125 + .00260 - .00002$$

$$= .87758.$$

The two results differ by only .00002, so that the first three terms give results which are accurate to the fourth decimal place. From a calculator, $\cos .5 = .8775826$.

Use the first three terms of the results given above for $\sin x$ and $\cos x$ to find the following. Compare your results with the values in Table 3, or with the approximation obtained by using the sine or cosine key on your calculator.

55. $\sin 1$ **56.** $\sin .1$ **57.** $\cos 1.4$ **58.** $\cos .8$

59. $\sin .01$ **60.** $\cos .01$

---

*Barbara Lando and Clifton Lando, "Is the Graph of Temperature Variation a Sine Curve?" *The Mathematics Teacher*, 70 (September, 1977): 534–37.

Find the lengths of the missing sides in each of the following right triangles.

61.
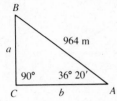

62.
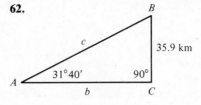

63.
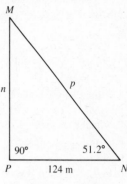

64.
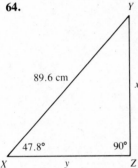

65.
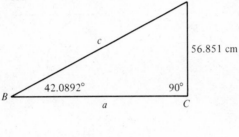

66.
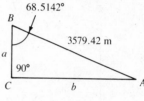

Solve each of the following problems.

67. A woman stands 59.7 m from the base of a tree. The angle her line of sight makes to the top of the tree is 39° 20′. If the woman's eyes are 1.74 m above the ground, find the height of the tree.

68. A ladder leans against a building, making an angle of 24° 10′ with the building. If the ladder is 4.88 m in length, find the distance between the bottom of the ladder and the building.

69. The length of the base of an isosceles triangle is 37.41 in. Each base angle is 49.74°. Find the length of each of the two equal sides of the triangle. (*Hint:* Divide the triangle into two right triangles.)

70. Find the altitude of an isosceles triangle having a base of 125.6 cm if the angle opposite the base is 72° 42′.

71. Find the angle of the sun above the horizon when a person 5.94 ft tall casts a shadow 9.74 ft long. (Find θ in the figure.)

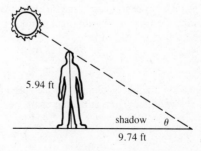

72. The angle of the sun above the horizon is 27.5°. Find the length of the shadow of a person 5.73 ft tall. (Find x in the figure at right.)

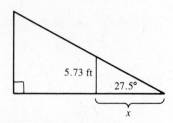

73. The solar panel shown in the figure below left must be tilted so that angle θ is 94° when the angle of elevation of the sun is 38°. (The sun's rays are assumed to be parallel, and they make an angle of 38° with the horizontal.) The panel is 4.5 ft long. Find h.

74. Find the minimum height h above the surface of the earth so that a pilot at point A in the figure below right can see an object on the horizon at C, 125 mi away. Assume that the radius of the earth is 4.00×10^3 mi.

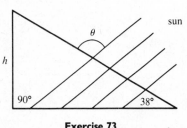

Exercise 73

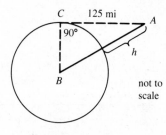

Exercise 74

75. A piece of land has the shape shown on the left below. Find x.

76. Find the value of x in the figure on the right below.

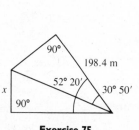

Exercise 75

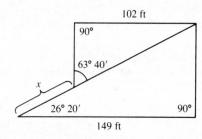

Exercise 76

When a light ray travels from one medium, such as air, to another medium, such as water or glass, the speed of the light changes, and the direction that the ray is traveling changes. (This is why a fish under water is in a different position than it appears to be.) These changes are given by Snell's law,

$$\frac{c_1}{c_2} = \frac{\sin \theta_1}{\sin \theta_2},$$

where c_1 is the speed in the first medium, c_2 is the speed in the second medium, and θ_1 and θ_2 are the angles shown in the figure at the top of the next page.

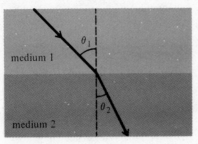

If this medium is less
dense, light travels at
a faster speed, c_1.

If this medium is more
dense, light travels at
a slower speed, c_2.

In the following exercises, assume that $c_1 = 3 \times 10^8$ m per sec. Find the speed of light in
the second medium.

77. $\theta_1 = 46°$, $\theta_2 = 31°$ **78.** $\theta_1 = 39°$, $\theta_2 = 28°$

Find θ_2 for the following values of θ_1 and c_2. Round to the nearest degree.

79. $\theta_1 = 40°$, $c_2 = 1.5 \times 10^8$ m per sec   **80.** $\theta_1 = 62°$, $c_2 = 2.6 \times 10^8$ m per sec

The figure below shows a fish's view of the world above the surface of the water.* Suppose
that a light ray comes from the horizon, enters the water, and strikes the fish's eye.

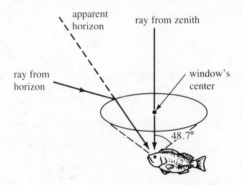

81. Let us assume that this ray gives a value of 90° for angle θ_1 in the formula for Snell's
law. (In a practical situation this angle would probably be a little less than 90°.) The
speed of light in water is about 2.254×10^8 m per sec. Find angle θ_2.

    (Your result should have been about 48.7°. This means that a fish sees the world
above the water as a cone, making an angle of 48.7° with the vertical.)

82. Suppose that an object is located at a true angle of 29.6° above the horizon. Find the
apparent angle above the horizon to a fish.

Chapter 2 Summary

Key Words

unit circle	cotangent	standard position
sine	secant	quadrantal angle
cosine	cosecant	coterminal angles
periodic function	reciprocal identities	degree measure
period	angle	radian measure
tangent	vertex	reference angle

Review Exercises

For each of the following arc lengths, find the coordinates of the corresponding point on the unit circle.

1. $-\pi/4$ **2.** $2\pi/3$ **3.** 3π **4.** $-\pi$

For each of the following values of s, find $\sin s$, $\cos s$, and $\tan s$.

5. $5\pi/4$ **6.** $5\pi/3$ **7.** $-\pi/3$ **8.** -4π

The point $(-2/3, \sqrt{5}/3)$ is the endpoint of arc s on the unit circle. Find the coordinates of the endpoints of the following arcs.

9. $-s$ **10.** $\pi + s$ **11.** $\pi - s$ **12.** $2\pi - s$

Give the quadrant in which θ terminates.

13. $\sin \theta > 0$, $\cos \theta < 0$ **14.** $\cos \theta > 0$, $\cot \theta < 0$ **15.** $\tan \theta > 0$, $\sin \theta < 0$ **16.** $\sec \theta < 0$, $\tan \theta < 0$

For each of the following, find the values of the other trigonometric functions.

17. $\sin s = 2/3$, s terminates in quadrant I   **18.** $\cos s = 4/5$, s terminates in quadrant IV

19. $\sec s = \sqrt{5}$, $\cot s = -1/2$ **20.** $\tan s = -4/3$, $\sec s = -5/3$

Work each of the following exercises.

21. A pulley is rotating 320 times per minute. Through how many degrees does a point on the edge of the pulley move in 2/3 sec?

22. The propeller of a speedboat rotates 650 times per minute. Through how many degrees will a point on the edge of the propeller rotate in 2.4 sec?

Convert decimal degrees to degrees, minutes, seconds, and convert degrees, minutes, seconds to decimal degrees. Round to the nearest second or the nearest thousandth of a degree, as appropriate.

23. $47° \ 25' \ 11''$ **24.** $119° \ 08' \ 03''$ **25.** $74.2983°$

26. $-61.5034°$ **27.** $183.0972°$ **28.** $275.1005°$

Convert radian measures to degrees, and degree measures to radians.

29. $3\pi/4$ **30.** $4\pi/5$ **31.** $31\pi/5$ **32.** $-11\pi/18$

33. $270°$ **34.** $480°$ **35.** $1020°$ **36.** $2000°$

Find the exact value of each of the following.

37. sin π/3   **38.** cos 2π/3   **39.** tan 4π/3

40. cot 390°   **41.** sec 900°   **42.** cot (− 1020°)

Find the sine, cosine, and tangent of each of the following angles.

43.
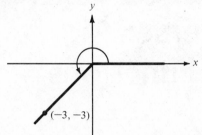

44.
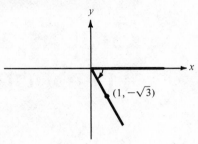

Find the value of each of the following. Use Table 3 or a calculator. Interpolate as necessary.

45. tan 235°   **46.** sec (− 87°)   **47.** sin 247° 10′

48. sec 28° 17′   **49.** cos 58° 4′   **50.** cos (− 3.1998)

Use a calculator or Table 3 to find a value of θ, where 0° ≤ θ ≤ 90°, for each of the following.

51. sin θ = .8258   **52.** cot θ = 1.124   **53.** cos θ = .9754   **54.** sec θ = 1.263

Find s in each of the following. Assume that 0 ≤ s ≤ π/2.

55. cos s = 0.9250   **56.** tan s = 4.011   **57.** sin s = 0.4924   **58.** csc s = 1.236

Find the lengths of the missing sides in each of the following right triangles.

59.
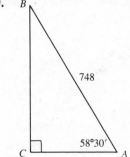

60.

61. A = 39.72°, b = 38.97 m   **62.** B = 47° 53′, b = 298.6 m

Solve each of the following problems.

63. Find t if θ = 5π/12 radians and ω = 8π/9 radians per sec.

64. Find θ if t = 12 sec and ω = 9 radians per sec.   **65.** Find ω if t = 8 sec and θ = 2π/5 radians.

66. Find ω if s = 12π/25 ft, r = 3/5 ft, and t = 15 sec.

67. Find s if r = 11.46 cm, ω = 4.283 radians per sec, and t = 5.813 sec.

68. Find the linear velocity of a point on the edge of a flywheel of radius 7 m if the flywheel is rotating 90 times per sec.

3

Graphs and Inverse Trigonometric Functions

Many things in daily life repeat with a predictable pattern: the daily newspaper is delivered at the same time each morning, in warm areas electricity use goes up in the summer and down in the winter, the price of fresh fruit goes down in the summer and up in the winter, and attendance at amusement parks increases in the summer and declines in autumn. There are many examples of these *periodic* phenomena. As this chapter will show, the trigonometric functions are periodic and very useful for describing periodic activities.

3.1 Graphs of Sine and Cosine

By the identities for coterminal angles,

$$\sin 0 = \sin 2\pi$$
$$\sin \pi/2 = \sin (\pi/2 + 2\pi)$$
$$\sin \pi = \sin (\pi + 2\pi),$$

and $\sin x = \sin (x + 2\pi)$ for any real number x. The value of the sine function is the same for x and for $x + 2\pi$, so that the function defined by $y = \sin x$ is a periodic function with period 2π. By definition,

Periodic Function

> a function f with the property that
>
> $$f(x) = f(x + p),$$
>
> for every real number x in the domain of f and for some positive real number p is a **periodic function.** The smallest possible positive value of p is the **period** of the function.

While it is true that $\sin x = \sin (x + 4\pi)$ and $\sin x = \sin (x + 6\pi)$, the *smallest* positive value of p making $\sin x = \sin (x + p)$ is $p = 2\pi$, so 2π is the period for the sine function.

We saw earlier that the sine function defined by $y = \sin x$ has the set of all real numbers as domain, with range $-1 \le y \le 1$, or $-1 \le \sin x \le 1$.* Since the sine function has period 2π, one period of $y = \sin x$ can be graphed by looking at values of x from 0 to 2π. To graph this period, look at Figure 1, which shows a unit circle with a point (p, q) marked on it. Based on the definitions for circular functions given earlier, for any angle x, $p = \cos x$ and $q = \sin x$. As x increases from 0 to $\pi/2$ (or 90°), q (or $\sin x$) increases from 0 to 1, while p (or $\cos x$) decreases from 1 to 0.

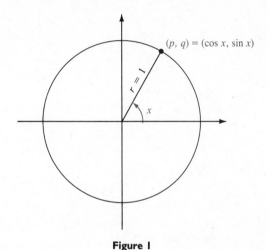

$(p, q) = (\cos x, \sin x)$

Figure 1

As x increases from $\pi/2$ to π (or 180°), q decreases from 1 to 0, while p decreases from 0 to -1. Similar results can be found for the other quadrants, as shown in the following table.

As x increases from	sin x	cos x
0 to $\pi/2$	Increases from 0 to 1	Decreases from 1 to 0
$\pi/2$ to π	Decreases from 1 to 0	Decreases from 0 to -1
π to $3\pi/2$	Decreases from 0 to -1	Increases from -1 to 0
$3\pi/2$ to 2π	Increases from -1 to 0	Increases from 0 to 1

Selecting key values of x and finding the corresponding values of $\sin x$ give the following results. (Decimals are rounded to the nearest tenth.)

x	0	$\pi/4$	$\pi/2$	$3\pi/4$	π	$5\pi/4$	$3\pi/2$	$7\pi/4$	2π
$\sin x$	0	.7	1	.7	0	$-.7$	-1	$-.7$	0

*In this chapter, x is used as the variable (as in $\sin x$) instead of s or θ to allow graphs to be drawn on the familiar xy-coordinate system.

Plotting the points from the table of values and connecting them with a smooth curve gives the solid portion of the graph in Figure 2. Since the function defined by $y = \sin x$ is periodic and has all real numbers as domain, the graph continues in both directions indefinitely, as indicated by the dashed lines. The graph is sometimes called a **sine wave** or **sinusoid.** It is a good idea to memorize the shape of this graph and be able to sketch it quickly. Some key points of the graph are $(0, 0)$, $(\pi/2, 1)$, $(\pi, 0)$, $(3\pi/2, -1)$, and $(2\pi, 0)$. Sketch the graph by plotting these five points and connecting them with the characteristic sine wave.

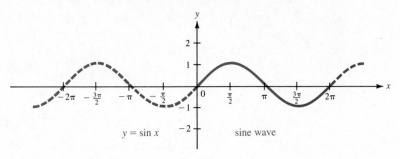

Figure 2

The same scales are used on both the x and y axes of Figure 2 so as not to distort the graph. Since the period of $y = \sin x$ is 2π, it is convenient to use subdivisions of 2π on the x-axis. Although the more familiar x-values, 1, 2, 3, 4, and so on, are still present, they are usually not shown to avoid cluttering the graph. These values are shown in Figure 3.

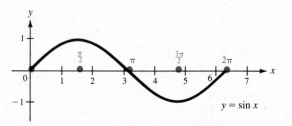

Figure 3

Sine graphs occur in many different practical applications. For one application, look back at Figure 1 and assume that the line from the origin to the point (p, q) is part of the pedal of a bicycle wheel, with a foot placed at (p, q). As mentioned above, q is equal to $\sin x$, showing that the height of the pedal from the horizontal axis in Figure 1 is given by $\sin x$. When various angles are chosen for the pedal and q is calculated for each angle, the height of the pedal leads to the sine curve shown in Figure 4.

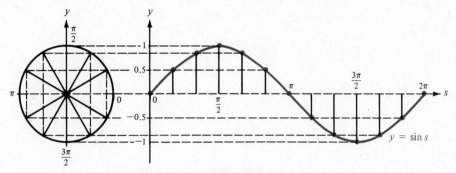

Figure 4

The graph of $y = \cos x$ can be found in much the same way as was the graph of $y = \sin x$. The domain of cosine is the set of all real numbers, and the range of the function defined by $y = \cos x$ is $-1 \le \cos x \le 1$. Here the key points are $(0, 1)$, $(\pi/2, 0)$, $(\pi, -1)$, $(3\pi/2, 0)$, and $(2\pi, 1)$. The graph of $y = \cos x$ has the same shape as the graph of $y = \sin x$. In fact, it is the sine wave, shifted $\pi/2$ units to the left. See Figure 5.

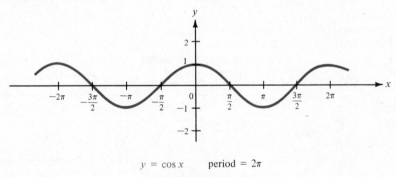

$y = \cos x$ period $= 2\pi$

Figure 5

The examples in the rest of this section show graphs that are "stretched" either vertically or horizontally, or both, when compared with the graphs of $y = \sin x$ or $y = \cos x$.

Example 1   Graph $y = 2 \sin x$.

For a given value of x, the value of y is twice as large as it would be for $y = \sin x$, as shown in the table of values. The only change in the graph is in the range, which becomes $-2 \le y \le 2$. See Figure 6 on the next page, which also shows a graph of $y = \sin x$ for comparison.   ■

x	0	$\pi/2$	π	$3\pi/2$	2π
$\sin x$	0	1	0	-1	0
$2 \sin x$	0	2	0	-2	0

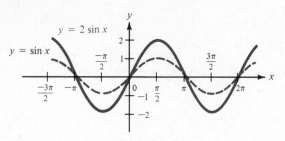

Figure 6

Example 1 suggests the following result.

Amplitude of Sine and Cosine

> For $a \neq 0$, the graph of $y = a \sin x$ or $y = a \cos x$ will have the same shape as $y = \sin x$ or $y = \cos x$, respectively, except with range $-|a| \leq y \leq |a|$. The number $|a|$ is called the **amplitude.**

No matter what the value of the amplitude, the period of both $y = a \sin x$ and $y = a \cos x$ is still 2π.

Example 2   Graph $y = \sin 2x$.

Start with a table of values.

x	0	$\pi/4$	$\pi/2$	$3\pi/4$	π	$5\pi/4$	$3\pi/2$	$7\pi/4$
$2x$	0	$\pi/2$	π	$3\pi/2$	2π	$5\pi/2$	3π	$7\pi/2$
$\sin 2x$	0	1	0	-1	0	1	0	-1

As the table shows, multiplying x by 2 shortens the period by half. The amplitude is not changed. Figure 7 shows the graph of $y = \sin 2x$. Again the graph of $y = \sin x$ is included for comparison.   ■

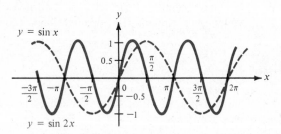

Figure 7

Example 2 suggests the following result. (See Exercise 47.)

Period of Sine and Cosine

> The graph of $y = \sin bx$ will look like that of $\sin x$, but with period $|2\pi/b|$. Also, the graph of $y = \cos bx$ looks like that of $y = \cos x$, but with period $|2\pi/b|$.

Example 3   Graph $y = \cos \dfrac{2}{3}x$.

The period here is $|2\pi/(2/3)| = 3\pi$. The amplitude is 1. The graph is shown in Figure 8.   ■

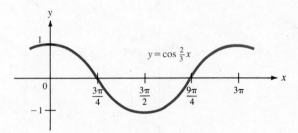

Figure 8

Throughout this chapter we assume $b > 0$. If the equation that defines a function has $b < 0$, the identities in the next chapter can be used to change the equation to one where $b > 0$. The steps used to graph $y = a \sin bx$ or $y = a \cos bx$, where $b > 0$, are given below.

Graphing Sine and Cosine

To graph $y = a \sin bx$ or $y = a \cos bx$, with $b > 0$:

1. Find the period, $2\pi/b$. Start at 0 on the x-axis and lay off a distance of $2\pi/b$.
2. Divide the interval from 0 to $2\pi/b$ into four equal parts.
3. Locate the points at which the graph crosses the x-axis.

Equation	Graph crosses x-axis at:
$y = a \sin bx$	$0, \dfrac{\pi}{b}, \dfrac{2\pi}{b}$ (beginning, middle, and end of interval)
$y = a \cos bx$	$\dfrac{\pi}{2b}, \dfrac{3\pi}{2b}$ (one fourth and three fourths points of interval)

4. Locate the points where the graph reaches maximum and minimum values.

Equation	Graph has a maximum when x is:	
$y = a \sin bx$	$\dfrac{\pi}{2b}$ (for $a > 0$) or	$\dfrac{3\pi}{2b}$ (for $a < 0$)
$y = a \cos bx$	0 and $\dfrac{2\pi}{b}$ (for $a > 0$) or	$\dfrac{\pi}{b}$ (for $a < 0$)

5. Use tables or a calculator to find as many additional points as needed. Then sketch the graph.
6. Draw additional periods of the graph, to the right and to the left, as needed.

Example 4 Graph $y = -2 \sin 3x$.

The period is $2\pi/3$. The amplitude is $|-2| = 2$. Sketch the graph using the steps shown in Figure 9. Notice how the minus sign affects the location of the maximum and minimum points. ∎

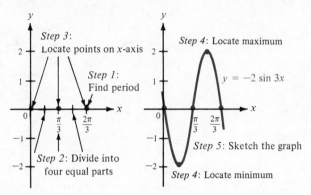

Figure 9

Example 5 Graph $y = 3 \cos \frac{1}{2}x$.

The period is $2\pi/(1/2) = 4\pi$. The amplitude is 3. Follow the steps shown in Figure 10. ∎

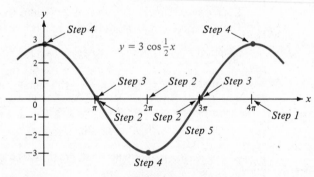

Figure 10

Vertical Translations If $c \neq 0$, a function defined by an equation of the form $y = c + a \sin bx$ or of the form $y = c + a \cos bx$ is shifted vertically when compared with $y = a \sin bx$ or $y = a \cos bx$, respectively. The next example shows how to draw a graph involving these **vertical translations.**

Example 6 Graph $y = 3 - 2 \cos 3x$.

The values of y will be 3 greater than the corresponding values of y in $y = -2 \cos 3x$. This means that the graph of $y = 3 - 2 \cos 3x$ is the same as the graph of $y = -2 \cos 3x$, except with a vertical translation of 3 units upward. See Figure 11. ■

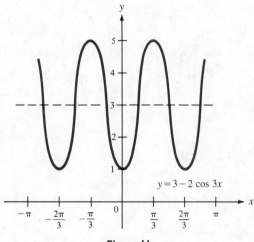

Figure 11

Radio stations send out a carrier signal in the form of a sine wave having equation

$$y = A_0 \sin (2\pi\omega_0 t),$$

where A_0 is the amplitude of the carrier signal, ω_0 is the number of periods the signal oscillates through in one second (its **frequency**), and t is time. A carrier signal received by a radio would be a pure tone. To transmit music and voices, the station might change or **modulate** A_0 according to

$$A_0(t) = A_0 + mA_0 \sin (2\pi\omega t),$$

where ω is the frequency of a pure tone and m is a constant called the **degree of modulation.** The transmitted signal has equation

$$y = A_0 \sin (2\pi\omega_0 t) + A_0 m \sin (2\pi\omega t) \sin (2\pi\omega_0 t)$$

$$= A_0[1 + m \sin (2\pi\omega t)] \sin (2\pi\omega_0 t).$$

A typical carrier signal and a typical graph of y are shown in Figure 12. This process of sending out a radio signal is called **amplitude modulation,** or AM, radio.

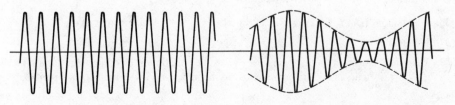

Carrier signal   Transmitted signal

Figure 12

Frequency modulation, or FM, radio involves altering the frequency of the carrier signal, rather than its amplitude. A typical graph is shown in Figure 13.

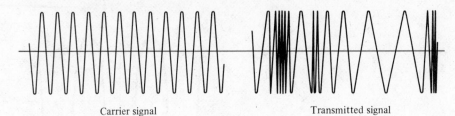

Carrier signal   Transmitted signal

Figure 13

3.1 Exercises

Graph the following over the interval $-2\pi \leq x \leq 2\pi$. Identify the amplitude.

1. $y = 2 \cos x$ **2.** $y = 3 \sin x$ **3.** $y = \frac{2}{3} \sin x$ **4.** $y = \frac{3}{4} \cos x$

5. $y = -\cos x$ **6.** $y = -\sin x$ **7.** $y = -2 \sin x$ **8.** $y = -3 \cos x$

Graph each of the following over a two-period interval. Give the period, the amplitude, and any vertical translations.

9. $y = \sin \frac{1}{2}x$ **10.** $y = \sin \frac{2}{3}x$ **11.** $y = \cos \frac{1}{3}x$ **12.** $y = \cos \frac{3}{4}x$

13. $y = \sin 3x$ **14.** $y = \sin 2x$ **15.** $y = \cos 2x$ **16.** $y = \cos 3x$

17. $y = -\sin 4x$ **18.** $y = -\cos 6x$ **19.** $y = 2 \sin \frac{1}{4}x$ **20.** $y = 3 \sin 2x$

21. $y = -2 \cos 3x$ **22.** $y = -5 \cos 2x$ **23.** $y = \frac{1}{2} \sin 3x$ **24.** $y = \frac{2}{3} \cos \frac{1}{2}x$

25. $y = -\frac{2}{3} \sin \frac{3}{4}x$ **26.** $y = -2 \cos 5x$ **27.** $y = 2 - \cos x$ **28.** $y = 1 + \sin x$

29. $y = -3 + 2 \sin x$ **30.** $y = 2 - 3 \cos x$ **31.** $y = 1 - 2 \cos \frac{1}{2}x$ **32.** $y = -3 + 3 \sin \frac{1}{2}x$

33. $y = \cos \pi x$ **34.** $y = -\sin \pi x$

Graph each of the following over two periods.

35. $y = (\sin x)^2$ [*Hint:* $(\sin x)^2 = \sin x \cdot \sin x$]   **36.** $y = (\cos x)^2$

37. $y = (\sin 2x)^2$ **38.** $y = (\cos 2x)^2$

39. The graph shown on the following page gives the variation in blood pressure for a typical person.* Systolic and diastolic pressures are the upper and lower limits of the periodic

*From *Calculus for the Life Sciences,* by Rodolfo De Sapio, W. H. Freeman and Company. Copyright © 1978 by W. H. Freeman and Company. Reprinted by permission.

changes in pressure that produce the pulse. The length of time between peaks is called
the period of the pulse.

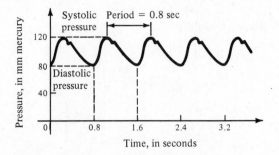

(a) Find the amplitude of the graph.
(b) Find the pulse rate (the number of pulse beats in one minute) for this person.

40. Scientists believe that the average annual temperature in a given location is periodic. The
overall temperature at a given place during a given season fluctuates as time goes on,
from colder to warmer, and back to colder. The graph shows an idealized description of
the temperature for the last few thousand years of a location at the same latitude as
Anchorage.

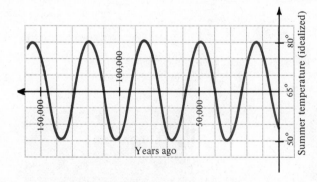

(a) Find the highest and lowest temperatures recorded.
(b) Use these two numbers to find the amplitude. (*Hint:* An alternative definition of the
amplitude is half the difference of the highest and lowest points on the graph.)
(c) Find the period of the graph.   (d) What is the trend of the temperature now?

41. Many of the activities of living organisms are periodic. For example, the graph below
shows the time that flying squirrels begin their evening activity.
(a) Find the amplitude of this graph.   (b) Find the period.

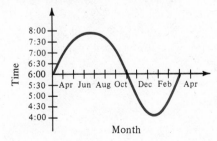

42. The figure shows schematic diagrams of a rhythmically moving arm. The upper arm RO rotates back and forth about the point R; the position of the arm is measured by the angle y between the actual position and the downward vertical position.*

    (a) Find an equation of the form $y = a \sin kt$ for the graph at the right.

    (b) How long does it take for a complete movement of the arm?

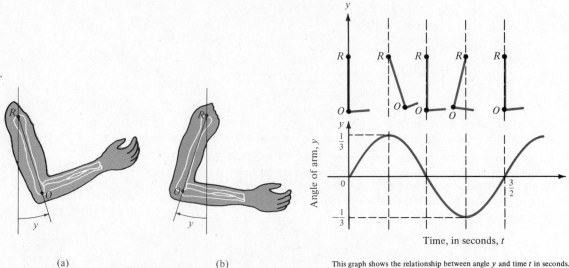

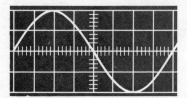

      (a)   (b)

This graph shows the relationship between angle y and time t in seconds.

Pure sounds produce single sine waves on an oscilloscope. Find the amplitude and period of each sine wave in the following photographs. On the vertical scale, each square represents .5, and on the horizontal scale each square represents 30°.

43. 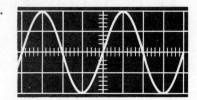   **44.**

45. The voltage E in an electrical circuit is given by

$$E = 5 \cos 120\pi t,$$

where t is time measured in seconds.

    (a) Find the amplitude and the period.

    (b) How many cycles are completed in one second? (The number of cycles completed in one second is the *frequency* of the function.)

    (c) Find E when $t = 0, .03, .06, .09, .12$.   **(d)** Graph E, for $0 \le t \le 1/30$.

---

*From *Calculus for the Life Sciences*, by Rodolfo De Sapio, W. H. Freeman and Company. Copyright © 1978 by W. H. Freeman and Company. Reprinted by permission.

46. For another electrical circuit, the voltage E is given by

$$E = 3.8 \cos 40\pi t,$$

where t is time measured in seconds.
(a) Find the amplitude and the period.
(b) Find the frequency. See Exercise 45(b).
(c) Find E when $t = .02, .04, .08, .12, .14$.
(d) Graph one period of E.

47. To find the period of $y = \sin bx$, where $b > 0$, first observe that as bx varies from 0 to 2π, we get one period of the graph of $y = \sin bx$. Show that x therefore must vary from 0 to $2\pi/b$, so that the period of $y = \sin bx$ is $2\pi/b$.

48. Sketch the graph of $y = \sin x$ for real number values of x from 0 to .2 in increments of .02. On the same axes, draw $y = x$. Use this sketch to argue that for small values of x, $\sin x \approx x$.

3.2 Graphs of the Other Trigonometric Functions

This section continues the discussion of graphs of the trigonometric functions, beginning with $y = \tan x$. As shown in the next chapter, the period of $y = \tan x$ is π, so that the tangent function need be graphed only within an interval of π units. A convenient interval for this purpose is $-\pi/2 < x < \pi/2$. Although the endpoints $-\pi/2$ and $\pi/2$ are not in the domain of tangent (why?), $\tan x$ exists for all other values in the interval.

There is no point on the graph of $y = \tan x$ at $x = -\pi/2$ or $x = \pi/2$. In the interval $0 < x < \pi/2$, however, the values of $\tan x$ are positive. As x goes from 0 to $\pi/2$, a calculator or Table 3 shows that $\tan x$ gets larger and larger without bound. As x goes from $-\pi/2$ up to 0, the values of $\tan x$ approach 0. These results, along with similar results for $y = \cot x$, are summarized in the following chart.

As x increases from	$\tan x$	$\cot x$
0 to $\pi/2$	Increases from 0, without bound	Decreases to 0
$-\pi/2$ to 0	Increases to 0	Decreases from 0 without bound

Based on these results, the graph of $y = \tan x$ will approach the vertical line $x = \pi/2$ but never touch it. The line $x = \pi/2$ is called a **vertical asymptote.** Since the period of tangent is π, the lines $x = \pi/2 + n\pi$, where n is any integer, are all vertical asymptotes. These asymptotes are indicated with a light dashed line on the graph in Figure 14 on the next page. In the interval $-\pi/2 < x < 0$, which corresponds to quadrant IV, $\tan x$ is negative, and as x goes from 0 to $-\pi/2$, the values of $\tan x$ get smaller and smaller. A table of values for $\tan x$, where $-\pi/2 < x < \pi/2$, is given below.

x	$-\pi/3$	$-\pi/4$	$-\pi/6$	0	$\pi/6$	$\pi/4$	$\pi/3$
$\tan x$	-1.7	-1	$-.6$	0	.6	1	1.7

Plotting the points from the table and letting the graph approach the asymptotes at $x = \pi/2$ and $x = -\pi/2$ gives the portion of the graph shown with a solid line in Figure 14. More of the graph can be sketched by repeating the same curve, as shown in the figure. This graph, like the graphs of the sine and cosine functions, should be memorized. Convenient key points are $(-\pi/4, -1)$, $(0, 0)$, and $(\pi/4, 1)$. The lines $x = \pi/2$ and $x = -\pi/2$ are vertical asymptotes. (The idea of *amplitude*, discussed earlier, applies only to the sine and cosine functions, and is not used here.)

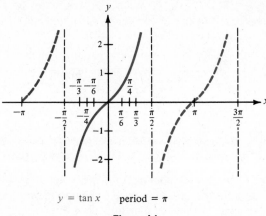

$y = \tan x$ period $= \pi$

Figure 14

Example 1   Graph $y = \tan 2x$.

Multiplying x by 2 changes the period from π to $\pi/2$. (Make a table of values to see this.) The effect on the graph is shown in Figure 15, where two periods of the function are graphed.   ■

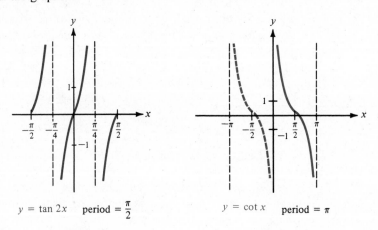

$y = \tan 2x$ period $= \dfrac{\pi}{2}$ $y = \cot x$ period $= \pi$

Figure 15   **Figure 16**

Period of Tangent

If $b > 0$, the graph of $y = \tan bx$ has period π/b.

The fact that cot $x = 1/(\tan x)$ can be used to find the graph of $y = \cot x$. The period of cotangent, like that of tangent, is π. The domain of $y = \cot x$ excludes $0 + n\pi$, where n is any integer (why?). Thus, the vertical lines $x = n\pi$ are asymptotes. The values of x that lead to asymptotes for $\tan x$ will make $\cot x = 0$, so $\cot (-\pi/2) = 0$, $\cot \pi/2 = 0$, $\cot 3\pi/2 = 0$, and so on. The values of $\tan x$ increase as x goes from $-\pi/2$ to $\pi/2$, so the values of $\cot x$ will *decrease* as x goes from $-\pi/2$ to $\pi/2$. Using these facts and plotting points as necessary gives the graph of $y = \cot x$ shown in Figure 16. (The graph shows two periods.)

The steps in graphing one period of $y = a \tan bx$, where $b > 0$, are summarized below. (The steps for graphing $y = a \cot bx$ are similar.)

Graphing Tangent

To graph $y = a \tan bx$, with $b > 0$:
1. Find the period, π/b.
2. Start at 0 on the x-axis and lay off two intervals, each with length half the period. One interval goes to the left and the other goes to the right of 0.
3. Draw the asymptotes as vertical dashed lines at the endpoints of the intervals of Step 2.
4. Locate a point at $(0, 0)$.
5. Sketch the graph, finding additional points as needed.
6. Draw additional periods, both to the right and to the left, as needed.

Example 2   Graph $y = -3 \tan \dfrac{1}{2} x$.

The period is $\pi/(1/2) = 2\pi$. Proceed as shown in Figure 17.   ∎

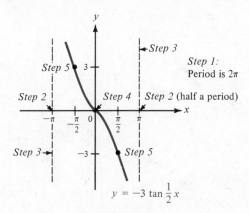

Figure 17

The graph of $y = \csc x$ is restricted to values of $x \neq n\pi$, where n is any integer (why?). This means that the lines $x = n\pi$ are asymptotes. Since $\csc x = 1/(\sin x)$, the period of $\csc x$ is 2π, the same as for $\sin x$. When $\sin x = 1$, the value of $\csc x$ is also 1, and when $0 < \sin x < 1$, then $\csc x > 1$. Also, if $-1 < \sin x < 0$, then csc $x < -1$. Using this information and plotting a few points shows that the graph takes the shape of the solid curve in Figure 18. To illustrate how the two graphs are related, the graph of $y = \sin x$ is also shown, as a dashed curve.

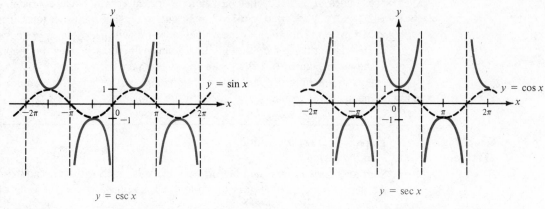

$y = \csc x$

Figure 18

$y = \sec x$

Figure 19

The graph of $y = \sec x$, shown in Figure 19, is related to the cosine graph in the same way that the graph of $y = \csc x$ is related to the sine graph.

Example 3   Graph $y = \dfrac{1}{2} \sec x$.

The amplitude of the corresponding cosine graph is 1/2 and the period is 2π. The graph of $y = (1/2) \sec x$ is shown in Figure 20.   ∎

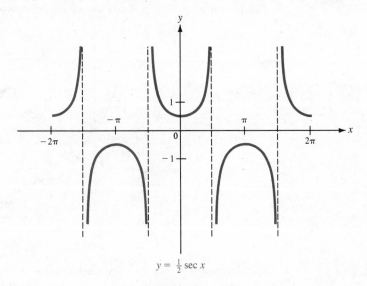

$y = \frac{1}{2} \sec x$

Figure 20

Example 4   Graph $y = 2 + \tan x$.

Every value of y will be 2 units more than the corresponding value of y in $y = \tan x$, causing the graph of $y = 2 + \tan x$ to be translated 2 units vertically compared with the graph of $y = \tan x$. See Figure 21.   ■

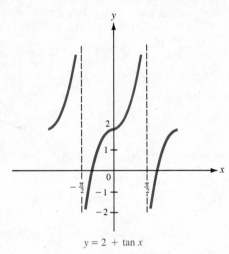

$y = 2 + \tan x$

Figure 21

3.2 Exercises

Graph the following over the interval $-2\pi \le x \le 2\pi$.

1. $y = 2 \tan x$
2. $y = -\tan x$
3. $y = -\cot x$
4. $y = \frac{1}{2} \cot x$

5. $y = 1 + \tan x$
6. $y = -2 + \tan x$
7. $y = -1 + 2 \tan x$
8. $y = 3 + \frac{1}{2} \tan x$

9. $y = 1 - \cot x$
10. $y = -2 - \cot x$
11. $y = -2 \csc x$
12. $y = -\frac{1}{2} \csc x$

13. $y = -\sec x$
14. $y = -3 \sec x$
15. $y = -2 - \csc x$
16. $y = 1 - \sec x$

Graph the following over a two-period interval. Identify the period.

17. $y = \tan 2x$
18. $y = 2 \tan \frac{1}{4} x$
19. $y = \cot 3x$
20. $y = -\cot \frac{1}{2} x$

21. $y = \csc 4x$
22. $y = \csc \frac{1}{4} x$
23. $y = \sec \frac{1}{2} x$
24. $y = -\sec 4x$

25. $y = 2 \csc \frac{1}{2} x$
26. $y = -2 \sec \frac{1}{4} x$

27. A rotating beacon is located at point A next to a long wall. (See the figure.) The beacon is 4 m from the wall. The distance d is given by

$$d = 4 \tan 2\pi t,$$

where t is time measured in seconds since the beacon started rotating. (When $t = 0$, the beacon is aimed at point R. When the beacon is aimed to the right of R, the value of d is positive; d is negative if the beacon is aimed to the left of R.) Find d for the following times.

(a) $t = 0$ (b) $t = .4$

(c) $t = .8$ (d) $t = 1.2$

(e) Why is .25 sec a meaningless value for t?

(f) What is a meaningful domain for t?

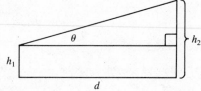

28. In the figure for Exercise 27, the distance a is given by

$$a = 4|\sec 2\pi t|.$$

Find a for the following times.

(a) $t = 0$ (b) $t = .86$ (c) $t = 1.24$

(d) Why are the absolute value bars needed here, but not in the function giving d?

29. Let a person h_1 ft tall stand d ft from an object h_2 ft tall, where $h_2 > h_1$. Let θ be the angle of elevation to the top of the object. (See the figure.)

(a) Show that $d = (h_2 - h_1) \cot \theta$.

(b) Let $h_2 = 55$ and $h_1 = 5$. Graph d for $0° < \theta < 90°$.

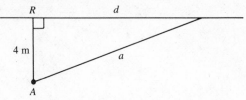

30. The quotient $y = \dfrac{\sin x}{\cos x}$ does not exist if $x = -\pi/2$ or if $x = \pi/2$. Start at $x = -1.4$ and evaluate the quotient with x increasing by .2 until $x = 1.4$ is reached. Plot the values obtained. What graph is suggested?

3.3 Horizontal Translations: Phase Shift

The previous sections showed the role played by the real numbers a, b, and c in the graph of a trigonometric function defined by $y = c + a \sin bx$. The value of a, in the case of sine and cosine, determines the maximum distance of the graph from the line $y = c$. (Recall: The amplitude of the graph is $|a|$.) The number b affects the period, so that if the usual period is p, the new period becomes p/b. (Recall: We assume $b > 0$.) Also, for $y = c + a \sin bx$, the graph is translated c units up from the x-axis if $c > 0$ and translated $|c|$ units down if $c < 0$. Now this section shows how the graph of $y = \sin(x - d)$ compares with that of $y = \sin x$. In the equation $y = \sin(x - d)$, the expression $x - d$ is called the **argument**.

Example 1 Graph $y = \sin\left(x - \dfrac{\pi}{3}\right)$.

Plotting points such as those given in the table of values below gives the graph in Figure 22. ■

x	0	$\pi/3$	π	$4\pi/3$	$11\pi/6$
$x - \pi/3$	$-\pi/3$	0	$2\pi/3$	π	$3\pi/2$
$\sin(x - \pi/3)$	$-.9$	0	$.9$	0	-1

The final result is a sine wave translated $\pi/3$ units to the *right*. This graph is said to have a *horizontal translation*, or **phase shift**, of $\pi/3$ to the right.

Phase Shift

> The graph of $y = \sin(x - d)$ has the shape of the basic graph $y = \sin x$, but with a translation of $|d|$ units—to the left if $d < 0$ and to the right if $d > 0$. The number d is the **phase shift** of the graph.
>
> Also, the graph of $y = \cos(x - d)$ has the shape of $y = \cos x$ but is translated $|d|$ units—to the left if $d < 0$ and to the right if $d > 0$. Again d is the phase shift.

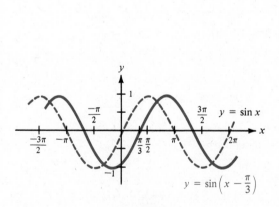

$$y = \sin\left(x - \frac{\pi}{3}\right)$$

Figure 22

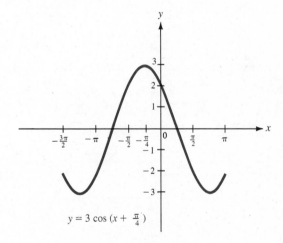

$$y = 3\cos\left(x + \frac{\pi}{4}\right)$$

Figure 23

Example 2 Graph $y = 3\cos\left(x + \dfrac{\pi}{4}\right)$.

Start by writing $3\cos(x + \pi/4)$ in the form $a\cos(x - d)$.

$$3\cos\left(x + \frac{\pi}{4}\right) = 3\cos\left[x - \left(-\frac{\pi}{4}\right)\right]$$

This result shows that $d = -\pi/4$. Since $-\pi/4$ is negative, the phase shift is $|-\pi/4| = \pi/4$ to the left. The graph is shown in Figure 23. ■

The next example shows the graph of a function defined by an expression of the form $y = a \cos b(x - d)$. The graphs of such functions have both a phase shift (if $d \neq 0$) and a period different from 2π (if $b \neq 1$).

Example 3   Graph $y = -2 \cos (3x + \pi)$.

The amplitude is $|-2| = 2$, and the period is $2\pi/3$. Find the phase shift by writing $-2 \cos (3x + \pi)$ in the form $a \cos b(x - d)$.

$$-2 \cos (3x + \pi) = -2 \cos 3\left(x + \frac{\pi}{3}\right) = -2 \cos 3\left[x - \left(-\frac{\pi}{3}\right)\right]$$

Since $d = -\pi/3$, the phase shift is $|-\pi/3| = \pi/3$ to the left. See the graph in Figure 24.   ■

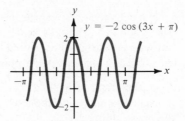

$$y = -2 \cos (3x + \pi)$$

Figure 24

A summary of sine and cosine graphs is given below. Assume $b > 0$.

Sine and Cosine Graphs		
Equation	$y = c + a \sin b(x - d)$	or
	$y = c + a \cos b(x - d)$	
Amplitude	$|a|$	
Period	$\dfrac{2\pi}{b}$	
Vertical translation	up c units if $c > 0$	
	down $|c|$ units if $c < 0$	
Phase shift (horizontal translation)	d units to the right if $d > 0$	
	$|d|$ units to the left if $d < 0$	
	(To find d, set the argument equal to 0.)	

The next example shows how phase shift applies to a tangent graph.

Example 4   Graph $y = \tan\left(2x + \frac{\pi}{2}\right)$.

Find the phase shift just as with sine and cosine: write the argument as $b(x - d)$.

$$\tan\left(2x + \frac{\pi}{2}\right) = \tan 2\left(x + \frac{\pi}{4}\right) = \tan 2\left[x - \left(-\frac{\pi}{4}\right)\right]$$

Since $d = -\pi/4$, which is negative, the phase shift is $|-\pi/4| = \pi/4$ to the left. The period is $\pi/2$, so one complete cycle of the graph is found between points $\pi/2$ units apart. The point at the origin is translated to $-\pi/4$, with other x-intercepts every $\pi/2$ units. As shown in Figure 25, the asymptotes are the vertical lines $x = 0 + \pi n/2$, where n is any integer. ■

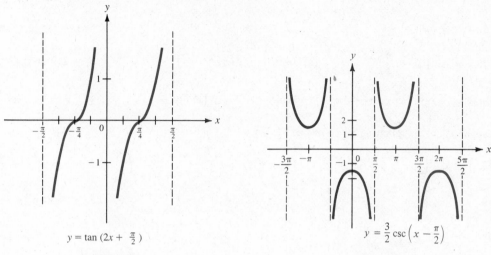

$y = \tan\left(2x + \frac{\pi}{2}\right)$

Figure 25

$y = \frac{3}{2}\csc\left(x - \frac{\pi}{2}\right)$

Figure 26

Example 5 Graph two periods of $y = \dfrac{3}{2}\csc\left(x - \dfrac{\pi}{2}\right)$.

Compared with $y = \csc x$, this graph has a phase shift of $\pi/2$ units to the right. Also, there are no values of y between $3/2$ and $-3/2$; note how this relates to the increased amplitude of $y = (3/2)\sin x$ as compared with $y = \sin x$. (Amplitude does not apply to the secant or cosecant functions; it enters only indirectly from the corresponding cosine or sine graphs.) See Figure 26.   ■

Example 6 One example of a phase shift occurs in work with electrical circuits. A simple alternating current circuit is shown in Figure 27. The relationship between voltage V and current I in the circuit is also shown in the figure.

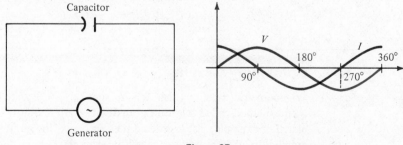

Capacitor

Generator

Figure 27

As this graph shows, current and voltage are *out of phase* by 90°. In this example, current *leads* the voltage by 90°, or voltage *lags* by 90°.   ■

3.3 Exercises

For each of the following, find the amplitude (if applicable), the period, any vertical translation, and any phase shift.

1. $y = 2 \sin x$

2. $y = \dfrac{2}{3} \sin 5x$

3. $y = 4 \cos \dfrac{x}{2}$

4. $y = -\cos \dfrac{2}{3} x$

5. $y = \tan (2x - \pi)$

6. $y = \tan \left(\dfrac{x}{2} + \pi \right)$

7. $y = \cot \left(3x + \dfrac{\pi}{4} \right)$

8. $y = \cot \left(2x - \dfrac{3\pi}{2} \right)$

9. $y = 2 - \sin \left(3x - \dfrac{\pi}{5} \right)$

10. $y = -1 + \dfrac{1}{2} \cos (2x - 3\pi)$

11. $y = -2 + 3 \tan (4x + \pi)$

12. $y = \dfrac{3}{2} - 2 \cot \left(\dfrac{1}{2} x - \pi \right)$

Graph each of the following over a one-period interval.

13. $y = \cos \left(x - \dfrac{\pi}{2} \right)$

14. $y = \sin \left(x + \dfrac{\pi}{4} \right)$

15. $y = \sin \left(x - \dfrac{\pi}{4} \right)$

16. $y = \cos \left(x + \dfrac{\pi}{3} \right)$

17. $y = 2 \cos \left(x - \dfrac{\pi}{3} \right)$

18. $y = 3 \sin \left(x + \dfrac{3\pi}{2} \right)$

19. $y = \dfrac{3}{2} \sin 2 \left(x - \dfrac{\pi}{4} \right)$

20. $y = -\dfrac{1}{2} \cos 4 \left(x + \dfrac{\pi}{2} \right)$

21. $y = -4 \sin (2x - \pi)$

22. $y = 3 \cos (4x + \pi)$

23. $y = \dfrac{1}{2} \cos \left(\dfrac{1}{2} x - \dfrac{\pi}{4} \right)$

24. $y = -\dfrac{1}{4} \sin \left(\dfrac{3}{4} x + \dfrac{\pi}{8} \right)$

25. $y = -3 + 2 \sin \left(x - \dfrac{\pi}{2} \right)$

26. $y = 4 - 3 \cos (x + \pi)$

27. $y = \dfrac{1}{2} + \sin 2 \left(x + \dfrac{\pi}{4} \right)$

28. $y = -\dfrac{5}{2} + \cos 3 \left(x - \dfrac{\pi}{6} \right)$

29. $y = \tan \left(x - \dfrac{\pi}{4} \right)$

30. $y = \cot \left(x + \dfrac{3\pi}{4} \right)$

31. $y = \sec \left(x + \dfrac{\pi}{4} \right)$

32. $y = \csc \left(x + \dfrac{\pi}{3} \right)$

33. $y = \tan \left(\dfrac{1}{2} x + \dfrac{\pi}{3} \right)$

34. $y = \csc \left(\dfrac{1}{2} x - \dfrac{\pi}{4} \right)$

35. $y = 2 + 3 \sec (2x - \pi)$

36. $y = 1 - 2 \cot \left(x + \dfrac{\pi}{2} \right)$

37. $y = 1 - \dfrac{1}{2} \csc \left(x - \dfrac{3\pi}{4} \right)$

38. $y = 2 + \dfrac{1}{4} \sec \left(\dfrac{1}{2} x + \pi \right)$

39. $y = \dfrac{2}{3} \tan \left(\dfrac{3}{4} x - \pi \right) - 2$

3.4   Combining Graphs

Trigonometric functions that are the sum of two or more functions can be graphed with a method called **addition of ordinates.** An **ordinate** is the y-value of an ordered pair. For example, in the ordered pair $(\pi, -1)$, the number -1 is the ordinate. This graphing method is best described by examples.

Example 1   Graph $y = x + \sin x$.

Begin by graphing $y = x$ and $y = \sin x$ separately on the same coordinate axes. Figure 28 shows the two graphs. Then select some x-values, and for these values add the two corresponding ordinates to get the ordinate of the sum, $x + \sin x$. For example, when $x = 0$, both ordinates are 0, so that $P_1 = (0, 0)$ is a point on the graph of $y = x + \sin x$. When $x = \pi/2$, the ordinates are $\pi/2$ and 1. Their sum is $\pi/2 + 1$, which is approximately 2.6, and $P_2 = (\pi/2, \pi/2 + 1)$, or approximately $(1.6, 2.6)$, is on the graph. At $x = 3\pi/2$, the sum $x + \sin x$ is $3\pi/2 + (-1)$, or approximately 3.7, with $P_3 = (3\pi/2, 3\pi/2 - 1)$, or approximately $(4.7, 3.7)$, on the graph. As many points as necessary can be located in this way. The graph is then completed by drawing a smooth curve through the points. The graph of $y = x + \sin x$ is shown in color in Figure 28.   ■

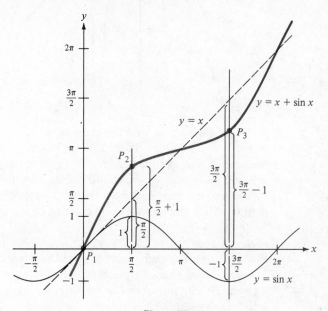

Figure 28

As shown in the graph in Figure 28, to get the graph of $y = x + \sin x$ the ordinates are actually treated as line segments. For example, the ordinate of P_2 is found by adding the lengths of the two line segments that represent the ordinates of x and $\sin x$ at $\pi/2$. The same is true for the ordinate of P_3 as well as for each of the other ordinates.

Example 2   Graph $y = \cos x - \tan x$.

Think of $y = \cos x - \tan x$ as $y = \cos x + (-\tan x)$. Start by graphing $y = \cos x$ and $y = -\tan x$ on the same axes, as in Figure 29. At $x = 0$, the ordinates are 1 and 0, so the ordinate of $y = \cos x + (-\tan x)$ is $1 + 0 = 1$ and the point $(0, 1)$ is on the graph. At any point where the graphs of $\cos x$ and $-\tan x$ cross, the ordinate is doubled. See, for example, P_1 and P_2 on the graph in Figure 29 on the next page.

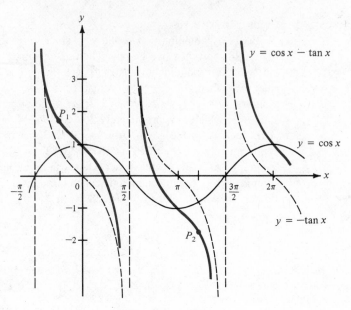

Figure 29

There are no points on the graph of $y = \cos x - \tan x$ when $x = \pi/2 + n\pi$, where n is an integer, because these numbers are not in the domain of tangent. Because of this, the lines $x = \pi/2 + n\pi$ are asymptotes, so that as x approaches $\pi/2$ and $3\pi/2$ from the right, the values of y get larger and larger. Also, when x approaches $\pi/2$ and $3\pi/2$ from the left, the values of y get smaller and smaller. A portion of this graph is shown in color in Figure 29. ∎

Figure A shows three sine waves, of amplitudes 1, $^1/_3$, and $^1/_5$, respectively. Figure B shows the wave that results from adding together the three waves in Figure A. Finally, Figure C shows the approximation to a square wave produced by adding together 19 sine waves similar to those in Figure A—sine waves with amplitudes 1, $^1/_3$, $^1/_5$, $^1/_7$, . . . , $^1/_{37}$.*

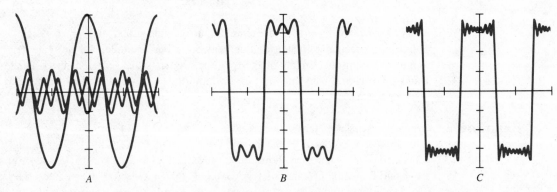

*From *The Science of Musical Sound,* by John R. Pierce. W. H. Freeman and Company. Copyright © 1983 by Scientific American Books. Reprinted by permission.

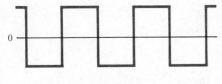

Square wave

A branch of advanced mathematics called Fourier analysis *shows how functions can be expressed as the sum of infinitely many sine waves. For example, the figure at the side shows a square wave, an idealized representation of a repetitive sound such as striking a piano key over and over.*

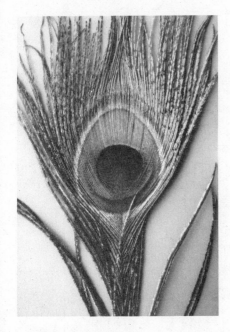

A biological application of the addition of ordinates of sine curves with equal frequency and amplitude is found in the colors of a peacock's feathers. Pure light waves (which are sine waves) are combined by interference to produce colors. There are no pigments in the feathers, which look clear under a microscope.

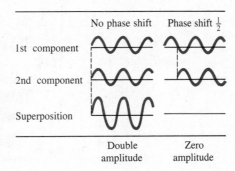

3.4 Exercises

Use the method of addition of ordinates to graph each of the following.

1. $y = x + \cos x$

2. $y = \sin x - 2x$

3. $y = 3x - \cos 2x$

4. $y = x + 2 - \sin x$

5. $y = \sin x + \sin 2x$

6. $y = \cos x - \cos \frac{1}{2} x$

7. $y = \sin x + \tan x$

8. $y = \sin x + \csc x$

9. $y = 2 \cos x - \sec x$

10. $y = 2 \sec x + \sin x$

11. $y = \cos x + \cot x$

12. $y = \sin x - 2 \cos x$

13. $y = -x + \sec x$

14. $y = x + \csc x$

Use methods similar to those in this section to graph the following.

15. $y = x \sin x$

16. $y = x \cos x$

17. $y = 2^{-x} \sin x$

18. $y = x^2 \sin x$

19. The function defined by $y = (6 \cos x)(\cos 8x)$ can occur in AM radio transmissions, as mentioned in Section 3.1. Graph this function on an interval from 0 to 2π. (*Hint:* First graph $y = 6 \cos x$ and $y = \cos 8x$ as dashed lines.)

20. Graph $y = \sin 8x + \sin 9x$. The period of this graph is very long. By placing two engines that are running at almost the same speed (as suggested by the $8x$ and $9x$) side by side, we get an effect of "beats." See the figure.

The top two sine waves represent pure tones, such as those put out by a tuning fork or an electronic oscillator. When two such pure tones, having slightly different periods, are played side by side, the amplitudes add algebraically, instant by instant, producing a result such as that shown in the bottom graph. The peaks are called beats. *Beats result, for example, from engines on an airplane that are running at almost, but not quite, the same speeds. Blowers in different apartments also can cause such beats; these can be quite annoying.*

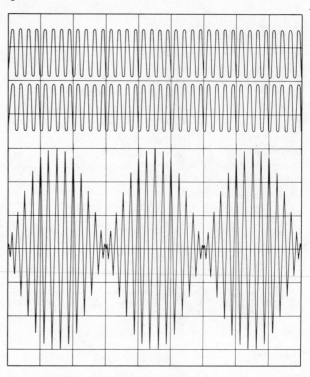

21. If a bumper of a car is given a firm downward push and then released, the shock absorbers of the car cause it to bounce back and then quickly return to stable position, producing an example of *damped oscillatory motion*. Such motion often can be represented by a function defined by

$$y = a \cdot e^{-kt} \cdot \sin b(t + c),$$

where a, k, b, and c are constants, with $k > 0$, t represents time, and e is the base of natural logarithms. (To six decimal places, $e = 2.718282$.) Graphs of $y = e^{-t}$ and $y = -e^{-t}$ for $t > 0$ are shown in the figure on the following page. Use these graphs to obtain the graph of $y = e^{-t} \cdot \sin t$.

22. Motion that gets out of control can be represented by a function defined by

$$y = a \cdot e^{kt} \cdot \sin b(t + c),$$

with variables as given in Exercise 21. Use the graphs of $y = e^t$ and $y = -e^t$ in the figure on the next page to graph $y = e^t \cdot \sin t$.

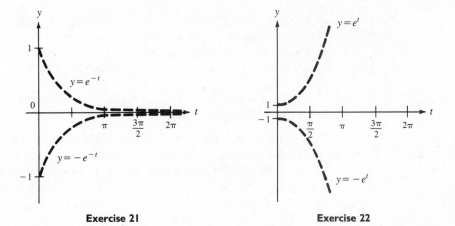

Exercise 21   Exercise 22

3.5 Simple Harmonic Motion (Optional)*

Case I: *Point P(x, y) is considered to move around the unit circle counterclockwise at a uniform speed.* This is the simplest case of harmonic motion, as illustrated in Figure 30. Let k be its **linear speed** (number of units P moves along the circle per unit of time). If we suppose that the point P is at $(1, 0)$ when $t = 0$, then the arc length u is given by the product of the rate and the time, or $u = kt$ after t units of time. (Recall that for a circle of radius 1, the arc length and the angle have the same measure.) So $y = \sin u = \sin kt$. If a horizontal line through P intersects the y-axis at Q with coordinates $(0, y)$, then the formula $y = \sin kt$ describes the up-and-down motion of Q along the y-axis as a function of time t. This oscillatory motion is called **simple harmonic motion.**

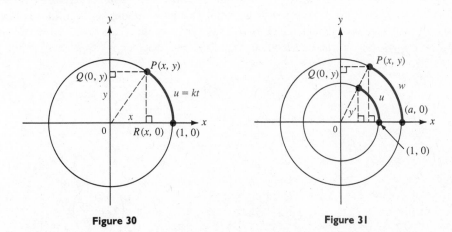

Figure 30   Figure 31

*From Floyd F. Helton and Margaret L. Lial, *Precalculus Mathematics: A Functions Approach* (Glenview, Ill.: Scott, Foresman and Company, 1983), pp. 211–15.

The amplitude of the motion is 1 and its period is $2\pi/k$. The moving points P and Q complete one cycle per period. The number of cycles per unit of time (called the **frequency**) is the reciprocal of the period, $k/(2\pi)$. The back-and-forth motion of a point R along the x-axis is another instance of simple harmonic motion, with equation $x = \cos kt$.

The basic notion is extended in Cases II and III below, to give a more general view of simple harmonic motion.

Case II: *The radius of the circle is allowed to be any number.* As in Figure 31, suppose that the outer circle has radius a and that P is at $(a, 0)$ when $t = 0$ and is moving counterclockwise at a uniform speed of k units per unit of time. After t units of time the arc length w is given by $w = kt$. Using geometric similarities in the figure,

$$\frac{w}{u} = \frac{a}{1} \quad \text{and} \quad \frac{y}{a} = \frac{y'}{1},$$

$$w = au \quad \text{and} \quad y = ay'.$$

From $y = ay'$,

$$y = a \sin u$$

$$= a \sin \frac{w}{a} \qquad \text{Since } w = au$$

$$= a \sin \frac{kt}{a} \qquad \text{Since } w = kt$$

$$= a \sin \left(\frac{k}{a}\right)t.$$

Let $b = k/a$ and write this as $y = a \sin bt$.

Case III: *The initial position of the point is anywhere on the circle.* Suppose that at time $t = 0$ the point P is at a distance w_0 from $(a, 0)$, measured along the circle counterclockwise. Then t units of time later it will have moved a distance kt from w_0 and so is along the circle at a distance $w = kt + w_0$ from $(a, 0)$. See Figure 32. Now

$$y = a \sin \frac{w}{a} \qquad \text{As in Case II}$$

$$= a \sin \frac{kt + w_0}{a}$$

$$= a \sin \left(\frac{kt}{a} + \frac{w_0}{a}\right)$$

$$= a \sin \frac{k}{a}\left(t + \frac{w_0}{k}\right).$$

The final equation may be written as $y = a \sin b(t + c)$, where $b = k/a$ and $c = w_0/k$. Note that $c = $ distance w_0 per speed $k = $ time for the particle to go from $(a, 0)$ to w_0 along the circle.

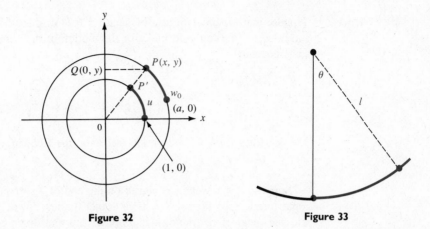

Figure 32   **Figure 33**

The three cases just described are represented by the following three formulas:

Simple Harmonic Motion	**Case I:** $y = \sin kt$, for motion on the unit circle, starting at the point (1, 0) and moving k units per unit of time

Case I: $y = \sin kt$, for motion on the unit circle, starting at the point (1, 0) and moving k units per unit of time

Case II: $y = a \sin bt$, where $b = k/a$, for motion on a circle of radius a, starting at the point $(a, 0)$ and moving k units per unit of time

Case III: $y = a \sin b(t + c)$, where $b = k/a$ and $c = w_0/k$, for motion on a circle of radius a, starting w_0 units from $(a, 0)$ and moving k units per unit of time

Example I   Write the equation of motion for a point moving at a linear speed of 5 around each of the following circles, and state the amplitude, period, and frequency.

(a) The unit circle, starting at the point (1, 0)

In this case, $y = \sin kt = \sin 5t$. The amplitude is 1, the period is $2\pi/5$, and the frequency is $5/(2\pi)$.

(b) A circle of radius 3, starting at the point (3, 0)

Here, the equation is $y = a \sin bt = 3 \sin (5/3)t$. The amplitude is 3, the period is $2\pi/(5/3) = 6\pi/5$, and the frequency is $5/(6\pi)$.   ■

The sine-wave $y = a \sin b(t + c)$ defines the general case of simple harmonic motion. Its graph is a sine wave of amplitude a (the radius of the circle), and its period is $2\pi/b$, or $2\pi a/k$ (the time for P to go once around the circle).

From the calculus it can be shown that for a pendulum, as shown in Figure 33, the angle θ is given by $\theta(t) = a \sin (\sqrt{g/l})t$, where a is the maximum angle of displacement of the pendulum arm from the vertical, g is the constant of gravitation (≈ 32), and l is the length of the pendulum arm. (Here, l is measured in feet and the angle θ in radians, and t is time in seconds.) This formula is actually an approximation for small values of θ. The period of motion is $2\pi/\sqrt{32/l}$, or $\pi\sqrt{2l}/4$.

Example 2   Suppose that a pendulum arm of length 2 ft is displaced 2 radians from the vertical and released. Write the formula for the motion of the pendulum and state the period and frequency.

In this case, $a = 2$ and $l = 2$.

$$\theta(t) = a \sin\left(\sqrt{\frac{g}{l}}\right)t = 2 \sin\left(\sqrt{\frac{32}{2}}\right)t = 2 \sin 4t$$

The period is $2\pi/4 = \pi/2 \approx 1.6$ (sec) and the frequency is $2/\pi \approx 0.6$ (cycles per sec).   ■

Suppose that a weight is placed on the end of a suspended spring and allowed to come to rest (see Figure 34). If it is then pulled down, stretching the spring, and released (neglecting friction), it oscillates up and down periodically in simple harmonic motion. The equation for this motion is $s(t) = a \sin (k/m)t$, where a is the maximum vertical displacement from the position of rest, k is the spring constant, and m is the mass of the weight. Here the spring constant k is determined by the fact (Hooke's law) that the force f required to stretch the spring a distance s is given by $f = ks$.

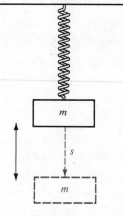

Figure 34

Mathematically, the formula for the motion of a weight on a spring is identical to that for the simple pendulum. Only the interpretations of the variables and constants are different. This is an example of the sort of situation that occurs quite frequently in mathematics.

3.5 Exercises

Solve the following problems.

1. Write the equation and then find the amplitude, period, and frequency of the simple harmonic motion determined by the uniform circular motion of a particle around a unit circle with the following linear speed.

   (a) 2 units per sec   (b) 3 units per sec   (c) 4 units per sec

2. Repeat Exercise 1 for a circle of radius 2 units.

3. Suppose that in Exercise 2 the point P is initially (when $t = 0$) at the point $(1, 3)$ and has a linear speed of 2 units per unit of time. Write the equation of the corresponding simple harmonic motion and determine its amplitude, period, and frequency.

4. What are the period and frequency of oscillation of a pendulum of length 1/2 ft?

5. How long should a pendulum be in order to have a period of one sec?

6. Suppose that a 4-lb force is required to stretch a spring 2 ft. (Since force $= ks$, $4 = k(2)$, so the spring constant is $k = 2$.) Let a mass of one unit be placed on the spring and allowed to come to rest. If the spring is then stretched 1/2 ft and released, what are the amplitude, period, and frequency of the resulting oscillatory motion?

7. The formula for the up-and-down motion of a weight on a spring is given by $s(t) = a \sin (k/m)t$.
(a) Write the formulas for the period and frequency of the motion.
(b) If the spring constant is $k = 4$, what mass m must be used to produce a period of 1 sec?

3.6 Inverse Trigonometric Functions

The definition of the inverse of a one-to-one function was given in Chapter 1. This section shows how to find the inverse of a trigonometric function. In looking for the inverse of $f(x) = \sin x$, recall that a function must be one-to-one before it can have an inverse. The graph of sine in Figure 35(a) shows that $f(x) = \sin x$ does not define a one-to-one function since different values of x can lead to the *same* value of y. However, by suitably restricting the domain of the sine function, a one-to-one function can be defined. It is customary to restrict the domain of $f(x) = \sin x$ to $-\pi/2 \le x \le \pi/2$, which is the solid portion of the graph in Figure 35(a). Reflecting that portion of the sine graph about the 45° line $y = x$ gives the graph of the inverse function, shown in Figure 35(b).

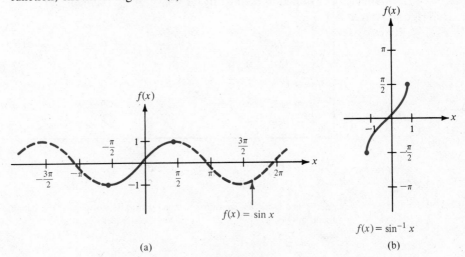

(a)

(b)

Figure 35

The equation of the inverse of $y = \sin x$ is found by first exchanging x and y to get $x = \sin y$. This equation then is solved for y by writing $y = \sin^{-1} x$. (Note that $\sin^{-1} x$ does not mean $1/\sin x$.) As suggested by Figure 35(b), the domain of $y = \sin^{-1} x$ is $-1 \le x \le 1$, while the restricted domain of $y = \sin x$ is the range of $y = \sin^{-1} x$, $-\pi/2 \le y \le \pi/2$. An alternative notation for $\sin^{-1} x$ is **arcsin x.** The name *arcsine* comes from the fact that if $y = \arcsin x$, then y is the arc length on the unit circle that corresponds to the number x. By definition,

Inverse Sine Function

$$y = \sin^{-1} x \quad \text{or} \quad y = \arcsin x \quad \text{if and only if} \quad x = \sin y,$$

where $-1 \le x \le 1$ and $-\pi/2 \le y \le \pi/2$.

Example 1   Find $\sin^{-1} 1/2$.

Let $y = \sin^{-1} 1/2$. Then by the definition of the inverse sine function, $\sin y = 1/2$. Since $\sin \pi/6 = 1/2$ and $\pi/6$ is in the range of $y = \sin^{-1} x$,

$$\sin^{-1} 1/2 = \pi/6. \quad \blacksquare$$

By the definition of the inverse sine function,

$$\sin (\sin^{-1} x) = x \quad \text{and} \quad \sin^{-1} (\sin y) = y$$

only for values of x satisfying $-1 \le x \le 1$ and y satisfying $-\pi/2 \le y \le \pi/2$. For example, since $7\pi/6$ is not in the interval $[-\pi/2, \pi/2]$, $\sin^{-1} (\sin 7\pi/6) \ne 7\pi/6$. (Show that $\sin^{-1} (\sin 7\pi/6) = -\pi/6$.)

For each of the other trigonometric functions, an inverse function can be defined by a suitable restriction on the domain, just as with sine. The *inverse trigonometric functions* and their ranges follow.

Inverse Trigonometric Functions

Function defined by	Range
$y = \sin^{-1} x$	$-\pi/2 \le y \le \pi/2$
$y = \cos^{-1} x$	$0 \le y \le \pi$
$y = \tan^{-1} x$	$-\pi/2 < y < \pi/2$
$y = \cot^{-1} x$	$0 < y < \pi$
$y = \sec^{-1} x$	$0 \le y < \pi/2$ or $\pi \le y < 3\pi/2$*
$y = \csc^{-1} x$	$0 < y \le \pi/2$ or $\pi < y \le 3\pi/2$*

The graphs of $y = \cos^{-1} x$ and $y = \tan^{-1} x$ are shown in Figures 36 and 37.

*Sec^{-1} and csc^{-1} are sometimes defined with different ranges. The definitions given here are the most useful in calculus.

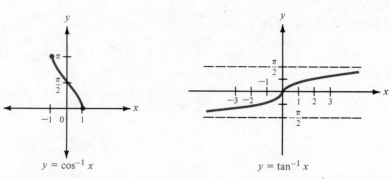

$y = \cos^{-1} x$
Figure 36

$y = \tan^{-1} x$

Figure 37

Example 2 Find $\cos^{-1}(-\sqrt{2}/2)$.

Because of the restriction $0 \le \cos^{-1} x \le \pi$, the angle represented by $\cos^{-1} x$ can terminate only in quadrants I and II. Since $-\sqrt{2}/2$ is negative, the angle is restricted to quadrant II. Let $y = \cos^{-1}(-\sqrt{2}/2)$. Then $\cos y = -\sqrt{2}/2$. In quadrant II, $\cos 3\pi/4 = -\sqrt{2}/2$, so

$$\cos^{-1}\left(-\frac{\sqrt{2}}{2}\right) = \frac{3\pi}{4}. \quad \blacksquare$$

Example 3 Find $\sin(\tan^{-1}(-3/2))$ without using tables or a calculator.

Let $y = \tan^{-1}(-3/2)$, so that $\tan y = -3/2$. Since the inverse tangent function is defined only in quadrants I and IV ($-\pi/2 < \tan^{-1} x < \pi/2$), and since $x = -3/2$, work in quadrant IV. Sketch y in quadrant IV and label a triangle as shown in Figure 38. The hypotenuse has a length of $\sqrt{13}$, so that $\sin y = -3/\sqrt{13}$, or

$$\sin\left(\tan^{-1}\left(-\frac{3}{2}\right)\right) = \frac{-3}{\sqrt{13}} = \frac{-3\sqrt{13}}{13}. \quad \blacksquare$$

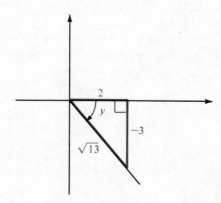

Figure 38

Example 4 Solve the equation $y = 2 \sin^{-1}(x + 4)$ for x.
First multiply by 1/2 to get

$$\frac{y}{2} = \sin^{-1}(x + 4).$$

Use the definition of the inverse sine function to get

$$\sin \frac{y}{2} = x + 4 \qquad \text{or} \qquad x = \left(\sin \frac{y}{2}\right) - 4. \quad \blacksquare$$

Example 5 Write $\sec(\tan^{-1} x)$ as an algebraic expression.
Let $\tan^{-1} x = u$. Then $\tan u = x$. Since $-\pi/2 < \tan^{-1} x < \pi/2$, u is in quadrant I or quadrant IV. As shown in Figure 39, in either case the value of secant is positive, so

$$\sec(\tan^{-1} x) = \sec u$$
$$= \sqrt{x^2 + 1}. \quad \blacksquare$$

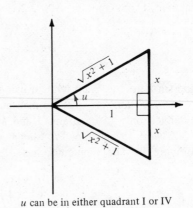

u can be in either quadrant I or IV

Figure 39

Alternate Notations for Inverse Trigonometric Functions

As mentioned earlier, there is an alternate notation for inverse trigonometric functions used in some books:

$$y = \sin^{-1} x \text{ is written } y = \arcsin x,$$
$$y = \cos^{-1} x \text{ is written } y = \arccos x,$$
$$y = \tan^{-1} x \text{ is written } y = \arctan x.$$
$$y = \cot^{-1} x \text{ is written } y = \text{arccot } x,$$
$$y = \sec^{-1} x \text{ is written } y = \text{arcsec } x,$$

and $\qquad\qquad y = \csc^{-1} x \text{ is written } y = \text{arccsc } x.$

In the following exercise set, these notations are used interchangeably.

3.6 Exercises

For the following, give the value of y in radians without using tables or a calculator.

1. $y = \sin^{-1}\left(-\dfrac{\sqrt{3}}{2}\right)$ **2.** $y = \cos^{-1}\dfrac{\sqrt{3}}{2}$ **3.** $y = \tan^{-1} 1$ **4.** $y = \tan^{-1}(-1)$

5. $y = \sin^{-1}(-1)$ **6.** $y = \cos^{-1}(-1)$ **7.** $y = \cos^{-1}\dfrac{1}{2}$ **8.** $y = \sin^{-1}\left(-\dfrac{\sqrt{2}}{2}\right)$

9. $y = \arccos\left(-\dfrac{\sqrt{2}}{2}\right)$ **10.** $y = \arctan\dfrac{\sqrt{3}}{3}$ **11.** $y = \arctan(-\sqrt{3})$ **12.** $y = \arccos\left(-\dfrac{1}{2}\right)$

For the following, give the value in degrees. (Round to the nearest ten minutes.)

13. $y = \sin^{-1}(-0.1334)$ **14.** $y = \cos^{-1}(-0.1334)$ **15.** $y = \cos^{-1}(-0.3987)$ **16.** $y = \sin^{-1} 0.7790$

17. $y = \cos^{-1} 0.9272$ **18.** $y = \tan^{-1} 1.767$ **19.** $y = \tan^{-1} 1.111$ **20.** $y = \sin^{-1} 0.8192$

21. $y = \arctan(-0.9217)$ **22.** $y = \arctan(-0.2867)$

Give the value of the following in radians to four significant digits.

23. $\sin^{-1} 0.7214$ **24.** $\cos^{-1} 0.3004$ **25.** $\tan^{-1}(-4.114)$ **26.** $\sin^{-1}(-0.9946)$

27. Enter 1.003 in your calculator and push the keys for inverse sine. The machine will tell you that something is wrong. What is wrong?

28. Enter 1.003 in your calculator and push the keys for inverse tangent. This time, unlike in Exercise 27, you get an answer easily. What is the difference?

Find each of the following to the nearest second.

29. $\sin^{-1}(-.443981)$ **30.** $\tan^{-1}(-.394511)$ **31.** $\cos^{-1} .91441$ **32.** $\tan^{-1} 14.76892$

Give the value of each of the following without using tables or a calculator.

33. $\tan\left(\arccos\dfrac{3}{4}\right)$ **34.** $\sin\left(\arccos\dfrac{1}{4}\right)$ **35.** $\cos(\tan^{-1}(-2))$ **36.** $\sec\left(\sin^{-1}\left(-\dfrac{1}{5}\right)\right)$

37. $\cot\left(\arcsin\left(-\dfrac{2}{3}\right)\right)$ **38.** $\cos\left(\arctan\dfrac{8}{3}\right)$ **39.** $\sec\left(\operatorname{arccot}\dfrac{3}{5}\right)$ **40.** $\cos\left(\arcsin\dfrac{12}{13}\right)$

41. $\cos\left(\arccos\dfrac{1}{2}\right)$ **42.** $\sin\left(\arcsin\dfrac{\sqrt{3}}{2}\right)$ **43.** $\tan(\tan^{-1}(-1))$ **44.** $\cot(\cot^{-1}(-\sqrt{3}))$

45. $\sec(\sec^{-1} 2)$ **46.** $\csc(\csc^{-1}\sqrt{2})$ **47.** $\arccos\left(\cos\dfrac{\pi}{4}\right)$

48. $\arctan\left(\tan\left(-\dfrac{\pi}{4}\right)\right)$ **49.** $\arcsin\left(\sin\dfrac{\pi}{3}\right)$ **50.** $\arccos(\cos 0)$

51. $\sin^{-1}(\sin \pi)$ **52.** $\tan^{-1}(\tan 2\pi/3)$ **53.** $\cos^{-1}(\cos 5\pi/4)$

54. $\sin^{-1}(\sin 3\pi/4)$ **55.** $\tan^{-1}(\tan 2\pi)$ **56.** $\cos^{-1}(\cos 5\pi/2)$

Use the inverse key of your calculator together with the trigonometric function keys to find each of the following to six decimal places.

57. $\cos(\arctan .3)$ **58.** $\sin(\arccos .75)$ **59.** $\tan(\arcsin .1225)$ **60.** $\cot(\arccos .5823)$

Solve each of the following equations for x.

61. $y = 4 \sin^{-1} x$ **62.** $3y = \cos^{-1} x$ **63.** $2y = \tan^{-1} 2x$ **64.** $y = 3 \sin^{-1} (x/2)$

65. $y = \sin^{-1} (x + 2)$ **66.** $y = \tan^{-1} (2x - 1)$

Write each of the following as an algebraic expression.

67. $\sin (\arccos u)$ **68.** $\tan (\arccos u)$ **69.** $\sec (\cot^{-1} u)$ **70.** $\csc (\sec^{-1} u)$

71. $\cot (\arcsin u)$ **72.** $\cos (\arcsin u)$ **73.** $\sin \left(\sec^{-1} \dfrac{u}{2} \right)$ **74.** $\cos \left(\tan^{-1} \dfrac{3}{u} \right)$

75. $\tan \left(\arcsin \dfrac{u}{\sqrt{u^2 + 2}} \right)$ **76.** $\cos \left(\arccos \dfrac{u}{\sqrt{u^2 + 5}} \right)$

77. $\sec \left(\operatorname{arccot} \dfrac{\sqrt{4 - u^2}}{u} \right)$ **78.** $\csc \left(\arctan \dfrac{\sqrt{9 - u^2}}{u} \right)$

Graph each of the following, and give the domain and range.

79. $y = \cot^{-1} x$ **80.** $y = \tan^{-1} 2x$ **81.** $y = \operatorname{arcsec} x$

82. $y = \operatorname{arccsc} x$ **83.** $y = 2 \cos^{-1} x$ **84.** $y = \sin^{-1} \dfrac{x}{2}$

85. Enter 1.74283 in your calculator (set for radians), and press the sine key. Then press the keys for inverse sine. You get 1.398763 instead of 1.74283. What happened?

The following were used by the mathematicians who computed the value of π to 100,000 decimal places. Use a calculator to verify that each is (approximately) correct.

86. $\pi = 16 \tan^{-1} \dfrac{1}{5} - 4 \tan^{-1} \dfrac{1}{239}$ **87.** $\pi = 24 \tan^{-1} \dfrac{1}{8} + 8 \tan^{-1} \dfrac{1}{57} + 4 \tan^{-1} \dfrac{1}{239}$

88. $\pi = 48 \tan^{-1} \dfrac{1}{18} + 32 \tan^{-1} \dfrac{1}{57} - 20 \tan^{-1} \dfrac{1}{239}$

Answer *true* or *false* for each of the following.

89. $2 \cos^{-1} x = \cos^{-1} 2x$ **90.** $\cot^{-1} x = 1/\tan^{-1} x$

91. $\tan^{-1} x = \sin^{-1} x / \cos^{-1} x$ **92.** $\sin^{-1} (-x) = -\sin^{-1} x$

93. $y = \sin^{-1} x$ defines an even function   **94.** $y = \cos^{-1} x$ defines an even function

95. $y = x \cdot \sin^{-1} x$ defines an odd function   **96.** $y = x^2 \cdot \cos^{-1} x$ defines an even function

Suppose that an airplane flying faster than sound goes directly over you. Assume that the plane is flying level. At the instant that you feel the sonic boom from the plane, the angle of elevation to the plane is given by

$$\alpha = 2 \arcsin \frac{1}{m},$$

where m is the Mach number of the plane's speed. Find α to the nearest degree for each of the following values of m:

97. $m = 1.2$ **98.** $m = 1.5$ **99.** $m = 2$ **100.** $m = 2.5$.

Chapter 3 Summary

Key Words   amplitude   addition of ordinates
vertical translation   harmonic motion
asymptote   inverse trigonometric functions
phase shift

Review Exercises

For each of the following, give the amplitude, period, vertical translation, and phase shift, as applicable.

1. $y = 2 \sin x$

2. $y = \tan 3x$

3. $y = -\dfrac{1}{2} \cos 3x$

4. $y = 2 \sin 5x$

5. $y = 1 + 2 \sin \dfrac{1}{4} x$

6. $y = 3 - \dfrac{1}{4} \cos \dfrac{2}{3} x$

7. $y = 3 \cos \left(x + \dfrac{\pi}{2} \right)$

8. $y = -\sin \left(x - \dfrac{3\pi}{4} \right)$

9. $y = \dfrac{1}{2} \csc \left(2x - \dfrac{\pi}{4} \right)$

10. $y = 2 \sec (\pi x - 2\pi)$

11. $y = \dfrac{1}{3} \cos \left(3x - \dfrac{\pi}{3} \right)$

12. $y = \cot \left(\dfrac{x}{2} + \dfrac{3\pi}{4} \right)$

Graph the following over a one-period interval.

13. $y = 3 \sin x$

14. $y = \dfrac{1}{2} \sec x$

15. $y = -\tan x$

16. $y = -2 \cos x$

17. $y = 2 + \cot x$

18. $y = -1 + \csc x$

19. $y = \sin 2x$

20. $y = \tan 3x$

21. $y = 3 \cos 2x$

22. $y = \dfrac{1}{2} \cot 3x$

23. $y = \cos \left(x - \dfrac{\pi}{4} \right)$

24. $y = \tan \left(x - \dfrac{\pi}{2} \right)$

25. $y = \sec \left(2x + \dfrac{\pi}{3} \right)$

26. $y = \sin \left(3x + \dfrac{\pi}{2} \right)$

27. $y = 1 + 2 \cos 3x$

28. $y = -1 - 3 \sin 2x$

29. $y = 2 \sin \pi x$

30. $y = -\dfrac{1}{2} \cos (\pi x - \pi)$

31. $y = 1 - 2 \sec \left(x - \dfrac{\pi}{4} \right)$

32. $y = -\csc (2x - \pi) + 1$

Graph each of the following using the method of addition of ordinates.

33. $y = \tan x - x$

34. $y = \cos x + \dfrac{1}{2} x$

35. $y = \sin x + \cos x$

36. $y = \tan x + \cot x$

37. The figure on the following page shows the population of lynx and hares in Canada for the years 1847–1903. The hares are food for the lynx. An increase in hare population causes an increase in lynx population some time later. The increasing lynx population then causes a decline in hare population.
(a) Estimate the length of one period.
(b) Estimate maximum and minimum hare population.

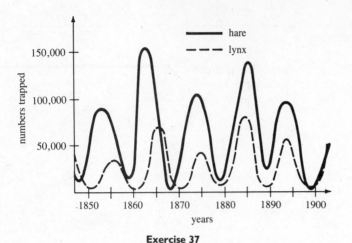

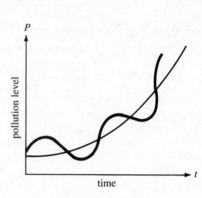

Exercise 37   Exercise 38

38. The amount of pollution in the air fluctuates with the seasons. It is lower after heavy spring rains and higher after periods of little rain. In addition to this seasonal fluctuation, the long-term trend is upward. An idealized graph of this situation is shown in the figure. Trigonometric functions can be used to describe the fluctuating part of the pollution levels. Powers of the number e (Recall: e is the base of natural logarithms; to six decimal places, $e = 2.718282$) can be used to show the long-term growth. In fact, the pollution level in a certain area might be given by

$$P(t) = 7(1 - \cos 2\pi t)(t + 10) + 100e^{.2t},$$

where t is time in years, with $t = 0$ representing January 1 of the base year. Thus, July 1 of the same year would be represented by $t = .5$, and October 1 of the following year would be represented by $t = 1.75$. Find the pollution levels on the following dates.
(a) January 1, base year   **(b)** July 1, base year
(c) January 1, following year   **(d)** July 1, following year

Find each of the following. Give answers in radians.

39. $y = \sin^{-1} \sqrt{2}/2$ **40.** $y = \cos^{-1}(-1/2)$ **41.** $y = \tan^{-1}(-\sqrt{3})$

42. $y = \arctan 1.780$ **43.** $y = \arcsin(-.6604)$ **44.** $y = \cos^{-1} .8039$

Find each of the following without using tables or a calculator.

45. $\sin\left(\sin^{-1}\dfrac{1}{2}\right)$ **46.** $\tan\left(\tan^{-1}\dfrac{2}{3}\right)$ **47.** $\cos(\arccos(-1))$

48. $\sin\left(\arcsin\left(-\dfrac{\sqrt{3}}{2}\right)\right)$ **49.** $\arccos\left(\cos\dfrac{3\pi}{4}\right)$ **50.** $\operatorname{arcsec}(\sec \pi)$

Write each of the following as an expression in u.

51. $\sin(\tan^{-1} u)$ **52.** $\cos\left(\arctan\dfrac{u}{\sqrt{1 - u^2}}\right)$ **53.** $\tan\left(\operatorname{arcsec}\dfrac{\sqrt{u^2 + 1}}{u}\right)$

Graph each of the following, and give the domain and range.

54. $y = \sin^{-1} x$ **55.** $y = \cos^{-1} x$ **56.** $y = \operatorname{arccot} x$

57. In the study of alternating current in electricity, instantaneous voltage is given by

$$e = E_{max} \sin 2\pi ft,$$

where f is the number of cycles per second, E_{max} is the maximum voltage, and t is time in seconds.

(a) Solve the equation for t.

(b) Find the smallest positive value of t in radians if $E_{max} = 12$, $e = 5$, and $f = 100$.

58. Many computer languages such as BASIC and FORTRAN have only the arctangent function available. To use the other inverse trigonometric functions, it is necessary to express them in terms of arctangent. This can be done as follows.

(a) Let $u = \arcsin x$. Solve the equation for x in terms of u.

(b) Use the result of part **(a)** to label the three sides of the triangle of the figure in terms of x.

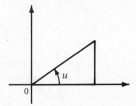

(c) Use the triangle from part **(b)** to write an equation for $\tan u$ in terms of x.

(d) Solve the equation from **(c)** for u.

(e) Use your equation from part **(d)** to calculate $\arcsin (1/2)$. Compare the answer with the actual value of $\arcsin (1/2)$.

4

Trigonometric Identities and Equations

A conditional equation, such as $2x + 1 = 9$ or $m^2 - 2m = 3$, is true for only certain values in the domain of its variable. For example, $2x + 1 = 9$ is true only for $x = 4$, and $m^2 - 2m = 3$ is true only for $m = 3$ and $m = -1$. On the other hand, an **identity** is an equation that is true for *every* value in the domain of its variable. Examples of identities include

$$5(x + 3) = 5x + 15 \qquad \text{and} \qquad (a + b)^2 = a^2 + 2ab + b^2.$$

This chapter discusses conditional equations and identities involving trigonometric functions. The domain of the variable is assumed to be all values for which a given function is defined.

4.1 Fundamental Identities

Let us begin by restating the fundamental identities introduced in Chapter 2.

Fundamental Identities

For all values of s for which denominators are not 0,

$$\sin s = \frac{1}{\csc s} \qquad \cos s = \frac{1}{\sec s} \qquad \tan s = \frac{1}{\cot s}$$

$$\cot s = \frac{1}{\tan s} \qquad \csc s = \frac{1}{\sin s} \qquad \sec s = \frac{1}{\cos s}$$

$$\tan s = \frac{\sin s}{\cos s} \qquad \cot s = \frac{\cos s}{\sin s}$$

$$\sin^2 s + \cos^2 s = 1 \qquad \tan^2 s + 1 = \sec^2 s$$

$$1 + \cot^2 s = \csc^2 s.$$

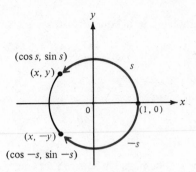

Figure 1

The unit circle in Figure 1 shows an arc of length s starting at $(1, 0)$, with endpoint (x, y). The same figure also shows an arc of length $-s$, again starting at $(1, 0)$. This arc has endpoint $(x, -y)$. Since $\sin s = y$,

$$\sin (-s) = -y = -\sin s,$$
$$\cos (-s) = x = \cos s$$

and
$$\tan (-s) = \frac{-y}{x} = -\tan s.$$

The reciprocal identities can be used to find $\csc (-s)$, $\sec (-s)$, and $\cot (-s)$.

Negative Angle Identities

$\sin (-s) = -\sin s$	$\cos (-s) = \cos s$	$\tan (-s) = -\tan s$
$\csc (-s) = -\csc s$	$\sec (-s) = \sec s$	$\cot (-s) = -\cot s$

These identities show that the sine and tangent functions are odd, while the cosine function is even.

Trigonometric identities are useful in several ways. One use of trigonometric identities is to find values of the other trigonometric functions from the value of a given trigonometric function. For example, given a value of $\tan s$, the value of $\cot s$ can be found by using the identity $\cot s = 1/\tan s$. In fact, given $\tan s$ and the quadrant in which s terminates, the values of all the other trigonometric functions can be found by using identities, as shown below.

Example 1   If $\tan s = -5/3$ and s is in quadrant II, find the values of the other trigonometric functions.

From the identity $\cot s = 1/\tan s$, $\cot s = -3/5$. Next, find $\sec s$ using the identity $\tan^2 s + 1 = \sec^2 s$.

$$\left(-\frac{5}{3} \right)^2 + 1 = \sec^2 s$$

$$\frac{25}{9} + 1 = \sec^2 s$$

Combining terms on the left,

$$\frac{34}{9} = \sec^2 s$$

$$-\sqrt{\frac{34}{9}} = \sec s$$

$$-\frac{\sqrt{34}}{3} = \sec s.$$

Choose the negative square root since the values of sec s are negative in quadrant II. Since cos s is the reciprocal of sec s,

$$\cos s = \frac{-3}{\sqrt{34}} = \frac{-3\sqrt{34}}{34},$$

after rationalizing the denominator. Now find sin s by using the identity $\sin^2 s + \cos^2 s = 1$.

$$\sin^2 s + \left(\frac{-3\sqrt{34}}{34}\right)^2 = 1$$

$$\sin^2 s = 1 - \frac{9}{34} = \frac{25}{34}$$

$$\sin s = \frac{5}{\sqrt{34}} \quad \text{or} \quad \sin s = \frac{5\sqrt{34}}{34}$$

Use the positive square root because the values of sin s are positive in quadrant II. Finally, csc $s = \sqrt{34}/5$ since csc s is the reciprocal of sin s.   ∎

Example 2   Express cos x in terms of tan x.

Since cos x and tan x are both related to sec x by an identity, start with $\tan^2 x + 1 = \sec^2 x$. Then

$$\frac{1}{\tan^2 x + 1} = \frac{1}{\sec^2 x} \quad \text{or} \quad \frac{1}{\tan^2 x + 1} = \cos^2 x.$$

Take the square root of both sides.

$$\pm\sqrt{\frac{1}{\tan^2 x + 1}} = \cos x$$

$$\cos x = \frac{\pm 1}{\sqrt{\tan^2 x + 1}}$$

Rationalize the denominator to get

$$\cos x = \frac{\pm\sqrt{\tan^2 x + 1}}{\tan^2 x + 1}.$$

Choose the + or the − sign, depending on the quadrant of x.   ∎

Another use of identities is to simplify trigonometric expressions by substituting one half of an identity for the other half. For example, the fact that $\sin^2 \theta + \cos^2 \theta$ equals 1 is used in the following example.

Example 3   Use the fundamental identities to write $\tan(-\theta) + \cot(-\theta)$ in terms of $\sin \theta$ and $\cos \theta$. Then simplify.

From the fundamental identities,

$$\tan(-\theta) + \cot(-\theta) = \frac{\sin(-\theta)}{\cos(-\theta)} + \frac{\cos(-\theta)}{\sin(-\theta)}.$$

Use the negative angle identities on the right, then simplify the expression by adding the two fractions on the right side, using the common denominator $\cos \theta \sin \theta$.

$$\tan(-\theta) + \cot(-\theta) = \frac{-\sin \theta}{\cos \theta} + \frac{\cos \theta}{-\sin \theta}$$

$$= \frac{-\sin \theta}{\cos \theta} - \frac{\cos \theta}{\sin \theta}$$

$$= \frac{-\sin^2 \theta - \cos^2 \theta}{\cos \theta \sin \theta}$$

$$= \frac{-(\sin^2 \theta + \cos^2 \theta)}{\cos \theta \sin \theta}$$

Now substitute 1 for $\sin^2 \theta + \cos^2 \theta$, to get

$$\tan(-\theta) + \cot(-\theta) = \frac{-1}{\cos \theta \sin \theta}. \quad \blacksquare$$

Example 4   Remove the radical in the expression $\sqrt{9 + x^2}$ by replacing x with $3 \tan \theta$, where $0 < \theta < \pi/2$.

Letting $x = 3 \tan \theta$ gives

$$\sqrt{9 + x^2} = \sqrt{9 + (3 \tan \theta)^2} = \sqrt{9 + 9 \tan^2 \theta}$$

$$= \sqrt{9(1 + \tan^2 \theta)} = 3\sqrt{1 + \tan^2 \theta}$$

$$= 3\sqrt{\sec^2 \theta}.$$

On the interval $0 < \theta < \pi/2$, the value of $\sec \theta$ is positive, giving

$$\sqrt{9 + x^2} = 3 \sec \theta. \quad \blacksquare$$

The result of Example 4 could be written as

$$\sec \theta = \frac{\sqrt{9 + x^2}}{3}.$$

In a right triangle, $\sec \theta$ is the ratio of the length of the hypotenuse to the side adjacent to the angle. This definition was used to label the right triangle in Figure 2 on the following page, with the Pythagorean theorem used to find the length of the side opposite angle θ. From the right triangle in Figure 2,

$$\sin \theta = \frac{x}{\sqrt{9 + x^2}}, \qquad \cos \theta = \frac{3}{\sqrt{9 + x^2}},$$

and so on.

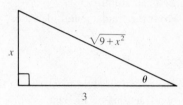

Figure 2

4.1 Exercises

Find sin s for each of the following.

1. $\cos s = 3/4$, s in quadrant I

2. $\cot s = -1/3$, s in quadrant IV

3. $\cos (-s) = \sqrt{5}/5$, $\tan s < 0$

4. $\tan s = -\sqrt{7}/2$, $\sec s > 0$

5. Find $\tan \theta$ if $\cos \theta = -2/5$, and $\sin \theta < 0$.

6. Find $\csc \alpha$ if $\tan \alpha = 6$, and $\cos \alpha > 0$.

Use the fundamental identities to find the values of the remaining five trigonometric functions of θ.

7. $\sin \theta = \dfrac{2}{3}$, θ in quadrant II

8. $\cos \theta = \dfrac{1}{5}$, θ in quadrant I

9. $\tan (-\theta) = \dfrac{1}{4}$, θ in quadrant IV

10. $\tan (-\theta) = -\dfrac{2}{3}$, θ in quadrant III

11. $\sec \theta = -3$, θ in quadrant II

12. $\csc \theta = -\dfrac{5}{2}$, θ in quadrant III

13. $\cot \theta = \dfrac{4}{3}$, $\sin \theta > 0$

14. $\sin \theta = -\dfrac{4}{5}$, $\cos \theta < 0$

15. $\sec \theta = \dfrac{4}{3}$, $\sin \theta < 0$

16. $\cos \theta = -\dfrac{1}{4}$, $\sin \theta > 0$

For each trigonometric expression in Column I, choose the expression from Column II which completes a fundamental identity.

Column I	Column II
17. $\dfrac{\cos x}{\sin x}$	**(a)** $\sin^2 x + \cos^2 x$
18. $\tan x$	**(b)** $\cot x$
19. $\cos (-x)$	**(c)** $\sec^2 x$
20. $\tan^2 x + 1$	**(d)** $\dfrac{\sin x}{\cos x}$
21. 1	**(e)** $\cos x$

For each expression in Column I, choose the expression from Column II which completes an identity. You will have to rewrite one or both expressions, using a fundamental identity, to recognize the matches.

Column I   Column II

22. $-\tan x \cos x$ **(a)** $\dfrac{\sin^2 x}{\cos^2 x}$

23. $\sec^2 x - 1$ **(b)** $\dfrac{1}{\sec^2 x}$

24. $\dfrac{\sec x}{\csc x}$ **(c)** $\sin(-x)$

25. $1 + \sin^2 x$ **(d)** $\csc^2 x - \cot^2 x + \sin^2 x$

26. $\cos^2 x$ **(e)** $\tan x$

In each of the following, use the fundamental identities to get an equivalent expression involving only sines and cosines and then simplify it.

27. $\csc^2 \beta - \cot^2 \beta$ **28.** $\dfrac{\tan(-\theta)}{\sec \theta}$ **29.** $\tan(-\alpha)\cos(-\alpha)$ **30.** $\cot^2 x(1 + \tan^2 x)$

31. $\tan^2 \theta - \dfrac{\sec^2 \theta}{\csc^2 \theta}$ **32.** $\dfrac{\tan x \csc x}{\sec x}$ **33.** $\sec \theta + \tan \theta$ **34.** $\dfrac{\sec \alpha}{\tan \alpha + \cot \alpha}$

35. $\sec^2 t - \tan^2 t$ **36.** $\csc^2 \gamma + \sec^2 \gamma$ **37.** $\cot^2 \beta - \csc^2 \beta$ **38.** $1 + \cot^2 \alpha$

39. $\dfrac{1 + \tan^2 \theta}{\cot^2 \theta}$ **40.** $\dfrac{1 - \sin^2 t}{\csc^2 t}$

41. $\cot^2 \beta \sin^2 \beta + \tan^2 \beta \cos^2 \beta$ **42.** $\sec^2 x + \cos^2 x$

43. $\dfrac{\cot^2 \alpha + \csc^2 \alpha}{\cos^2 \alpha}$ **44.** $1 - \tan^4 \theta$ **45.** $1 - \cot^4 s$ **46.** $\tan^4 \gamma - \cot^4 \gamma$

Complete this chart so that each expression in the column at the left is written in terms of the expressions given across the top.

	$\sin \theta$	$\cos \theta$	$\tan \theta$	$\cot \theta$	$\sec \theta$	$\csc \theta$
47. $\sin \theta$	$\sin \theta$	$\pm\sqrt{1 - \cos^2 \theta}$	$\dfrac{\pm\tan \theta}{\sqrt{1 + \tan^2 \theta}}$			$\dfrac{1}{\csc \theta}$
48. $\cos \theta$		$\cos \theta$	$\dfrac{\pm\sqrt{\tan^2 \theta + 1}}{\tan^2 \theta + 1}$		$\dfrac{1}{\sec \theta}$	
49. $\tan \theta$			$\tan \theta$	$\dfrac{1}{\cot \theta}$		
50. $\cot \theta$			$\dfrac{1}{\tan \theta}$	$\cot \theta$	$\dfrac{\pm\sqrt{\sec^2 \theta - 1}}{\sec^2 \theta - 1}$	
51. $\sec \theta$		$\dfrac{1}{\cos \theta}$			$\sec \theta$	
52. $\csc \theta$	$\dfrac{1}{\sin \theta}$					$\csc \theta$

53. Suppose that $\cos \theta = x/(x + 1)$. Find $\sin \theta$.   **54.** Find $\tan \alpha$ if $\sec \alpha = (p + 4)/p$.

Show that each of the following is not an identity by replacing the variables with numbers that show the result to be false.

55. $(\sin s + \cos s)^2 = 1$ **56.** $(\tan s + 1)^2 = \sec^2 s$ **57.** $2 \sin s = \sin 2s$

58. $\sin x = \sqrt{1 - \cos^2 x}$ **59.** $\sin^3 x + \cos^3 x = 1$ **60.** $\sin x + \sin y = \sin (x + y)$

Use the indicated substitution to remove the radical in the given expression in Exercises 61–66. Assume $0 < \theta < \pi/2$. Then find the indicated values.

61. $\sqrt{16 + 9x^2}$, let $x = \dfrac{4}{3} \tan \theta$; find $\sin \theta$ and $\cos \theta$

62. $\sqrt{x^2 - 25}$, let $x = 5 \sec \theta$; find $\sin \theta$ and $\tan \theta$

63. $\sqrt{(1 - x^2)^3}$, let $x = \cos \theta$; find $\sin \theta$ and $\tan \theta$

64. $\dfrac{\sqrt{x^2 - 9}}{x}$, let $x = 3 \sec \theta$; find $\sin \theta$ and $\tan \theta$

65. $x^2\sqrt{1 + 16x^2}$, let $x = \dfrac{1}{4} \tan \theta$; find $\sin \theta$ and $\cos \theta$

66. $x^2\sqrt{9 + x^2}$, let $x = 3 \tan \theta$; find $\sin \theta$ and $\cos \theta$

67. Let $\cos x = 1/5$. Find all possible values for

$$\frac{\sec x - \tan x}{\sin x}.$$

68. Let $\csc x = -3$. Find all possible values for

$$\frac{\sin x + \cos x}{\sec x}.$$

Prove the following for first-quadrant values of s.

69. $\log \sin s = -\log \csc s$

70. $\log \tan s = \log \sin s - \log \cos s$

4.2 Verifying Trigonometric Identities

One of the skills required for more advanced work in mathematics (and especially in calculus) is the ability to use the trigonometric identities to write trigonometric expressions in alternate forms. This skill is developed by using the fundamental identities to verify that a trigonometric equation is an identity (for those values of the variable for which it is defined). Here are some hints that may help you get started.

Verifying Identities

1. Memorize the fundamental identities given in the last section. Whenever you see either half of a fundamental identity, the other half should come to mind.
2. Try to rewrite the more complicated side of the equation so that it is identical to the simpler side.
3. It is often helpful to express all other trigonometric functions in the equation in terms of sine and cosine and then simplify the result.
4. You should usually perform any factoring or indicated algebraic operations. For example, the expression $\sin^2 x + 2 \sin x + 1$ can be factored as $(\sin x + 1)^2$. The sum or difference of two trigonometric expressions, such as

$$\frac{1}{\sin \theta} + \frac{1}{\cos \theta}$$

can be added or subtracted in the same way as any other rational expression. In this example,

$$\frac{1}{\sin \theta} + \frac{1}{\cos \theta} = \frac{\cos \theta}{\sin \theta \cos \theta} + \frac{\sin \theta}{\sin \theta \cos \theta} = \frac{\cos \theta + \sin \theta}{\sin \theta \cos \theta}.$$

5. Keep in mind the side you are not changing as you select substitutions. It represents your goal. For example, to verify the identity

$$\tan^2 x + 1 = \frac{1}{\cos^2 x},$$

try to think of an identity that relates $\tan x$ to $\cos x$. Here, since $\sec x = 1/\cos x$ and $\sec^2 x = \tan^2 x + 1$, the secant function is the best link between the two sides of the equation.
6. If an expression contains $1 + \sin x$, multiplying both numerator and denominator by $1 - \sin x$ would give $1 - \sin^2 x$, which could be replaced with $\cos^2 x$.

These hints are used in the following examples. A word of warning: Verifying identities is not the same as solving equations. Techniques used in solving equations, such as adding the same terms to both sides, or multiplying both sides by the same term, are not valid when working with identities.

Example 1 Verify that

$$\cot s + 1 = \csc s (\cos s + \sin s)$$

is an identity.

Use the fundamental identities to rewrite one side of the equation so that it is identical to the other side. Since the right side is more complicated, it is a good idea to start with it. Begin by changing each trigonometric function to sine or cosine.

$$\csc s (\cos s + \sin s) = \frac{1}{\sin s} (\cos s + \sin s) = \frac{\cos s}{\sin s} + \frac{\sin s}{\sin s} = \cot s + 1$$

The equation is an identity because the right side equals the left side, for all values of s for which denominators are not zero. ∎

Example 2 Verify that

$$\tan^2 \alpha(1 + \cot^2 \alpha) = \frac{1}{1 - \sin^2 \alpha}$$

is an identity.

Working with the left side gives

$$\tan^2 \alpha(1 + \cot^2 \alpha) = \tan^2 \alpha + \tan^2 \alpha \cot^2 \alpha$$

$$= \tan^2 \alpha + \tan^2 \alpha \cdot \frac{1}{\tan^2 \alpha}$$

$$= \tan^2 \alpha + 1 = \sec^2 \alpha$$

$$= \frac{1}{\cos^2 \alpha} = \frac{1}{1 - \sin^2 \alpha}. \quad \blacksquare$$

Example 3 Show that

$$\frac{\tan t - \cot t}{\sin t \cos t} = \sec^2 t - \csc^2 t.$$

Work with the left side.

$$\frac{\tan t - \cot t}{\sin t \cos t} = \frac{\tan t}{\sin t \cos t} - \frac{\cot t}{\sin t \cos t}$$

$$= \tan t \cdot \frac{1}{\sin t \cos t} - \cot t \cdot \frac{1}{\sin t \cos t}$$

$$= \frac{\sin t}{\cos t} \cdot \frac{1}{\sin t \cdot \cos t} - \frac{\cos t}{\sin t} \cdot \frac{1}{\sin t \cos t}$$

$$= \frac{1}{\cos^2 t} - \frac{1}{\sin^2 t} = \sec^2 t - \csc^2 t \quad \blacksquare$$

Example 4 Show that $\dfrac{\sin \theta}{1 + \cos \theta} = \dfrac{1 - \cos \theta}{\sin \theta}$.

Multiplying numerator and denominator on the left by $1 - \cos \theta$ gives useful results.

$$\frac{\sin \theta}{1 + \cos \theta} = \frac{\sin \theta}{1 + \cos \theta} \cdot \frac{1 - \cos \theta}{1 - \cos \theta}$$

$$= \frac{\sin \theta (1 - \cos \theta)}{1 - \cos^2 \theta}$$

Now use the identity $1 - \cos^2 \theta = \sin^2 \theta$ to get

$$\frac{\sin \theta}{1 + \cos \theta} = \frac{\sin \theta (1 - \cos \theta)}{\sin^2 \theta}$$

$$= \frac{1 - \cos \theta}{\sin \theta}. \quad \blacksquare$$

4.2 Exercises

For each of the following, perform the indicated operations and simplify the result.

1. $\tan \theta + \dfrac{1}{\tan \theta}$

2. $\dfrac{\cos x}{\sin x} + \dfrac{\sin x}{\cos x}$

3. $\cot s(\tan s + \sin s)$

4. $\sec \beta (\cos \beta + \sin \beta)$

5. $\dfrac{1}{\csc^2 \theta} + \dfrac{1}{\sec^2 \theta}$

6. $\dfrac{1}{\sin \alpha - 1} - \dfrac{1}{\sin \alpha + 1}$

7. $\dfrac{\cos x}{\sec x} + \dfrac{\sin x}{\csc x}$

8. $\dfrac{\cos \gamma}{\sin \gamma} + \dfrac{\sin \gamma}{1 + \cos \gamma}$

9. $(1 + \sin t)^2 + \cos^2 t$

10. $(1 + \tan s)^2 - 2 \tan s$

Factor each of the following trigonometric expressions.

11. $\sin^2 \gamma - 1$

12. $9 \sec^2 \theta - 25$

13. $(\sin x + 1)^2 - (\sin x - 1)^2$

14. $(\tan x + \cot x)^2 - (\tan x - \cot x)^2$

15. $2 \sin^2 x + 3 \sin x + 1$

16. $4 \tan^2 \beta + \tan \beta - 3$

17. $4 \sec^2 x + 3 \sec x - 1$

18. $2 \csc^2 x + 7 \csc x - 30$

19. $\cos^4 x + 2 \cos^2 x + 1$

20. $\cot^4 x + 3 \cot^2 x + 2$

Use the fundamental identities to simplify each of the given expressions.

21. $\tan \theta \cos \theta$

22. $\cot \alpha \sin \alpha$

23. $\sec r \cos r$

24. $\cot t \tan t$

25. $\dfrac{\sin \beta \tan \beta}{\cos \beta}$

26. $\dfrac{\csc \theta \sec \theta}{\cot \theta}$

27. $\sec^2 x - 1$

28. $\csc^2 t - 1$

29. $\dfrac{\sin^2 x}{\cos^2 x} + \sin x \csc x$

30. $\dfrac{1}{\tan^2 \alpha} + \cot \alpha \tan \alpha$

Verify each of the following trigonometric identities.

31. $1 - \sec \alpha \cos \alpha = \tan \alpha \cot \alpha - 1$

32. $\csc^4 \theta = \cot^4 \theta + 2 \cot^2 \theta + 1$

33. $\dfrac{\sin^2 \theta}{\cos^2 \theta} = \sec^2 \theta - 1$

34. $\cot \beta \sin \beta = \cos \beta$

35. $\sin^2 \alpha + \tan^2 \alpha + \cos^2 \alpha = \sec^2 \alpha$

36. $\sin^2 s - 1 = -\cos^2 s$

37. $\dfrac{\sin^2 \gamma}{\cos \gamma} = \sec \gamma - \cos \gamma$

38. $(1 + \tan^2 x) \cos^2 x = 1$

39. $\cot s + \tan s = \sec s \csc s$

40. $\dfrac{\cos \alpha}{\sec \alpha} + \dfrac{\sin \alpha}{\csc \alpha} = \sec^2 \alpha - \tan^2 \alpha$

41. $\dfrac{\cos \alpha}{\sin \alpha \cot \alpha} = 1$

42. $\sin^4 \theta - \cos^4 \theta = 2 \sin^2 \theta - 1$

43. $\dfrac{1 + \sin x}{\cos x} = \dfrac{\cos x}{1 - \sin x}$

44. $(1 - \cos^2 \alpha)(1 + \cos^2 \alpha) = 2 \sin^2 \alpha - \sin^4 \alpha$

45. $\dfrac{(\sec \theta - \tan \theta)^2 + 1}{\sec \theta \csc \theta - \tan \theta \csc \theta} = 2 \tan \theta$

46. $\dfrac{\cos \theta + 1}{\tan^2 \theta} = \dfrac{\cos \theta}{\sec \theta - 1}$

47. $\dfrac{1}{\sec \alpha - \tan \alpha} = \sec \alpha + \tan \alpha$

48. $\dfrac{1}{1 - \sin \theta} + \dfrac{1}{1 + \sin \theta} = 2 \sec^2 \theta$

49. $\dfrac{1 - \cos x}{1 + \cos x} = (\cot x - \csc x)^2$

50. $\dfrac{\tan s}{1 + \cos s} + \dfrac{\sin s}{1 - \cos s} = \cot s + \sec s \csc s$

51. $\dfrac{1}{\tan \alpha - \sec \alpha} + \dfrac{1}{\tan \alpha + \sec \alpha} = -2 \tan \alpha$

52. $\dfrac{\cot \alpha + 1}{\cot \alpha - 1} = \dfrac{1 + \tan \alpha}{1 - \tan \alpha}$

53. $\dfrac{\csc \theta + \cot \theta}{\tan \theta + \sin \theta} = \cot \theta \csc \theta$

54. $\sin^2 \alpha \sec^2 \alpha + \sin^2 \alpha \csc^2 \alpha = \sec^2 \alpha$

55. $\sec^4 x - \sec^2 x = \tan^4 x + \tan^2 x$

56. $\dfrac{1 - \sin \theta}{1 + \sin \theta} = \sec^2 \theta - 2 \sec \theta \tan \theta + \tan^2 \theta$

57. $\sin \theta + \cos \theta = \dfrac{\sin \theta}{1 - \dfrac{\cos \theta}{\sin \theta}} + \dfrac{\cos \theta}{1 - \dfrac{\sin \theta}{\cos \theta}}$

58. $\dfrac{\sin \theta}{1 - \cos \theta} - \dfrac{\sin \theta \cos \theta}{1 + \cos \theta} = \csc \theta + \csc \theta \cos^2 \theta$

59. $\dfrac{\sec^4 s - \tan^4 s}{\sec^2 s + \tan^2 s} = \sec^2 s - \tan^2 s$

60. $\dfrac{\cot^2 t - 1}{1 + \cot^2 t} = 1 - 2 \sin^2 t$

61. $\dfrac{\tan^2 t - 1}{\sec^2 t} = \dfrac{\tan t - \cot t}{\tan t + \cot t}$

62. $(1 + \sin x + \cos x)^2 = 2(1 + \sin x)(1 + \cos x)$

63. $(\sin s + \cos s)^2 \cdot \csc s = 2 \cos s + \dfrac{1}{\sin s}$

64. $\dfrac{\sin^3 t - \cos^3 t}{\sin t - \cos t} = 1 + \sin t \cos t$

65. $\dfrac{1 + \cos x}{1 - \cos x} - \dfrac{1 - \cos x}{1 + \cos x} = 4 \cot x \csc x$

66. $(\sec \alpha - \tan \alpha)^2 = \dfrac{1 - \sin \alpha}{1 + \sin \alpha}$

67. $(\sec \alpha + \csc \alpha)(\cos \alpha - \sin \alpha) = \cot \alpha - \tan \alpha$

68. $\dfrac{\sin^4 \alpha - \cos^4 \alpha}{\sin^2 \alpha - \cos^2 \alpha} = 1$

69. $\dfrac{\cot^2 x + \sec^2 x + 1}{\cot^2 x} = \sec^4 x$

70. $\dfrac{\cos x - (\sin x - 1)}{\cos x + (\sin x - 1)} = \dfrac{\sin x}{1 - \cos x}$

71. $\ln e^{|\sin x|} = |\sin x|$

72. $\ln |\tan x| = -\ln |\cot x|$

73. $\ln |\sec x - \tan x| = -\ln |\sec x + \tan x|$

74. $-\ln |\csc t - \cot t| = \ln |\csc t + \cot t|$

Given a complicated equation involving trigonometric functions, it is a good idea to decide whether it is likely to be an identity before trying to prove that it is. Substitute $s = 1$ and $s = 2$ into each of the following. (Be sure to use radians.) If you get the same results on both sides of the equation, it *may* be an identity. Then prove that it is.

75. $\dfrac{2 + 5 \cos s}{\sin s} = 2 \csc s + 5 \cot s$

76. $1 + \cot^2 s = \dfrac{\sec^2 s}{\sec^2 s - 1}$

77. $\dfrac{\tan s - \cot s}{\tan s + \cot s} = 2 \sin^2 s$

78. $\dfrac{1}{1 + \sin s} + \dfrac{1}{1 - \sin s} = \sec^2 s$

79. $\dfrac{1 - \tan^2 s}{1 + \tan^2 s} = \cos^2 s - \sin s$

80. $\dfrac{\sin^3 s - \cos^3 s}{\sin s - \cos s} = \sin^2 s + 2 \sin s \cos s + \cos^2 s$

81. $\sin^2 s + \cos^2 s = \dfrac{1}{2}(1 - \cos 4s)$

82. $\cos 3s = 3 \cos s + 4 \cos^3 s$

Show that the following are *not* identities for real numbers s and t.

83. $\sin(\csc s) = 1$

84. $\sqrt{\cos^2 s} = \cos s$

85. $\csc t = \sqrt{1 + \cot^2 t}$

86. $\sin t = \sqrt{1 - \cos^2 t}$

87. Let $\tan \theta = t$ and show that $\sin \theta \cos \theta = \dfrac{t}{t^2 + 1}$.

88. When does $\sin x = \sqrt{1 - \cos^2 x}$?

4.3 Sum and Difference Identities for Cosine

Several examples presented throughout this book have shown that $\cos(A - B)$ does *not* equal $\cos A - \cos B$. For example, if $A = \pi/2$ and $B = 0$,

$$\cos(A - B) = \cos(\pi/2 - 0) = \cos \pi/2 = 0,$$

while

$$\cos A - \cos B = \cos \pi/2 - \cos 0 = 0 - 1 = -1.$$

The actual formula for $\cos(A - B)$ is derived in this section. Start by letting $A(1, 0)$, $B(x_1, y_1)$, $C(x_2, y_2)$, and $D(x_3, y_3)$ be points on the unit circle such that the length of arc AD is s, the length of arc AB is t, and the length of arc AC is $s - t$.

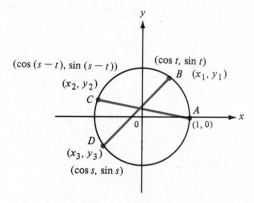

Figure 3

(See Figure 3.) Here we assume that $0 < t < s < 2\pi$ although the results are valid for any values of s and t. Arc BD also has length $s - t$, so that arcs AC and BD are equal in length. Since these arcs are equal in length, line segments AC and BD must also have equal length. By the distance formula,

$$\sqrt{(x_2 - 1)^2 + (y_2 - 0)^2} = \sqrt{(x_3 - x_1)^2 + (y_3 - y_1)^2}.$$

Squaring both sides and simplifying gives

$$x_2^2 - 2x_2 + 1 + y_2^2 = x_3^2 - 2x_3x_1 + x_1^2 + y_3^2 - 2y_3y_1 + y_1^2. \qquad (*)$$

Since points B, C, and D are on the unit circle,

$$x_1{}^2 + y_1{}^2 = 1, \qquad x_2{}^2 + y_2{}^2 = 1, \qquad \text{and} \qquad x_3{}^2 + y_3{}^2 = 1.$$

Substituting these results into equation $(*)$ gives

$$2 - 2x_2 = 2 - 2x_3x_1 - 2y_3y_1$$

or

$$x_2 = x_3x_1 + y_3y_1.$$

Since $x_2 = \cos(s - t)$, $x_3 = \cos s$, $x_1 = \cos t$, $y_3 = \sin s$, and $y_1 = \sin t$,

Cosine of Difference of Two Angles

$$\cos(s - t) = \cos s \cos t + \sin s \sin t.$$

Example 1   Find the value of $\cos 15°$.

To find $\cos 15°$, write $15°$ as the difference of two angles which have known function values. Since the trigonometric function values of both $45°$ and $30°$ are known, write $15°$ as $45° - 30°$. Then use the identity for the cosine of the difference of two angles.

$$\cos 15° = \cos(45° - 30°)$$
$$= \cos 45° \cos 30° + \sin 45° \sin 30°$$
$$= \frac{\sqrt{2}}{2} \cdot \frac{\sqrt{3}}{2} + \frac{\sqrt{2}}{2} \cdot \frac{1}{2}$$
$$\cos 15° = \frac{\sqrt{6} + \sqrt{2}}{4} \qquad \blacksquare$$

The formula for $\cos(s - t)$ can be used to find a formula for $\cos(s + t)$. Do this by using the result for $\cos(s - t)$ and writing $s + t$ as $s - (-t)$.

$$\cos(s + t) = \cos[s - (-t)]$$
$$= \cos s \cos(-t) + \sin s \sin(-t)$$

From earlier work, $\cos(-t) = \cos t$, and $\sin(-t) = -\sin t$. Making these substitutions gives

Cosine of Sum of Two Angles

$$\cos(s + t) = \cos s \cos t - \sin s \sin t.$$

Example 2   Find $\cos \dfrac{5}{12}\pi$.

$$\cos \frac{5}{12}\pi = \cos\left(\frac{\pi}{6} + \frac{\pi}{4}\right) = \cos \frac{\pi}{6} \cos \frac{\pi}{4} - \sin \frac{\pi}{6} \sin \frac{\pi}{4}$$
$$= \frac{\sqrt{3}}{2} \cdot \frac{\sqrt{2}}{2} - \frac{1}{2} \cdot \frac{\sqrt{2}}{2} = \frac{\sqrt{6} - \sqrt{2}}{4} \qquad \blacksquare$$

Example 3   Suppose $\sin x = 1/2$, $\cos y = -12/13$, and x and y are both in quadrant II. Find $\cos (x + y)$.

By the identity above, $\cos (x + y) = \cos x \cos y - \sin x \sin y$. The values of $\sin x$ and $\cos y$ are given, so that $\cos (x + y)$ can be found if $\cos x$ and $\sin y$ are known. To find $\cos x$, use the fact that $\sin^2 x + \cos^2 x = 1$, and substitute $1/2$ for $\sin x$.

$$\sin^2 x + \cos^2 x = 1$$
$$\left(\frac{1}{2}\right)^2 + \cos^2 x = 1$$
$$\frac{1}{4} + \cos^2 x = 1$$
$$\cos^2 x = \frac{3}{4}$$

Since x is in quadrant II, $\cos x$ is negative, so $\cos x = -\sqrt{3}/2$. Find $\sin y$ similarly.

$$\sin^2 y + \cos^2 y = 1$$
$$\sin^2 y + \left(-\frac{12}{13}\right)^2 = 1$$
$$\sin^2 y + \frac{144}{169} = 1$$
$$\sin^2 y = \frac{25}{169}$$

Since y is in quadrant II, $\sin y = 5/13$. Now find $\cos (x + y)$.

$$\cos (x + y) = \cos x \cos y - \sin x \sin y$$
$$= -\frac{\sqrt{3}}{2} \cdot \left(-\frac{12}{13}\right) - \frac{1}{2} \cdot \frac{5}{13}$$
$$= \frac{12\sqrt{3}}{26} - \frac{5}{26}$$
$$\cos (x + y) = \frac{12\sqrt{3} - 5}{26} \quad \blacksquare$$

The identities for the cosine of the sum and difference of two angles can be used to derive other identities. For example, substituting $\pi/2$ for s in the identity for $\cos (s - t)$ gives

$$\cos (\pi/2 - t) = \cos \pi/2 \cdot \cos t + \sin \pi/2 \cdot \sin t$$
$$= 0 \cdot \cos t + 1 \cdot \sin t$$
$$\cos (\pi/2 - t) = \sin t.$$

This result is true for any value of t since the identity for $\cos (s - t)$ is true for any value of s and t. The identity $\cos (\pi/2 - t) = \sin t$ is a **cofunction identity.** The cofunction identities are listed on the next page.

Cofunction Identities	$\cos (\pi/2 - t) = \sin t$	$\sin (\pi/2 - t) = \cos t$
	$\tan (\pi/2 - t) = \cot t$	$\cot (\pi/2 - t) = \tan t$
	$\sec (\pi/2 - t) = \csc t$	$\csc (\pi/2 - t) = \sec t$

The derivation of some of the cofunction identities is included in Exercises 72 and 73. Because of these identities, sine and cosine are called **cofunctions,** as are tangent and cotangent, and secant and cosecant.

Example 4 Find a number s which satisfies each of the following.

(a) $\cot s = \tan \pi/12$

Since tangent and cotangent are cofunctions,

$$\cot s = \tan (\pi/2 - s) = \tan \pi/12.$$

One solution to this equation is found by letting $\pi/2 - s = \pi/12$. Then $s = 5\pi/12$.

(b) $\sin \theta = \cos (-30°)$

In a similar way,

$$\sin \theta = \cos (90° - \theta) = \cos (-30°).$$
$$90° - \theta = -30°$$
$$\theta = 120°. \quad \blacksquare$$

Example 5 Write $\cos (\pi - t)$ as a function of t.

Use the identity for $\cos (s - t)$. Replace s with π.

$$\cos (\pi - t) = \cos \pi \cdot \cos t + \sin \pi \cdot \sin t$$
$$= (-1) \cdot \cos t + (0) \cdot \sin t = -\cos t \quad \blacksquare$$

4.3 Exercises

Write each of the following in terms of the cofunction of a complementary angle.

1. $\tan 87°$ **2.** $\sin 15°$ **3.** $\cos \pi/12$ **4.** $\sin 2\pi/5$

5. $\csc (-14° 24')$ **6.** $\sin 142° 14'$ **7.** $\sin 5\pi/8$ **8.** $\cot 9\pi/10$

9. $\sec 146° 42'$ **10.** $\tan 174° 3'$ **11.** $\cot 176.9814°$ **12.** $\sin 98.0142°$

Use the cofunction identities to fill in each of the following blanks with the appropriate trigonometric function name.

13. $\cot \dfrac{\pi}{3} = \underline{\hspace{1cm}} \dfrac{\pi}{6}$ **14.** $\sin \dfrac{2\pi}{3} = \underline{\hspace{1cm}} \left(-\dfrac{\pi}{6}\right)$ **15.** $\underline{\hspace{1cm}} 33° = \sin 57°$

16. $\underline{\hspace{1cm}} 72° = \cot 18°$ **17.** $\cos 70° = \dfrac{1}{\underline{\hspace{1cm}} 20°}$ **18.** $\tan 24° = \dfrac{1}{\underline{\hspace{1cm}} 66°}$

Tell whether each of the following is *true* or *false*.

19. $\cos 42° = \cos (30° + 12°)$ **20.** $\cos (-24°) = \cos 16° - \cos 40°$

21. $\cos 74° = \cos 60° \cos 14° + \sin 60° \sin 14°$ **22.** $\cos 140° = \cos 60° \cos 80° - \sin 60° \sin 80°$

23. $\cos \pi/3 = \cos \pi/12 \cos \pi/4 - \sin \pi/12 \sin \pi/4$

24. $\cos 2\pi/3 = \cos 11\pi/12 \cos \pi/4 + \sin 11\pi/12 \sin \pi/4$

25. $\cos 70° \cos 20° - \sin 70° \sin 20° = 0$ **26.** $\cos 85° \cos 40° + \sin 85° \sin 40° = \sqrt{2}/2$

Use the cofunction identities to find a value of θ that makes each of the following true.

27. $\tan \theta = \cot (45° + 2\theta)$ **28.** $\sin \theta = \cos (2\theta - 10°)$

29. $\sec \theta = \csc (\theta/2 + 20°)$ **30.** $\cos \theta = \sin (3\theta + 10°)$

31. $\sin (3\theta - 15°) = \cos (\theta + 25°)$ **32.** $\cot (\theta - 10°) = \tan (2\theta + 20°)$

Use the sum and difference identities for cosine to find the value of each of the following without using calculators or tables.

33. $\cos 285°$ **34.** $\cos (-15°)$ **35.** $\cos (-105°)$ **36.** $\cos 75°$

37. $\cos 7\pi/12$ **38.** $\cos (-5\pi/12)$

39. $\cos 40° \cos 50° - \sin 40° \sin 50°$ **40.** $\cos 80° \cos 35° + \sin 80° \sin 35°$

41. $\cos (-10°) \cos 35° + \sin (-10°) \sin 35°$

42. $\cos 112.146° \cos 67.854° - \sin 112.146° \sin 67.854°$

43. $\cos 174.983° \cos 95.017° - \sin 174.983° \sin 95.017°$

44. $\cos 348.502° \cos 78.502° + \sin 348.502° \sin 78.502°$

45. $\cos 2\pi/5 \cos \pi/10 - \sin 2\pi/5 \sin \pi/10$ **46.** $\cos 7\pi/9 \cos 2\pi/9 - \sin 7\pi/9 \sin 2\pi/9$

Write each of the following as a function of θ or x.

47. $\cos (30° + \theta)$ **48.** $\cos (45° - \theta)$ **49.** $\cos (60° + \theta)$

50. $\cos (\theta - 30°)$ **51.** $\cos (3\pi/2 - x)$ **52.** $\cos (x + \pi/4)$

For each of the following, find $\cos (s + t)$ and $\cos (s - t)$.

53. $\cos s = -1/5$ and $\sin t = 3/5$, s and t in quadrant II

54. $\sin s = 2/3$ and $\sin t = -1/3$, s in quadrant II and t in quadrant IV

55. $\sin s = 3/5$ and $\sin t = -12/13$, s in quadrant I and t in quadrant III

56. $\cos s = -8/17$ and $\cos t = -3/5$, s and t in quadrant III

57. $\cos s = -15/17$ and $\sin t = 4/5$, s in quadrant II and t in quadrant I

58. $\sin s = -8/17$ and $\cos t = -8/17$, s and t in quadrant III

59. $\sin s = \sqrt{5}/7$ and $\sin t = \sqrt{6}/8$, s and t in quadrant I

60. $\cos s = \sqrt{2}/4$ and $\sin t = -\sqrt{5}/6$, s and t in quadrant IV

Verify each of the following identities.

61. $\cos (\pi/2 + x) = -\sin x$ **62.** $\sec (\pi - x) = -\sec x$

63. $\cos 2x = \cos^2 x - \sin^2 x$ (*Hint:* $\cos 2x = \cos (x + x)$.)

64. $\cos (x + y) + \cos (x - y) = 2 \cos x \cos y$ **65.** $\dfrac{\cos (\alpha - \theta) - \cos (\alpha + \theta)}{\cos (\alpha - \theta) + \cos (\alpha + \theta)} = \tan \theta \tan \alpha$

66. $1 + \cos 2x - \cos^2 x = \cos^2 x$ (*Hint:* Use the result in Exercise 63.)

67. $\cos (\pi + s - t) = -\sin s \sin t - \cos s \cos t$ **68.** $\cos (\pi/2 + s - t) = \sin t \cos s - \cos t \sin s$

69. $\cos (\alpha + \beta) \cos (\alpha - \beta) = 1 - \sin^2 \alpha - \sin^2 \beta$

70. $\cos 4x \cos 7x - \sin 4x \sin 7x = \cos 11x$

71. Use the identities for the cosine of the sum and difference of two angles to complete each of the following.

$\cos (0 - t) =$ $\cos (0 + t) =$ $\cos (\pi/2 - t) =$ $\cos (\pi/2 + t) =$
$\cos (\pi - t) =$ $\cos (\pi + t) =$ $\cos (3\pi/2 - t) =$ $\cos (3\pi/2 + t) =$

72. Use the identity $\cos (\pi/2 - t) = \sin t$; replace t with $\pi/2 - t$, and derive the identity $\cos t = \sin (\pi/2 - t)$.

73. Derive each identity.
  (a) $\tan t = \cot (\pi/2 - t)$
  (b) $\csc t = \sec (\pi/2 - t)$

74. Let $f(x) = \cos x$. Prove that $\dfrac{f(x + h) - f(x)}{h} = \cos x \left(\dfrac{\cos h - 1}{h} \right) - \sin x \left(\dfrac{\sin h}{h} \right)$.

Use the identities of this section to find each of the following.

75. $\cos (\sin^{-1} 8/17 + \tan^{-1} 3/4)$

76. $\cos (\tan^{-1} 5/12 - \cos^{-1} 4/5)$

77. $\cos (\cos^{-1} 1/4 + \tan^{-1} 5/8)$

78. $\cos (\sin^{-1} 1/3 - \cos^{-1} 2/5)$

4.4 Sum and Difference Identities for Sine and Tangent

Formulas for $\sin (s + t)$ and $\sin (s - t)$ can be developed from the results of the previous section. From a cofunction identity,

$$\sin (s + t) = \cos \left[\frac{\pi}{2} - (s + t) \right]$$

$$= \cos \left[\left(\frac{\pi}{2} - s \right) - t \right].$$

Using the identity for $\cos (s - t)$ from the previous section gives

$$\sin (s + t) = \cos \left(\frac{\pi}{2} - s \right) \cdot \cos t + \sin \left(\frac{\pi}{2} - s \right) \cdot \sin t.$$

Substitute from the cofunction identities again to get

Sine of Sum or Difference of Two Angles

$$\sin (s + t) = \sin s \cos t + \cos s \sin t$$
$$\sin (s - t) = \sin s \cos t - \cos s \sin t.$$

The second identity comes from the first by writing $\sin (s - t)$ as $\sin [s + (-t)]$.

Using the identities for $\sin (s + t)$, $\cos (s + t)$, $\sin (s - t)$, and $\cos (s - t)$, and the identity $\tan x = \sin x/\cos x$, gives the following new identities.

**Tangent of Sum
or Difference of
Two Angles**

$$\tan (s + t) = \frac{\tan s + \tan t}{1 - \tan s \tan t}$$

$$\tan (s - t) = \frac{\tan s - \tan t}{1 + \tan s \tan t}$$

We show the proof for the first of these two identities. The proof of the other is very similar. Start with

$$\tan (s + t) = \frac{\sin (s + t)}{\cos (s + t)}$$

$$= \frac{\sin s \cos t + \cos s \sin t}{\cos s \cos t - \sin s \sin t}.$$

To express this result in terms of the tangent function, multiply both numerator and denominator by $1/(\cos s \cos t)$.

$$\tan (s + t) = \frac{\dfrac{\sin s \cos t + \cos s \sin t}{1}}{\dfrac{\cos s \cos t - \sin s \sin t}{1}} \cdot \frac{\dfrac{1}{\cos s \cos t}}{\dfrac{1}{\cos s \cos t}}$$

$$\tan (s + t) = \frac{\dfrac{\sin s \cos t}{\cos s \cos t} + \dfrac{\cos s \sin t}{\cos s \cos t}}{\dfrac{\cos s \cos t}{\cos s \cos t} - \dfrac{\sin s \sin t}{\cos s \cos t}}$$

Using the identity $\tan x = \sin x/\cos x$ gives

$$\tan (s + t) = \frac{\tan s + \tan t}{1 - \tan s \tan t}.$$

Example 1 Use identities to find the exact value of each of the following.

(a) $\sin 75°$

$$\sin 75° = \sin (45° + 30°)$$
$$= \sin 45° \cos 30° + \cos 45° \sin 30°$$
$$= \frac{\sqrt{2}}{2} \cdot \frac{\sqrt{3}}{2} + \frac{\sqrt{2}}{2} \cdot \frac{1}{2} = \frac{\sqrt{6} + \sqrt{2}}{4}$$

(b) $\tan \dfrac{7}{12} \pi = \tan \left(\dfrac{\pi}{3} + \dfrac{\pi}{4} \right)$

$$= \frac{\tan \dfrac{\pi}{3} + \tan \dfrac{\pi}{4}}{1 - \tan \dfrac{\pi}{3} \tan \dfrac{\pi}{4}} = \frac{\sqrt{3} + 1}{1 - \sqrt{3} \cdot 1}$$

To simplify this result, rationalize the denominator by multiplying numerator and denominator by $1 + \sqrt{3}$.

$$\tan \frac{7}{12} \pi = \frac{\sqrt{3} + 1}{1 - \sqrt{3}} \cdot \frac{1 + \sqrt{3}}{1 + \sqrt{3}}$$

$$= \frac{3 + 2\sqrt{3} + 1}{1 - 3}$$

$$\tan \frac{7}{12} \pi = -2 - \sqrt{3} \quad \blacksquare$$

Example 2   If $\sin s = 4/5$ and $\cos t = -5/13$, where s is in quadrant II and t is in quadrant III, find each of the following.

(a) $\sin (s + t)$

Use the identity for the sine of the sum of two angles,

$$\sin (s + t) = \sin s \cos t + \cos s \sin t.$$

In addition to the given values of $\sin s$ and $\cos t$, this identity requires values of $\cos s$ and $\sin t$. The first of these can be found by using $\sin^2 s + \cos^2 s = 1$.

$$\sin^2 s + \cos^2 s = 1$$

$$\frac{16}{25} + \cos^2 s = 1$$

$$\cos^2 s = \frac{9}{25}$$

$$\cos s = -\frac{3}{5} \quad (s \text{ is in quadrant II})$$

In the same way, find $\sin t = -12/13$. Now use the formula for $\sin (s + t)$.

$$\sin (s + t) = \frac{4}{5} \left(-\frac{5}{13} \right) + \left(-\frac{3}{5} \right) \left(-\frac{12}{13} \right)$$

$$= -\frac{20}{65} + \frac{36}{65} = \frac{16}{65}.$$

(b) $\tan (s + t)$

Use the results of part (a) to get $\tan s = -4/3$ and $\tan t = 12/5$. Then use the identity for $\tan (s + t)$.

$$\tan (s + t) = \frac{\tan s + \tan t}{1 - \tan s \tan t}$$

$$\tan (s + t) = \frac{-\frac{4}{3} + \frac{12}{5}}{1 - \left(-\frac{4}{3} \right) \left(\frac{12}{5} \right)} = \frac{\frac{16}{15}}{1 + \frac{48}{15}} = \frac{\frac{16}{15}}{\frac{63}{15}} = \frac{16}{63}. \quad \blacksquare$$

Example 3   Write each of the following as a function of θ.

(a) $\sin (30° + \theta)$

Using the identity for $\sin (s + t)$,

$$\sin (30° + \theta) = \sin 30° \cos \theta + \cos 30° \sin \theta$$

$$\sin (30° + \theta) = \frac{1}{2} \cos \theta + \frac{\sqrt{3}}{2} \sin \theta.$$

(b) $\tan (45° - \theta) = \dfrac{\tan 45° - \tan \theta}{1 + \tan 45° \tan \theta}$

$$\tan (45° - \theta) = \frac{1 - \tan \theta}{1 + \tan \theta} \quad \blacksquare$$

Reduction Identity   Equations of the form $y = a \sin x + b \cos x$ occur so often in mathematics that it is useful to know how to rewrite them in simpler form. In particular, these equations must often be graphed, which can be done with the procedure developed here. Figure 4 shows a circle of radius r with the point (a, b) on the circle and on the terminal side of angle α. The circle has equation $x^2 + y^2 = r^2$, so that

$$a^2 + b^2 = r^2, \quad \text{or} \quad r = \sqrt{a^2 + b^2}.$$

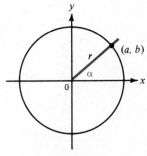

Figure 4

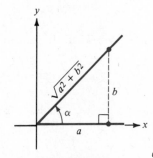

Figure 5

From Figure 5,

$$\sin \alpha = \frac{b}{r} = \frac{b}{\sqrt{a^2 + b^2}} \quad \text{and} \quad \cos \alpha = \frac{a}{r} = \frac{a}{\sqrt{a^2 + b^2}}.$$

Now rewrite $a \sin x + b \cos x$ as follows.

$$a \sin x + b \cos x = \frac{a}{1} \cdot \frac{\sqrt{a^2 + b^2}}{\sqrt{a^2 + b^2}} \sin x + \frac{b}{1} \cdot \frac{\sqrt{a^2 + b^2}}{\sqrt{a^2 + b^2}} \cos x$$

$$= \sqrt{a^2 + b^2} \left(\frac{a}{\sqrt{a^2 + b^2}} \sin x + \frac{b}{\sqrt{a^2 + b^2}} \cos x \right)$$

$$= \sqrt{a^2 + b^2} \left(\sin x \frac{a}{\sqrt{a^2 + b^2}} + \cos x \frac{b}{\sqrt{a^2 + b^2}} \right).$$

Substitute $\sin \alpha$ and $\cos \alpha$ from above.

$$= \sqrt{a^2 + b^2} \, (\sin x \cos \alpha + \cos x \sin \alpha)$$

Using the identity for $\sin (s + t)$ produces

$$a \sin x + b \cos x = \sqrt{a^2 + b^2} \, \sin (x + \alpha).$$

This result, the **reduction identity,** is summarized as follows.

Reduction Identity	$a \sin x + b \cos x = \sqrt{a^2 + b^2} \, \sin (x + \alpha),$ where $\quad \sin \alpha = \dfrac{b}{\sqrt{a^2 + b^2}} \quad$ and $\quad \cos \alpha = \dfrac{a}{\sqrt{a^2 + b^2}}.$

Example 4 Rewrite $\dfrac{1}{2} \sin \theta + \dfrac{\sqrt{3}}{2} \cos \theta$ using the reduction identity.

From the identity above, $a = \dfrac{1}{2}$ and $b = \dfrac{\sqrt{3}}{2}$, so that

$$a \sin \theta + b \cos \theta = \sqrt{a^2 + b^2} \, \sin (\theta + \alpha)$$

becomes

$$\frac{1}{2} \sin \theta + \frac{\sqrt{3}}{2} \cos \theta = 1 \cdot \sin (\theta + \alpha),$$

where angle α satisfies the conditions

$$\sin \alpha = \frac{b}{\sqrt{a^2 + b^2}} = \frac{\sqrt{3}}{2} \quad \text{and} \quad \cos \alpha = \frac{a}{\sqrt{a^2 + b^2}} = \frac{1}{2}.$$

The smallest possible positive value of α that satisfies both of these conditions is $\alpha = 60°$. Thus

$$\frac{1}{2} \sin \theta + \frac{\sqrt{3}}{2} \cos \theta = \sin (\theta + 60°). \quad \blacksquare$$

The reduction identity of this section is useful for graphs which involve sums of sines and cosines. It can be used instead of the method of addition of ordinates discussed in Chapter 3.

Example 5 Graph $y = \sin x + \cos x$.

Rewrite $\sin x + \cos x$ using the reduction identity. Since $a = b = 1$, $\sqrt{a^2 + b^2} = \sqrt{2}$, and

$$\sin x + \cos x = \sqrt{2} \, \sin (x + \alpha).$$

To find α, let

$$\sin \alpha = \frac{b}{\sqrt{a^2 + b^2}} = \frac{1}{\sqrt{2}} \quad \text{and} \quad \cos \alpha = \frac{a}{\sqrt{a^2 + b^2}} = \frac{1}{\sqrt{2}}.$$

The smallest positive angle satisfying these conditions is $\pi/4$, so that

$$y = \sin x + \cos x = \sqrt{2} \sin\left(x + \frac{\pi}{4}\right).$$

The last expression shows that the graph has amplitude $\sqrt{2}$, a period of 2π, and a phase shift of $\pi/4$ to the left, as shown in Figure 6. ■

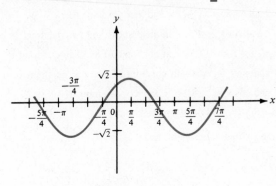

$y = \sin x + \cos x$

Figure 6

4.4 Exercises

Use the identities of this section to find the value of each of the following without using calculators or tables.

1. $\sin 15°$

2. $\sin 105°$

3. $\tan 15°$

4. $\tan (-105°)$

5. $\sin (-105°)$

6. $\tan \dfrac{5\pi}{12}$

7. $\sin \dfrac{5\pi}{12}$

8. $\sin 285°$

9. $\sin 76° \cos 31° - \cos 76° \sin 31°$

10. $\sin 40° \cos 50° + \cos 40° \sin 50°$

11. $\dfrac{\tan 80° + \tan 55°}{1 - \tan 80° \tan 55°}$

12. $\dfrac{\tan 80° - \tan (-55°)}{1 + \tan 80° \tan (-55°)}$

13. $\dfrac{\tan 100° + \tan 80°}{1 - \tan 100° \tan 80°}$

14. $\sin 100° \cos 10° - \cos 100° \sin 10°$

15. $\sin \pi/5 \cos 3\pi/10 + \cos \pi/5 \sin 3\pi/10$

16. $\dfrac{\tan 5\pi/12 + \tan \pi/4}{1 - \tan 5\pi/12 \tan \pi/4}$

17. $\sin 79.802° \cos 100.198° + \cos 79.802° \sin 100.198°$

18. $\sin 296.4372° \cos 26.4372° - \cos 296.4372° \sin 26.4372°$

19. $\dfrac{\tan 151.9063° + \tan 28.0937°}{1 - \tan 151.9063° \tan 28.0937°}$

20. $\dfrac{\tan 214.91° + \tan 145.09°}{1 - \tan 214.91° \tan 145.09°}$

Write each of the following as a function of θ or s.

21. $\sin (45° + \theta)$

22. $\sin (\theta - 30°)$

23. $\tan (\theta + 30°)$

24. $\tan (60° - \theta)$

25. $\tan (\pi/4 + s)$

26. $\sin (\pi/4 + s)$

27. $\sin (180° - \theta)$

28. $\sin (270° - \theta)$

29. $\tan (180° + \theta)$

30. $\tan (360° - \theta)$

31. $\sin (\pi + \theta)$

32. $\tan (\pi - \theta)$

For each of the following, find sin $(s + t)$, sin $(s - t)$, tan $(s + t)$, and tan $(s - t)$.

33. cos $s = 3/5$ and sin $t = 5/13$, s and t in quadrant I

34. cos $s = -1/5$ and sin $t = 3/5$, s and t in quadrant II

35. sin $s = 2/3$ and sin $t = -1/3$, s in quadrant II and t in quadrant IV

36. sin $s = 3/5$ and sin $t = -12/13$, s in quadrant I and t in quadrant III

37. cos $s = -8/17$ and cos $t = -3/5$, s and t in quadrant III

38. cos $s = -15/17$ and sin $t = 4/5$, s in quadrant II and t in quadrant I

39. sin $s = -4/5$ and cos $t = 12/13$, s in quadrant III and t in quadrant IV

40. sin $s = -5/13$ and sin $t = 3/5$, s in quadrant III and t in quadrant II

41. sin $s = -8/17$ and cos $t = -8/17$, s and t in quadrant III

42. sin $s = 2/3$ and sin $t = 2/5$, s and t in quadrant I

Verify that each of the following are identities.

43. $\sin\left(\dfrac{\pi}{2} + x\right) = \cos x$

44. $\sin\left(\dfrac{3\pi}{2} + x\right) = -\cos x$

45. $\tan\left(\dfrac{\pi}{2} + x\right) = -\cot x$ $\left(\textit{Hint: } \tan \theta = \dfrac{\sin \theta}{\cos \theta}\right)$

46. $\tan\left(\dfrac{\pi}{4} + x\right) = \dfrac{1 + \tan x}{1 - \tan x}$

47. $\sin 2x = 2 \sin x \cos x$ (*Hint:* $\sin 2x = \sin (x + x)$)

48. $\sin (x + y) + \sin (x - y) = 2 \sin x \cos y$

49. $\tan (x - y) - \tan (y - x) = \dfrac{2(\tan x - \tan y)}{1 + \tan x \tan y}$

50. $\sin (210° + x) - \cos (120° + x) = 0$

51. $\dfrac{\cos (\alpha - \beta)}{\cos \alpha \sin \beta} = \tan \alpha + \cot \beta$

52. $\dfrac{\sin (s + t)}{\cos s \cos t} = \tan s + \tan t$

53. $\dfrac{\sin (x - y)}{\sin (x + y)} = \dfrac{\tan x - \tan y}{\tan x + \tan y}$

54. $\dfrac{\tan (\alpha + \beta) - \tan \beta}{1 + \tan (\alpha + \beta) \tan \beta} = \tan \alpha$

Let sin $s = 0.599832$, where s terminates in quadrant II. Let sin $t = -0.845992$, where t terminates in quadrant III. Find each of the following.

55. sin $(s - t)$ 　　　　　 **56.** tan $(s + t)$ 　　　　　 **57.** sin $2s$

Use the reduction identity to simplify each of the following for angles between 0° and 360°. Use a calculator or Table 3 to find angles to the nearest degree. Choose the smallest possible positive value of α.

58. $-\sin x + \cos x$ 　　 **59.** $\sqrt{3} \sin x - \cos x$ 　　 **60.** $5 \sin \theta - 12 \cos \theta$ 　　 **61.** $12 \sin A + 5 \cos A$

62. $-15 \sin x + 8 \cos x$ 　　 **63.** $15 \sin B - 8 \cos B$ 　　 **64.** $-7 \sin \theta - 24 \cos \theta$

65. $24 \cos t - 7 \sin t$ 　　 **66.** $3 \sin x + 4 \cos x$ 　　 **67.** $-4 \sin x + 3 \cos x$

Graph each of the following by first changing to the form $y = a \sin (x + \alpha)$.

68. $y = \sqrt{3} \sin x + \cos x$ 　　　　　 **69.** $y = \sin x - \sqrt{3} \cos x$

70. $y = -\sin x + \cos x$ 　　　　　 **71.** $y = -\sin x - \cos x$

72. Why is it not possible to follow Example 3 and find a formula for tan $(270° - \theta)$?

73. What happens when you try to evaluate

$$\frac{\tan 65.902° + \tan 24.098°}{1 - \tan 65.902° \tan 24.098°} ?$$

Derive a formula for each of the following.

74. $\sin (A + B + C)$ **75.** $\cos (A + B + C)$

76. Let $f(x) = \sin x$. Show that

$$\frac{f(x + h) - f(x)}{h} = \sin x \left(\frac{\cos h - 1}{h}\right) + \cos x \left(\frac{\sin h}{h}\right).$$

77. The slope of a line is defined as the ratio of the vertical change and the horizontal change. As shown in the sketch on the left, the tangent of the *angle of inclination* θ is given by the ratio of the side opposite and the side adjacent. This ratio is the same as that used in finding the slope, m, so that $m = \tan \theta$.

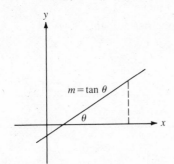 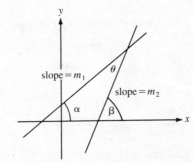

In the figure on the right, let the two lines have angles of inclination α and β, and slopes m_1 and m_2, respectively. Let θ be the smallest positive angle between the lines. Show that

$$\tan \theta = \frac{m_2 - m_1}{1 + m_1 m_2}.$$

Use the results from Exercise 77 to find the angle between the following pairs of lines. Round to the nearest tenth of a degree.

78. $x + y = 9$, $2x + y = -1$ **79.** $5x - 2y + 4 = 0$, $3x + 5y = 6$

Use the identities of this section to find each of the following.

80. $\sin (\cos^{-1} 5/13 + \tan^{-1} 3/4)$ **81.** $\sin (\sin^{-1} 2/3 + \tan^{-1} 1/4)$

82. $\sin (\cos^{-1} 1/2 - \tan^{-1} (- 3))$ **83.** $\tan (\tan^{-1} 4/5 - \sin^{-1} 5/13)$

84. Use the results of Exercise 29 to help show that each of the following has a period of π: **(a)** tangent   **(b)** cotangent.

4.5   Multiple-Angle Identities

Some special cases of the identities for the sum of two angles are used often enough to be expressed as separate identities. These are the identities that result from the addition identities when $s = t$, so that $s + t = 2s$. These identities are called

double-angle identities. For example, in the identity for $\cos(s + t)$, let $t = s$ to derive an expression for $\cos 2s$.

$$\begin{aligned}
\cos 2s &= \cos(s + s) \\
&= \cos s \cos s - \sin s \sin s \\
&= \cos^2 s - \sin^2 s
\end{aligned}$$

Substitution from either $\cos^2 s = 1 - \sin^2 s$ or $\sin^2 s = 1 - \cos^2 s$ leads to two alternate forms for this identity, as shown below.

Cosine of Double Angle

$$\begin{aligned}
\cos 2s &= \cos^2 s - \sin^2 s \\
\cos 2s &= 1 - 2 \sin^2 s \\
\cos 2s &= 2 \cos^2 s - 1
\end{aligned}$$

Now find an identity for $\sin 2s$ by starting with the one for $\sin(s + t)$.

$$\begin{aligned}
\sin 2s &= \sin(s + s) \\
&= \sin s \cos s + \cos s \sin s
\end{aligned}$$

Sine of Double Angle

$$\sin 2s = 2 \sin s \cos s.$$

Finally, find $\tan 2s$ from the identity for $\tan(s + t)$.

$$\begin{aligned}
\tan 2s &= \tan(s + s) \\
&= \frac{\tan s + \tan s}{1 - \tan s \tan s}
\end{aligned}$$

Tangent of Double Angle

$$\tan 2s = \frac{2 \tan s}{1 - \tan^2 s}$$

Example 1 Simplify each of the following.

(a) $\sin 15° \cos 15°$

The product of the sine and cosine of the same angle suggests the identity for $\sin 2s$: $\sin 2s = 2 \sin s \cos s$. With this identity,

$$\begin{aligned}
\sin 15° \cos 15° &= \left(\frac{1}{2}\right)(2) \sin 15° \cos 15° \\
&= \frac{1}{2}(2 \sin 15° \cos 15°) = \frac{1}{2}(\sin 2 \cdot 15°) \\
&= \frac{1}{2} \sin 30° = \frac{1}{2} \cdot \frac{1}{2} = \frac{1}{4}.
\end{aligned}$$

(b) $\cos^2 7x - \sin^2 7x$

This expression suggests an identity for $\cos 2s$: $\cos 2s = \cos^2 s - \sin^2 s$. Substituting $7x$ for s gives

$$\cos^2 7x - \sin^2 7x = \cos 2(7x) = \cos 14x. \quad \blacksquare$$

The double-angle identities for $2s$ can be used to find values of the trigonometric functions of s as shown in the following example.

Example 2   Find the values of the six trigonometric functions of θ if $\cos 2\theta = 4/5$ and θ terminates in quadrant II.

Use one of the double-angle identities for cosine.

$$\cos 2\theta = 1 - 2 \sin^2 \theta$$

$$\frac{4}{5} = 1 - 2 \sin^2 \theta$$

$$-\frac{1}{5} = -2 \sin^2 \theta$$

$$\frac{1}{10} = \sin^2 \theta$$

$$\sin \theta = \frac{\sqrt{10}}{10}$$

The positive square root was chosen since θ terminates in quadrant II. The values of $\cos \theta$ and $\tan \theta$ can now be found using the fundamental identities.

$$\sin^2 \theta + \cos^2 \theta = 1$$

$$\frac{1}{10} + \cos^2 \theta = 1$$

$$\cos^2 \theta = \frac{9}{10}$$

$$\cos \theta = \frac{-3}{\sqrt{10}} \qquad \text{since } \theta \text{ is in quadrant II}$$

$$\cos \theta = \frac{-3\sqrt{10}}{10}$$

Verify that $\tan \theta = \sin \theta / \cos \theta = -1/3$. Use reciprocals to find that $\csc \theta = \sqrt{10}$, $\sec \theta = -\sqrt{10}/3$, and $\cot \theta = -3$. $\quad \blacksquare$

Example 3   Given $\cos \theta = 3/5$, where $3\pi/2 < \theta < 2\pi$, find $\cos 2\theta$, $\sin 2\theta$, and $\tan 2\theta$.

From $\cos \theta = 3/5$, the identity $\sin^2 \theta + \cos^2 \theta = 1$ leads to $\sin \theta = \pm 4/5$. Since θ terminates in quadrant IV, $\sin \theta = -4/5$. Then, using the double-angle identities,

$$\sin 2\theta = 2 \sin \theta \cos \theta = 2 \left(-\frac{4}{5}\right)\left(\frac{3}{5}\right) = -\frac{24}{25}.$$

Also,

$$\cos 2\theta = \cos^2 \theta - \sin^2 \theta = \frac{9}{25} - \frac{16}{25} = -\frac{7}{25}$$

$$\tan 2\theta = \frac{\sin 2\theta}{\cos 2\theta} = \frac{-24/25}{-7/25} = \frac{24}{7}.$$

As an alternative way of finding $\tan 2\theta$, start with $\sin \theta = -4/5$ and $\cos \theta = 3/5$, to get $\tan \theta = -4/3$, with

$$\tan 2\theta = \frac{2 \tan \theta}{1 - \tan^2 \theta} = \frac{2\left(-\frac{4}{3}\right)}{1 - \frac{16}{9}} = \frac{-\frac{8}{3}}{-\frac{7}{9}} = \frac{24}{7}. \qquad \blacksquare$$

These identities, together with the addition identities, allow trigonometric functions of multiple values of s to be rewritten in terms of s.

Example 4   Write $\sin 3s$ in terms of $\sin s$.

$$\begin{aligned}
\sin 3s &= \sin (2s + s) \\
&= \sin 2s \cos s + \cos 2s \sin s \\
&= (2 \sin s \cos s) \cos s + (\cos^2 s - \sin^2 s) \sin s \\
&= 2 \sin s \cos^2 s + \cos^2 s \sin s - \sin^3 s \\
&= 2 \sin s(1 - \sin^2 s) + (1 - \sin^2 s) \sin s - \sin^3 s \\
&= 2 \sin s - 2 \sin^3 s + \sin s - \sin^3 s - \sin^3 s \\
\sin 3s &= 3 \sin s - 4 \sin^3 s \qquad \blacksquare
\end{aligned}$$

From the alternate forms of the double-angle identity for cosine, three additional identities can be derived. These **half-angle identities,** listed below, are used in the study of calculus.

Half-angle Identities

$$\cos \frac{s}{2} = \pm\sqrt{\frac{1 + \cos s}{2}} \qquad \sin \frac{s}{2} = \pm\sqrt{\frac{1 - \cos s}{2}}$$

$$\tan \frac{s}{2} = \pm\sqrt{\frac{1 - \cos s}{1 + \cos s}} \qquad \tan \frac{s}{2} = \frac{\sin s}{1 + \cos s}$$

$$\text{and} \qquad \tan \frac{s}{2} = \frac{1 - \cos s}{\sin s}$$

In these identities, the plus or minus sign is selected according to the quadrant in which $s/2$ terminates. For example, if s represents an angle of $324°$, then $s/2 = 162°$, which lies in quadrant II. In quadrant II, $\cos s/2$ and $\tan s/2$ are negative, and $\sin s/2$ is positive.

To derive the identity for sin $s/2$, start with the identity

$$\cos 2x = 1 - 2\sin^2 x.$$

Now solve for sin x.

$$2\sin^2 x = 1 - \cos 2x$$

$$\sin x = \pm\sqrt{\frac{1 - \cos 2x}{2}}$$

Let $2x = s$, so that $x = s/2$, and substitute into this last expression.

$$\sin\frac{s}{2} = \pm\sqrt{\frac{1 - \cos s}{2}}$$

The identity for cos $s/2$ from the box above is derived in a similar way, by starting with the double-angle identity $\cos 2x = 2\cos^2 x - 1$. One identity for tan $s/2$ comes from the half-angle identities for sine and cosine.

$$\tan\frac{s}{2} = \frac{\pm\sqrt{\dfrac{1 - \cos s}{2}}}{\pm\sqrt{\dfrac{1 + \cos s}{2}}} = \pm\sqrt{\frac{1 - \cos s}{1 + \cos s}}$$

The other two identities for tan $s/2$ given above are proven in Exercise 96 together with results of Example 4, Section 4.2.

Example 5   Find cos 112.5°.

Since $112.5° = 225°/2$, use the identity for cos $s/2$ with $s = 225°$. Since 112.5° is in quadrant II, where cosine is negative, the minus sign must be used on the radical.

$$\cos 112.5° = \cos\frac{225°}{2} = -\sqrt{\frac{1 + \cos 225°}{2}}$$

$$= -\sqrt{\frac{1 - \dfrac{\sqrt{2}}{2}}{2}} = -\sqrt{\frac{2 - \sqrt{2}}{4}} = -\frac{\sqrt{2 - \sqrt{2}}}{2} \quad\blacksquare$$

Example 6   Find tan 22.5°.

Use the identity $\tan\dfrac{s}{2} = \dfrac{\sin s}{1 + \cos s}$, with $s = 45°$.

$$\tan 22.5° = \tan\frac{45°}{2} = \frac{\sin 45°}{1 + \cos 45°}$$

$$= \frac{\dfrac{\sqrt{2}}{2}}{1 + \dfrac{\sqrt{2}}{2}} = \frac{\sqrt{2}}{2 + \sqrt{2}} = \sqrt{2} - 1$$

(Here the denominator was rationalized.) See also Exercise 95. $\blacksquare$

Example 7 Simplify each of the following.

(a) $\pm\sqrt{\dfrac{1 + \cos 12x}{2}}$

Start with the identity for cos $s/2$,

$$\cos s/2 = \pm\sqrt{\dfrac{1 + \cos s}{2}},$$

and replace s with $12x$ to get

$$\pm\sqrt{\dfrac{1 + \cos 12x}{2}} = \cos\dfrac{12x}{2} = \cos 6x.$$

(b) $\dfrac{1 - \cos 5\alpha}{\sin 5\alpha}$

Use the third identity for tan $s/2$ given above to get

$$\dfrac{1 - \cos 5\alpha}{\sin 5\alpha} = \tan\dfrac{5\alpha}{2}. \quad \blacksquare$$

Example 8 Given cos $s = 2/3$, with $3\pi/2 < s < 2\pi$, find cos $s/2$, sin $s/2$ and tan $s/2$.

Since $3\pi/2 < s < 2\pi$, $3\pi/4 < s/2 < \pi$. Thus, $s/2$ terminates in quadrant II and cos $s/2$ and tan $s/2$ are negative, while sin $s/2$ is positive. From the half-angle identities,

$$\sin\dfrac{s}{2} = \sqrt{\dfrac{1 - \dfrac{2}{3}}{2}} = \sqrt{\dfrac{1}{6}} = \dfrac{\sqrt{6}}{6}$$

$$\cos\dfrac{s}{2} = -\sqrt{\dfrac{1 + \dfrac{2}{3}}{2}} = -\sqrt{\dfrac{5}{6}} = -\dfrac{\sqrt{30}}{6}$$

$$\tan\dfrac{s}{2} = \dfrac{\dfrac{\sqrt{6}}{6}}{-\dfrac{\sqrt{30}}{6}} = \dfrac{-\sqrt{5}}{5}. \quad \blacksquare$$

4.5 Exercises

Write each of the following as a single trigonometric function value.

1. $\sqrt{\dfrac{1 - \cos 40°}{2}}$

2. $\sqrt{\dfrac{1 + \cos 76°}{2}}$

3. $\sqrt{\dfrac{1 - \cos 147°}{1 + \cos 147°}}$

4. $\sqrt{\dfrac{1 + \cos 165°}{1 - \cos 165°}}$

5. $\dfrac{1 - \cos 59.74°}{\sin 59.74°}$

6. $\dfrac{\sin 158.2°}{1 + \cos 158.2°}$

7. $\pm\sqrt{\dfrac{1 + \cos 18x}{2}}$

8. $\pm\sqrt{\dfrac{1 + \cos 20\alpha}{2}}$

9. $\pm\sqrt{\dfrac{1 - \cos 8\theta}{1 + \cos 8\theta}}$

10. $\pm\sqrt{\dfrac{1 - \cos 5A}{1 + \cos 5A}}$

11. $\pm\sqrt{\dfrac{1 + \cos x/4}{2}}$

12. $\pm\sqrt{\dfrac{1 - \cos 3\theta/5}{2}}$

Write each of the following as a single trigonometric function value or as a single number.

13. $1 - 2 \sin^2 15°$

14. $\dfrac{2 \tan 15°}{1 - \tan^2 15°}$

15. $2 \sin \pi/3 \cos \pi/3$

16. $\dfrac{2 \tan \pi/3}{1 - \tan^2 \pi/3}$

17. $\sin \pi/8 \cos \pi/8$

18. $\cos^2 \pi/8 - 1/2$

19. $\dfrac{\tan 51°}{1 - \tan^2 51°}$

20. $\dfrac{1}{4} - \dfrac{1}{2} \sin^2 47.1°$

21. $\dfrac{1}{8} \sin 29.5° \cos 29.5°$

22. $\sin^2 2\pi/5 - \cos^2 2\pi/5$

23. $\cos^2 2\alpha - \sin^2 2\alpha$

24. $\dfrac{\tan 2y/5}{1 - \tan^2 2y/5}$

Determine whether the positive or negative square root should be selected.

25. $\sin 195° = \pm \sqrt{\dfrac{1 - \cos 390°}{2}}$

26. $\cos 58° = \pm \sqrt{\dfrac{1 + \cos 116°}{2}}$

27. $\tan 225° = \pm \sqrt{\dfrac{1 - \cos 450°}{1 + \cos 450°}}$

28. $\sin (-10°) = \pm \sqrt{\dfrac{1 - \cos (-20°)}{2}}$

Use the identities of this section to find the values of sine, cosine, and tangent for each of the following.

29. $\theta = 22.5°$

30. $\theta = 15°$

31. $\theta = 195°$

32. $x = -\pi/8$

33. $x = 5\pi/2$

34. $x = 3\pi/2$

Find values of k and t to make the following statements true.

35. $8 \sin 3\theta \cos 3\theta = k \sin t\theta$

36. $2 \cos^2 3\theta - 2 \sin^2 3\theta = k \cos t\theta$

37. $3 - 3 \cos 4\theta = k \sin^2 t\theta$

38. $\dfrac{\tan 2\theta}{1 - \tan^2 2\theta} = k \tan t\theta$

Find each of the following.

39. $\cos x$, if $\cos 2x = -5/12$ and $\pi/2 < x < \pi$

40. $\sin x$, if $\cos 2x = 2/3$ and $\pi < x < 3\pi/2$

41. $\cos \alpha/2$, if $\cos \alpha = -1/4$ and $\pi < \alpha < 3\pi/2$

42. $\sin \beta/2$, if $\cos \beta = 3/4$ and $3\pi/2 < \beta < 2\pi$

Use the identities of this section to find values of the six trigonometric functions for each of the following.

43. x, given $\cos 2x = -5/12$ and $\pi/2 < x < \pi$

44. t, given $\cos 2t = 2/3$ and $\pi/2 < t < \pi$

45. 2θ, given $\sin \theta = 2/5$ and $\cos \theta < 0$

46. 2β, given $\cos \beta = -12/13$ and $\sin \beta > 0$

47. $2x$, given $\tan x = 2$ and $\cos x > 0$

48. $2x$, given $\tan x = 5/3$ and $\sin x < 0$

49. $\alpha/2$, given $\cos \alpha = 1/3$ and $\sin \alpha < 0$

50. $\theta/2$, given $\cos \theta = -2/3$ and $\sin \theta > 0$

Verify that each of the following equations is an identity.

51. $(\sin \gamma + \cos \gamma)^2 = \sin 2\gamma + 1$

52. $\cos 2s = \cos^4 s - \sin^4 s$

53. $\sec 2x = \dfrac{\sec^2 x + \sec^4 x}{2 + \sec^2 x - \sec^4 x}$

54. $\cot s + \tan s = 2 \csc 2s$

55. $\sin 4\alpha = 4 \sin \alpha \cos \alpha \cos 2\alpha$

56. $\dfrac{1 + \cos 2x}{\sin 2x} = \cot x$

57. $\sec^2 \dfrac{x}{2} = \dfrac{2}{1 + \cos x}$ **58.** $\cot^2 \dfrac{x}{2} = \dfrac{(1 + \cos x)^2}{\sin^2 x}$ **59.** $\sin^2 \dfrac{x}{2} = \dfrac{\tan x - \sin x}{2 \tan x}$

60. $\dfrac{\sin 2x}{2 \sin x} = \cos^2 \dfrac{x}{2} - \sin^2 \dfrac{x}{2}$ **61.** $\dfrac{2}{1 + \cos x} - \tan^2 \dfrac{x}{2} = 1$ **62.** $\tan \dfrac{\gamma}{2} = \csc \gamma - \cot \gamma$

63. $\tan 8k - \tan 8k \cdot \tan^2 4k = 2 \tan 4k$ **64.** $\sin 2\gamma = \dfrac{2 \tan \gamma}{1 + \tan^2 \gamma}$

65. $-\tan 2\theta = \dfrac{2 \tan \theta}{\sec^2 \theta - 2}$ **66.** $\cos 2y = \dfrac{2 - \sec^2 y}{\sec^2 y}$

67. $\dfrac{2 \cos 2\alpha}{\sin 2\alpha} = \cot \alpha - \tan \alpha$ **68.** $\sin 2\alpha \cos 2\alpha = \sin 2\alpha - 4 \sin^3 \alpha \cos \alpha$

69. $\dfrac{\tan \dfrac{x}{2} + \cot \dfrac{x}{2}}{\cot \dfrac{x}{2} - \tan \dfrac{x}{2}} = \sec x$ **70.** $1 - \tan^2 \dfrac{\theta}{2} = \dfrac{2 \cos \theta}{1 + \cos \theta}$

71. $\cos x = \dfrac{1 - \tan^2 \dfrac{x}{2}}{1 + \tan^2 \dfrac{x}{2}}$ **72.** $\dfrac{\sin 2\alpha - 2 \sin \alpha}{2 \sin \alpha + \sin 2\alpha} = -\tan^2 \dfrac{\alpha}{2}$

Express each of the following as trigonometric functions of x.

73. $\tan^2 2x$ **74.** $\cos^2 2x$ **75.** $\cos 3x$

76. $\sin 4x$ **77.** $\cos 4x$ **78.** $\tan 4x$

An airplane flying faster than sound sends out sound waves that form a cone, as shown in the figure. The cone intersects the ground to form a hyperbola. As this hyperbola passes over a particular point on the ground, a sonic boom is heard at that point.

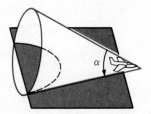

If α is the angle at the vertex of the cone, then

$$\sin \frac{\alpha}{2} = \frac{1}{m}$$

where m is the Mach number of the plane. (We assume $m > 1$.) The Mach number is the ratio of the speed of the plane and the speed of sound. For example, a speed of Mach 1.4 means that the plane is flying 1.4 times the speed of sound. Find α or m, as necessary, for each of the following.

79. $m = 3/2$ **80.** $m = 5/4$ **81.** $m = 2$

82. $m = 5/2$ **83.** $\alpha = 30°$ **84.** $\alpha = 60°$

Let $\sin s = -0.481143$, with $7\pi/4 < s < 2\pi$. Find each of the following.

85. $\sin 2s$

86. $\sin \frac{1}{2}s$

87. $\cos 2s$

88. $\cos \frac{1}{2}s$

89. $\tan 2s$

90. $\tan \frac{1}{2}s$

Let $\cos s = -0.592147$, with $\pi < s < 3\pi/2$. Use the identities of this section to find each of the following.

91. $\sin \frac{1}{2}s$

92. $\cos \frac{1}{2}s$

93. $\tan \frac{1}{2}s$

94. $\csc \frac{1}{2}s$

95. In Example 6 the identity

$$\tan \frac{s}{2} = \frac{\sin s}{1 + \cos s}$$

was used to find that $\tan 22.5° = \sqrt{2} - 1$.
  (a) Find $\tan 22.5°$ with the identity $\tan s/2 = \pm\sqrt{(1 - \cos s)/(1 + \cos s)}$.
  (b) Show that both answers are the same.

96. Go through the following steps to prove that

$$\tan \frac{A}{2} = \frac{\sin A}{1 + \cos A}.$$

  (a) Start with $\tan A/2 = \pm\sqrt{(1 - \cos A)/(1 + \cos A)}$, and multiply numerator and denominator by $\sqrt{1 + \cos A}$ to show that

$$\tan \frac{A}{2} = \pm\left|\frac{\sin A}{1 + \cos A}\right|.$$

  (b) Show that $1 + \cos A \geq 0$, giving

$$\tan \frac{A}{2} = \frac{\pm|\sin A|}{1 + \cos A}.$$

  (c) By considering the quadrant in which A lies, show that $\tan A/2$ and $\sin A$ have the same sign, with

$$\tan \frac{A}{2} = \frac{\sin A}{1 + \cos A}.$$

Simplify each of the following.

97. $\sin^2 2x + \cos^2 2x$

98. $\sin^2 \frac{x}{2} + \cos^2 \frac{x}{2}$

99. $1 + \tan^2 4x$

100. $\cot^2 \frac{x}{3} + 1$

4.6 Sum and Product Identities

One group of identities in this section can be used to rewrite a product of two functions as a sum or difference. The other group can be used to rewrite a sum or difference of two functions as a product. Some of these identities can also be used to rewrite an expression involving both sine and cosine functions as one with only one of these functions. In the next section on conditional equations, the need for this kind of change will become clear. These identities are also useful in graphing and in calculus.

The identities of this section all result from the sum and difference identities for sine and cosine. Adding the identities for $\sin (s + t)$ and $\sin (s - t)$ gives

$$\sin (s + t) = \sin s \cos t + \cos s \sin t$$
$$\sin (s - t) = \sin s \cos t - \cos s \sin t$$
$$\overline{\sin (s + t) + \sin (s - t) = 2 \sin s \cos t,}$$

or

$$\sin s \cos t = \frac{1}{2} [\sin (s + t) + \sin (s - t)].$$

Subtract $\sin (s - t)$ from $\sin (s + t)$ to get

$$\cos s \sin t = \frac{1}{2} [\sin (s + t) - \sin (s - t)].$$

Using the identities for $\cos (s + t)$ and $\cos (s - t)$ in a similar way gives

$$\cos s \cos t = \frac{1}{2} [\cos (s + t) + \cos (s - t)]$$

$$\sin s \sin t = \frac{1}{2} [\cos (s - t) - \cos (s + t)].$$

Example 1   Rewrite $\cos 2\theta \sin \theta$ as the sum or difference of two functions.
Using the identity for $\cos s \sin t$ gives

$$\cos 2\theta \sin \theta = \frac{1}{2} (\sin 3\theta - \sin \theta)$$

$$\cos 2\theta \sin \theta = \frac{1}{2} \sin 3\theta - \frac{1}{2} \sin \theta. \quad \blacksquare$$

Example 2   Evaluate $\cos 15° \cos 45°$.
Use the identity for $\cos s \cos t$.

$$\cos 15° \cos 45° = \frac{1}{2} [\cos (15° + 45°) + \cos (15° - 45°)]$$

$$= \frac{1}{2} [\cos 60° + \cos (-30°)] = \frac{1}{2} (\cos 60° + \cos 30°)$$

$$= \frac{1}{2} \left(\frac{1}{2} + \frac{\sqrt{3}}{2} \right) = \frac{1 + \sqrt{3}}{4} \quad \blacksquare$$

From these new identities further identities can be found which are used in calculus to rewrite a sum of trigonometric functions as a product. To begin, let $s + t = x$, and let $s - t = y$. Then adding x and y gives

$$x + y = (s + t) + (s - t) = 2s$$

or

$$s = \frac{x + y}{2}.$$

Subtracting y from x produces

$$x - y = (s + t) - (s - t) = 2t$$

from which

$$t = \frac{x - y}{2}.$$

With these results, the identity

$$\sin s \cos t = \frac{1}{2} [\sin (s + t) + \sin (s - t)]$$

becomes

$$\sin \left(\frac{x + y}{2} \right) \cos \left(\frac{x - y}{2} \right) = \frac{1}{2} (\sin x + \sin y),$$

or, by reversing the equality and multiplying on both sides by 2,

$$\sin x + \sin y = 2 \sin \left(\frac{x + y}{2} \right) \cos \left(\frac{x - y}{2} \right).$$

The following three identities can be obtained in a similar way.

$$\sin x - \sin y = 2 \cos \left(\frac{x + y}{2} \right) \sin \left(\frac{x - y}{2} \right)$$

$$\cos x + \cos y = 2 \cos \left(\frac{x + y}{2} \right) \cos \left(\frac{x - y}{2} \right)$$

$$\cos x - \cos y = -2 \sin \left(\frac{x + y}{2} \right) \sin \left(\frac{x - y}{2} \right)$$

Example 3   Write $\sin 2\gamma - \sin 4\gamma$ as a product of two functions.
Use the identity for $\sin x - \sin y$.

$$\sin 2\gamma - \sin 4\gamma = 2 \cos \left(\frac{2\gamma + 4\gamma}{2} \right) \sin \left(\frac{2\gamma - 4\gamma}{2} \right)$$

$$= 2 \cos \frac{6\gamma}{2} \sin \frac{-2\gamma}{2}$$

$$= 2 \cos 3\gamma \sin (-\gamma)$$

$$\sin 2\gamma - \sin 4\gamma = -2 \cos 3\gamma \sin \gamma \quad \blacksquare$$

Example 4 Verify that $\dfrac{\sin 3s + \sin s}{\cos s + \cos 3s} = \tan 2s$ is an identity.

Work on the left side as follows.

$$\frac{\sin 3s + \sin s}{\cos s + \cos 3s} = \frac{2 \sin\left(\dfrac{3s + s}{2}\right) \cos\left(\dfrac{3s - s}{2}\right)}{2 \cos\left(\dfrac{s + 3s}{2}\right) \cos\left(\dfrac{s - 3s}{2}\right)}$$

$$= \frac{\sin 2s \cos s}{\cos 2s \cos (-s)} = \frac{\sin 2s}{\cos 2s} = \tan 2s \quad \blacksquare$$

4.6 Exercises

Rewrite each of the following as a sum or difference of trigonometric functions.

1. $\cos 35° \sin 25°$ **2.** $2 \sin 2x \sin 4x$ **3.** $3 \cos 5x \cos 3x$ **4.** $2 \sin 74° \cos 114°$

5. $\sin (-\theta) \sin (-3\theta)$ **6.** $4 \cos (-32°) \sin 15°$ **7.** $-8 \cos 4y \cos 5y$ **8.** $2 \sin 3k \sin 14k$

Rewrite each of the following as a product of trigonometric functions.

9. $\sin 60° - \sin 30°$ **10.** $\sin 28° + \sin (-18°)$ **11.** $\cos 42° + \cos 148°$ **12.** $\cos 2x - \cos 8x$

13. $\sin 12\beta - \sin 3\beta$ **14.** $\cos 5x + \cos 10x$ **15.** $-3 \sin 2x + 3 \sin 5x$ **16.** $-\cos 8s + \cos 14s$

Verify that each of the following is an identity.

17. $\tan x = \dfrac{\sin 3x - \sin x}{\cos 3x + \cos x}$ **18.** $\dfrac{\sin 5t + \sin 3t}{\cos 3t - \cos 5t} = \cot t$ **19.** $\dfrac{\cot 2\theta}{\tan 3\theta} = \dfrac{\cos 5\theta + \cos \theta}{\cos \theta - \cos 5\theta}$

20. $\dfrac{\cos \alpha + \cos \beta}{\cos \alpha - \cos \beta} = -\cot\left(\dfrac{\alpha + \beta}{2}\right) \cot\left(\dfrac{\alpha - \beta}{2}\right)$ **21.** $\dfrac{1}{\tan 2s} = \dfrac{\sin 3s - \sin s}{\cos s - \cos 3s}$

22. $\dfrac{\sin^2 5\alpha - 2 \sin 5\alpha \sin 3\alpha + \sin^2 3\alpha}{\sin^2 5\alpha - \sin^2 3\alpha} = \dfrac{\tan \alpha}{\tan 4\alpha}$ **23.** $\sin 6\theta \cos 4\theta - \sin 3\theta \cos 7\theta = \sin 3\theta \cos \theta$

24. $\sin 8\beta \sin 4\beta + \cos 10\beta \cos 2\beta = \cos 6\beta \cos 2\beta$ **25.** $\sin^2 u - \sin^2 v = \sin (u + v) \sin (u - v)$

26. $\cos^2 u - \cos^2 v = -\sin (u + v) \sin (u - v)$

27. Show that the double-angle identity for sine can be considered a special case of the identity $\sin s \cos t = (1/2) [\sin (s + t) + \sin (s - t)]$.

28. Show that the double-angle identity $\cos 2s = 2 \cos^2 s - 1$ is a special case of the identity $\cos s \cos t = \dfrac{1}{2} [\cos (s + t) + \cos (s - t)]$.

4.7 Trigonometric Equations

So far in this chapter trigonometric identities, statements which are true for every value in the domain of the variable, have been discussed. This section discusses conditional equations which involve trigonometric functions. As mentioned in the introduction to this chapter, a **conditional equation** is an equation in which some replacements for the variable make the statement true, while others make it false. For example, $2x + 3 = 5$, $x^2 - 5x = 10$, and $2^x = 8$ are conditional equations. Conditional equations with trigonometric functions can usually be solved by using algebraic methods and trigonometric identities to simplify the equations. The next examples show methods for solving trigonometric equations.

Example 1   Solve $3 \sin x = \sqrt{3} + \sin x$ for x in degrees.

First, solve for $\sin x$. Collect all terms with $\sin x$ on one side of the equation.

$$3 \sin x = \sqrt{3} + \sin x$$
$$2 \sin x = \sqrt{3}$$
$$\sin x = \frac{\sqrt{3}}{2}$$

There are many possible solutions for x: $\sin 60° = \sqrt{3}/2$; so do $\sin 120°$, $\sin 420°$, and so on. There are an infinite number of values of x that satisfy this equation. This infinite number of solutions gives the solution set

$$\{x | x = 60° + 360° \cdot n \text{ or } x = 120° + 360° \cdot n, \, n \text{ any integer}\}.$$

Often the solutions of a trigonometric equation are required only for some particular interval. For example, if the solutions of this equation were restricted to the interval $0° \leq x < 360°$, just the solutions $60°$ and $120°$ would be given.   ∎

When an equation involves more than one trigonometric function, it is often helpful to use a suitable identity to rewrite the equation in terms of just one trigonometric function, as in the following example.

Example 2   Find all solutions for $\sin x + \cos x = 0$ in the interval $[0°, 360°)$.

Since $\sin x/\cos x = \tan x$ and $\cos x/\cos x = 1$, divide both sides of the equation by $\cos x$ (assuming $\cos x \neq 0$) to get

$$\sin x + \cos x = 0$$
$$\frac{\sin x}{\cos x} + \frac{\cos x}{\cos x} = \frac{0}{\cos x}$$
$$\tan x + 1 = 0$$
$$\tan x = -1.$$

For the last equation the solutions in the given interval are

$$x = 135° \qquad \text{and} \qquad x = 315°,$$

with solution set $\{135°, 315°\}$.

It was assumed here that $\cos x \neq 0$. If $\cos x = 0$, the given equation becomes $\sin x + 0 = 0$, or $\sin x = 0$. If $\cos x = 0$, it is not possible for $\sin x = 0$, so that no solutions were missed by assuming $\cos x \neq 0$.   ∎

The next example shows how factoring may be used to solve a trigonometric equation.

Example 3   Find all solutions for $\sin x \tan x = \sin x$ in the interval $[0°, 360°)$.
Subtract $\sin x$ from both sides, then factor on the left.

$$\sin x \tan x = \sin x$$
$$\sin x \tan x - \sin x = 0$$
$$\sin x \, (\tan x - 1) = 0$$

Now set each factor equal to 0.

$$\sin x = 0 \qquad\qquad \tan x - 1 = 0$$
$$\tan x = 1$$
$$x = 0° \quad \text{or} \quad x = 180° \qquad x = 45° \quad \text{or} \quad x = 225°$$

The solution set is $\{0°, 45°, 180°, 225°\}$.   ∎

There are four solutions for Example 3. Trying to solve the equation by dividing both sides by $\sin x$ would result in $\tan x = 1$, which would give $x = 45°$ or $x = 225°$. The other two solutions would not appear. The missing solutions are the ones that make the divisor, $\sin x$, equal 0. For this reason, it is best to avoid dividing by a variable expression. However, in an equation like that in Example 2, dividing both sides by a variable simplified the solution considerably.

It is important to remember that dividing by a variable expression requires checking to see whether the numbers that make that expression equal to 0 are solutions of the original equation.

Sometimes a trigonometric equation can be solved by first squaring both sides, and then using a trigonometric identity. This works for those identities that involve squares, like $\sin^2 x + \cos^2 x = 1$ or $\tan^2 x + 1 = \sec^2 x$. When squaring both sides of an equation, be sure to check for any numbers that satisfy the squared equation but not the given equation, by substituting potential solutions into the given equation.

Example 4   Find all solutions for $\tan x + \sqrt{3} = \sec x$ in the interval $[0, 2\pi)$.
Square both sides, then express $\sec^2 x$ in terms of $\tan^2 x$.

$$\tan x + \sqrt{3} = \sec x$$
$$\tan^2 x + 2\sqrt{3} \tan x + 3 = \sec^2 x$$
$$\tan^2 x + 2\sqrt{3} \tan x + 3 = 1 + \tan^2 x$$
$$2\sqrt{3} \tan x = -2$$
$$\tan x = -\frac{1}{\sqrt{3}}$$

The possible solutions in the given interval are $5\pi/6$ and $11\pi/6$. Now check the possible solutions. Try $5\pi/6$ first.

$$\tan x + \sqrt{3} = \tan \frac{5\pi}{6} + \sqrt{3} = \frac{-\sqrt{3}}{3} + \sqrt{3} = \frac{2\sqrt{3}}{3}$$

$$\sec x = \sec \frac{5\pi}{6} = \frac{-2\sqrt{3}}{3}$$

By this check, $5\pi/6$ is not a solution. Now try $11\pi/6$.

$$\tan \frac{11\pi}{6} + \sqrt{3} = \frac{-\sqrt{3}}{3} + \sqrt{3} = \frac{2\sqrt{3}}{3}$$

$$\sec \frac{11\pi}{6} = \frac{2\sqrt{3}}{3}$$

So $11\pi/6$ is a solution to the given equation, and the solution set, $\{11\pi/6\}$, contains only one element.   ■

Some trigonometric equations are quadratic in form and can be solved by the methods used to solve quadratic equations.

Example 5   Find all solutions of $\tan^2 x + \tan x - 2 = 0$ in the interval $[0, 2\pi)$.

Let $y = \tan x$, so that the equation becomes $y^2 + y - 2 = 0$, with the left side factorable as $(y - 1)(y + 2) = 0$. Substituting $\tan x$ back for y gives

$$(\mathbf{tan}\ \mathbf{x} - 1)(\mathbf{tan}\ \mathbf{x} + 2) = 0.$$

Set each factor equal to 0.

$$\tan x - 1 = 0 \qquad \text{or} \qquad \tan x + 2 = 0$$
$$\tan x = 1 \qquad \text{or} \qquad \tan x = -2$$

If $\tan x = 1$, then for $[0, 2\pi)$, $x = \pi/4$ or $x = 5\pi/4$. If $x = -2$, then a calculator or Table 3 gives $x = 2.0344$ or 5.1760 (approximately). The solution set for $[0, 2\pi)$ is $\{\pi/4, 5\pi/4, 2.0344, 5.1760\}$.   ■

When a trigonometric equation which is quadratic in form cannot be factored, the quadratic theorem can be used to solve the equation.

Example 6   Find all solutions for $\cot^2 x + 3 \cot x = 1$ in the interval $[0°, 360°)$

Write the equation with 0 on one side.

$$\cot^2 x + 3 \cot x - 1 = 0$$

Since the expression on the left cannot be readily factored, use the quadratic formula with $a = 1$, $b = 3$, $c = -1$, and $\cot x$ as the variable.

$$\cot x = \frac{-3 \pm \sqrt{9 + 4}}{2} = \frac{-3 \pm \sqrt{13}}{2} = \frac{-3 \pm 3.6056}{2}$$

$$\cot x = .3028 \quad \text{or} \quad \cot x = -3.3028$$

From cot $x = .3028$, use a calculator or Table 3 to find $x = 73.2°$ or $253.2°$ (since cotangent has period $180°$). Also, cot $x = -3.3028$ leads to $x = 163.2°$ or $343.2°$ giving the solution set

$$\{73.2°, 163.2°, 253.2°, 343.2°\}. \quad \blacksquare$$

The methods for solving trigonometric equations illustrated in the examples are summarized as follows.

Solving Trigonometric Equations

1. If only one trigonometric function is present, first solve the equation for that function.
2. If more than one trigonometric function is present, rearrange the equation so that one side equals 0. Then try to factor.
3. If method 2 does not work, try using identities to change the form of the equation. It may be helpful to square both sides of the equation first. Be sure to check all proposed solutions.
4. If the equation is quadratic in form, but not easily factorable, use the quadratic formula.

Conditional trigonometric equations where a half angle or multiple angle is given, such as $2 \sin (x/2) = 1$, often require an additional step to solve. This extra step is shown in the following example.

Example 7   Find all solutions for $2 \sin x/2 = 1$ in the interval $[0°, 360°)$.

Dividing the inequality $0° \le x < 360°$ through by 2 gives

$$0° \le \frac{x}{2} < 180°,$$

so that the first step is to find all values of $x/2$ in the interval $[0°, 180°)$. Begin as before by solving for the trigonometric function.

$$2 \sin \frac{x}{2} = 1$$

$$\sin \frac{x}{2} = \frac{1}{2}$$

Both $\sin 30° = 1/2$ and $\sin 150° = 1/2$, and both $30°$ and $150°$ are in the interval $[0°, 180°)$, so

$$\frac{x}{2} = 30° \quad \text{or} \quad \frac{x}{2} = 150°,$$

from which $x = 60° \quad$ or $\quad x = 300°,$

giving the solution set $\{60°, 300°\}. \quad \blacksquare$

Example 8 Find all solutions for $4 \sin x \cos x = \sqrt{3}$ in the interval $[0°, 360°)$.
The identity $2 \sin x \cos x = \sin 2x$ is useful here.

$$4 \sin x \cos x = \sqrt{3}$$
$$2(2 \sin x \cos x) = \sqrt{3}$$
$$2 \sin 2x = \sqrt{3}$$
$$\sin 2x = \frac{\sqrt{3}}{2}$$

The interval $0° \le x < 360°$ implies $0° \le 2x < 720°$, so

$$2x = 60°, 120°, 420°, \text{ or } 480°$$
$$x = 30°, 60°, 210°, \text{ or } 240°,$$

and there are four solutions. The solution set is $\{30°, 60°, 210°, 240°\}$. ∎

Example 9 Find all solutions of $\cos 6x - \cos 2x = -\sin 4x$ in the interval $[0, 2\pi)$.
Use the identity

$$\cos x - \cos y = -2 \sin \left(\frac{x + y}{2}\right) \sin \left(\frac{x - y}{2}\right)$$

to rewrite the given equation as

$$-2 \sin \left(\frac{6x + 2x}{2}\right) \sin \left(\frac{6x - 2x}{2}\right) = -\sin 4x$$
$$-2 \sin 4x \sin 2x = -\sin 4x$$

or $\qquad\qquad\qquad 2 \sin 4x \sin 2x - \sin 4x = 0.$

Factor to get $\qquad\qquad \sin 4x (2 \sin 2x - 1) = 0$

$$\sin 4x = 0 \qquad \text{or} \qquad 2 \sin 2x - 1 = 0$$

$$\sin 4x = 0 \qquad \text{or} \qquad \sin 2x = \frac{1}{2}.$$

The solutions of $\sin 4x = 0$ are given by $4x = 0 + n \cdot \pi$, so

$$x = 0 + n \cdot \frac{\pi}{4}.$$

Letting $n = 0, 1, 2, 3, 4, 5, 6, 7$ gives the solutions

$$0, \ \pi/4, \ \pi/2, \ 3\pi/4, \ \pi, \ 5\pi/4, \ 3\pi/2, \text{ and } 7\pi/4.$$

If $\sin 2x = 1/2$, then $2x = \pi/6 + n \cdot 2\pi$, and $2x = 5\pi/6 + n \cdot 2\pi$. The first of these produces the solutions $\pi/12$ and $13\pi/12$, while the second produces $5\pi/12$ and $17\pi/12$. In summary, the solution set of the original equation is

$$\{0, \ \pi/4, \ \pi/2, \ 3\pi/4, \ \pi, \ 5\pi/4, \ 3\pi/2, \ 7\pi/4, \ \pi/12, \ 13\pi/12, \ 5\pi/12, \ 17\pi/12\}. \quad ∎$$

The next example shows how to solve equations involving inverse trigonometric functions.

Example 10   Solve $\sin^{-1} x - \cos^{-1} x = \pi/6$.

Begin by adding $\cos^{-1} x$ to both sides of the equation to get

$$\sin^{-1} x = \cos^{-1} x + \frac{\pi}{6}.$$

By the definition of the inverse sine, this becomes

$$\sin\left(\cos^{-1} x + \frac{\pi}{6}\right) = x.$$

Let $u = \cos^{-1} x$. Then

$$\sin\left(u + \frac{\pi}{6}\right) = x.$$

Use the identity for $\sin(s + t)$, which gives

$$\sin u \cos \frac{\pi}{6} + \cos u \sin \frac{\pi}{6} = x. \qquad (*)$$

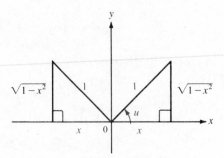

u can be in either quadrant I or II

Figure 7

Since $u = \cos^{-1} x$, we have $\cos u = x$. Because of the domain of $\cos^{-1} x$, u can be in either quadrant I or II. Sketch triangles in each of these quadrants and label them as shown in Figure 7. In either quadrant I or II, $\sin u$ is positive, with

$$\sin u = \sqrt{1 - x^2}.$$

Replace $\sin u$ with $\sqrt{1 - x^2}$, $\sin \pi/6$ with $1/2$, $\cos \pi/6$ with $\sqrt{3}/2$, and $\cos u$ with x, so that equation $(*)$ becomes

$$\sqrt{1 - x^2} \cdot \frac{\sqrt{3}}{2} + x \cdot \frac{1}{2} = x$$
$$\sqrt{1 - x^2} \cdot \sqrt{3} + x = 2x$$
$$\sqrt{3} \cdot \sqrt{1 - x^2} = x.$$

Squaring both sides gives

$$3(1 - x^2) = x^2,$$

from which

$$3 - 3x^2 = x^2$$
$$3 = 4x^2$$
$$x = \pm\sqrt{\frac{3}{4}} = \pm\frac{\sqrt{3}}{2}.$$

Check by substitution in the original equation that $\sqrt{3}/2$ is a solution, while $-\sqrt{3}/2$ is not. The solution set is $\{\sqrt{3}/2\}$. ∎

4.7 Exercises

Find all solutions for the following equations in the interval $[0, 2\pi)$. Use 3.1416 as an approximation for π when you need values from Table 3.

1. $3 \tan x + 5 = 2$

2. $\tan x + 1 = 2$

3. $2 \sec x + 1 = \sec x + 3$

4. $\tan^2 x - 1 = 0$

5. $(\cot x - \sqrt{3})(2 \sin x + \sqrt{3}) = 0$

6. $(\tan x - 1)(\cos x - 1) = 0$

7. $(\sec x - 2)(\sqrt{3} \sec x - 2) = 0$

8. $(2 \sin x + 1)(\sqrt{2} \cos x + 1) = 0$

9. $\cos^2 x + 2 \cos x + 1 = 0$

10. $2 \cos^2 x - \sqrt{3} \cos x = 0$

11. $-2 \sin^2 x = 3 \sin x + 1$

12. $3 \sin^2 x - \sin x = 2$

13. $\cos^2 x - \sin^2 x = 0$

14. $\dfrac{2 \tan x}{3 - \tan^2 x} = 1$

15. $\sin 2x = 0$

16. $\cos 2x = 1$

17. $3 \tan 2x = \sqrt{3}$

18. $\cot 2x = \sqrt{3}$

19. $\sqrt{2} \cos 2x = -1$

20. $2\sqrt{3} \sin 2x = -3$

21. $\sin \dfrac{x}{2} = \sqrt{2} - \sin \dfrac{x}{2}$

22. $\cos 2x - \cos x = 0$

23. $4\sqrt{2} \sin \dfrac{x}{2} = -4$

24. $2 + \cos \dfrac{x}{2} = 2 - \sqrt{3} - \cos \dfrac{x}{2}$

Find all solutions for the following equations in the interval $[0°, 360°)$. Find θ to the nearest tenth of a degree.

25. $\tan \theta + 6 \cot \theta = 5$

26. $\csc \theta = 2 \sin \theta + 1$

27. $\sec^2 \theta = 2 \tan \theta + 4$

28. $2 \tan^2 \theta \sin \theta - \tan^2 \theta = 0$

29. $5 \sec^2 \theta = 6 \sec \theta$

30. $\cos^2 \theta = \sin^2 \theta + 1$

31. $\csc^2 \theta - 2 \cot \theta = 0$

32. $3 \cot^3 \theta = \cot \theta$

33. $\sin^2 \theta \cos^2 \theta = 0$

34. $\sec^2 \theta \tan \theta = 2 \tan \theta$

35. $2 \sin 2\theta = \sqrt{3}$

36. $2 \cos 2\theta = \sqrt{2}$

37. $\cos \dfrac{\theta}{2} = 1$

38. $\sin \dfrac{\theta}{2} = 1$

39. $2\sqrt{3} \sin \dfrac{\theta}{2} = 3$

40. $2\sqrt{3} \cos \dfrac{\theta}{2} = -3$

41. $2 \sin \theta = 2 \cos 2\theta$

42. $\cos \theta - 1 = \cos 2\theta$

43. $\sin 2\theta = 2 \cos^2 \theta$

44. $\csc^2 \dfrac{\theta}{2} = 2 \sec \theta$

45. $\cos \theta = \sin^2 \dfrac{\theta}{2}$

46. $4 \cos 2\theta = 8 \sin \theta \cos \theta$

47. $2 \cos^2 2\theta = 1 - \cos 2\theta$

48. $\sin \theta = \cos \dfrac{\theta}{2}$

Give all solutions (to the nearest tenth of a degree) for each of the following. Write the solutions in the form used in Example 1.

49. $\sin x - \cos x = 1$ **50.** $\cos 2x = 1$ **51.** $\cos x \left(\sin x - \dfrac{1}{2} \right) = 0$

52. $(2 \sin x - \sqrt{3})(\cos x + 1) = 0$ **53.** $\tan^2 x + 2 \tan x = 3$

54. $\cot^2 x - 4 \cot x - 5 = 0$ **55.** $\sin 2x = \cos 2x$ **56.** $\tan \dfrac{1}{2} x = \cot \dfrac{1}{2} x$

To solve the following equations, you will need the quadratic formula. Find all solutions in the interval $[0°, 360°)$. Give solutions to the nearest tenth of a degree.

57. $9 \sin^2 x - 6 \sin x = 1$ **58.** $4 \cos^2 x + 4 \cos x = 1$

59. $\tan^2 x + 4 \tan x + 2 = 0$ **60.** $3 \cot^2 x - 3 \cot x - 1 = 0$

61. $\sin^2 x - 2 \sin x + 3 = 0$ **62.** $2 \cos^2 x + 2 \cos x - 1 = 0$

63. $\cot x + 2 \csc x = 3$ **64.** $2 \sin x = 1 - 2 \cos x$

For the following equations, use the sum and product identities of Section 7.6. Give all solutions in the interval $[0, 2\pi)$.

65. $\sin x + \sin 3x = \cos x$ **66.** $\cos 4x - \cos 2x = \sin x$ **67.** $\sin 3x - \sin x = 0$

68. $\cos 2x + \cos x = 0$ **69.** $\sin 4x + \sin 2x = 2 \cos x$ **70.** $\cos 5x + \cos 3x = 2 \cos 4x$

In an electric circuit, let V represent the electromotive force in volts at t seconds. Assume $V = \cos 2\pi t$. Find the smallest positive value of t where $0 \le t \le 1/2$ for each of the following values of V.

71. $V = 0$ **72.** $V = .5$

A coil of wire rotating in a magnetic field induces a voltage given by

$$e = 20 \sin \left(\frac{\pi t}{4} - \frac{\pi}{2} \right),$$

where t is time in seconds. Find the smallest positive time to produce the following voltages.

73. 0 **74.** $10\sqrt{3}$

75. The equation

$$.342D \cos \theta + h \cos^2 \theta = \frac{16D^2}{V_0^2}$$

is used in reconstructing accidents in which a vehicle vaults into the air after hitting an obstruction. V_0 is the velocity in feet per second of the vehicle when it hits, D is the distance (in feet) from the obstruction to the landing point, and h is the difference in height (in feet) between the landing point and the takeoff point. Angle θ is the takeoff angle, the angle between the horizontal and the path of the vehicle. Find θ to the nearest degree if $V_0 = 60$, $D = 80$, and $h = 2$.

76. The seasonal variation in the length of daylight can be represented by a sine function. For example, the daily number of hours of daylight in New Orleans is given by

$$h = \frac{35}{3} + \frac{7}{3} \sin \frac{2\pi x}{365},$$

where x is the number of days after March 21 (disregarding leap year).*
(a) On what date will there be about 14 hours of daylight?
(b) What date has the least number of hours of daylight?
(c) When will there be about 10 hours of daylight?

77. The British nautical mile is defined as the length of a minute of arc of a meridian. Since the earth is flat at its poles, the nautical mile, in feet, is given by

$$L = 6{,}077 - 31 \cos 2\theta,$$

where θ is the latitude in degrees. (See the figure.)
(a) Find the latitude(s) at which the nautical mile is 6,074 feet.
(b) At what latitude(s) is the nautical mile 6,108 feet?
(c) In the United States the nautical mile is defined everywhere as 6,080.2 feet. At what latitude(s) does this agree with the British nautical mile?*

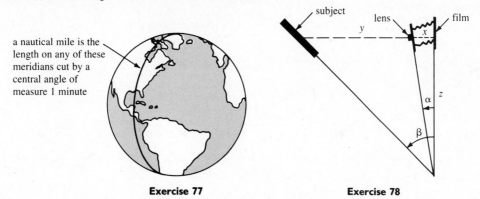

a nautical mile is the length on any of these meridians cut by a central angle of measure 1 minute

Exercise 77   **Exercise 78**

78. When a large view camera is used to take a picture of an object that is not parallel to the film, the lens board should be tilted so that the planes containing the subject, the lens board, and the film intersect in a line (see the figure). This gives the best "depth of field."*
(a) Write two equations, one relating α, x, and z, and the other relating β, x, y, and z.
(b) Eliminate z from the equations in part (a) to get one equation relating α, β, x, and y.
(c) Solve the equation from part (b) for α.
(d) Solve the equation from part (b) for β.

Solve each of the following equations.

79. $\sin^{-1} x = \tan^{-1} 3/4$

80. $\tan^{-1} x = \cos^{-1} 5/13$

81. $\arccos x = \arcsin 3/5$

82. $\arctan x = \arcsin (-4/5)$

83. $\sin^{-1} x - \tan^{-1} 1 = -\pi/4$

84. $\cos^{-1} x + \cos^{-1} 1 = \pi/2$

85. $\cos^{-1} x + 2 \sin^{-1} \sqrt{3}/2 = \pi$

86. $\sin^{-1} x + \tan^{-1} \sqrt{3} = 2\pi/3$

87. $\sin^{-1} 2x + \cos^{-1} x = \pi/6$

88. $\sin^{-1} 2x + \sin^{-1} x = \pi/2$

89. $\cos^{-1} x + \tan^{-1} x = \pi/2$

90. $\tan^{-1} x + \cos^{-1} x = \pi/4$

*From *A Sourcebook of Applications of School Mathematics* by Donald Bushaw et al. Copyright © 1980 by The Mathematical Association of America. Reprinted by permission. The material was prepared with the support of National Science Foundation Grant No. SED72-01123 A05. However, any opinions, findings, conclusions, or recommendations expressed herein are those of the authors and do not necessarily reflect the views of NSF.

Chapter 4 Summary

Key Words   identity   cofunction identities   half-angle identities
                fundamental identities   reduction identity   sum and product identities
                sum and difference identities double-angle identities   conditional equation

Review Exercises

1. Use the trigonometric identities to find the remaining five trigonometric function values of
   x, given that $\cos x = 3/5$ and x is in quadrant IV.

2. Given $\tan x = -5/4$, where x is in the interval $(\pi/2, \pi)$, use trigonometric identities to
   find the other trigonometric function values of x.

3. Given $\sin x = -1/4$, $\cos y = -4/5$, and both x and y are in quadrant III, find $\sin (x + y)$
   and $\cos (x - y)$.

4. Given $\sin 2\theta = \sqrt{3}/2$ and 2θ terminates in quadrant II, use trigonometric identities to
   find $\tan \theta$.

5. Given $x = \pi/8$, use trigonometric identities to find $\sin x$, $\cos x$, and $\tan x$.

For each item in List I, give the letter of the item in List II which completes an identity.

List I		List II
6. $\sin 35°$	**(a)** $\sin (-35°)$	**(e)** $\cos 150° \cos 60° - \sin 150° \sin 60°$
7. $\tan (-35°)$	**(b)** $\cos 55°$	**(f)** $\cot (-35°)$
8. $\cos 35°$		**(g)** $\cos^2 150° - \sin^2 150°$
9. $\cos 75°$	**(c)** $\sqrt{\dfrac{1 + \cos 150°}{2}}$	**(h)** $\sin 15° \cos 60° + \cos 15° \sin 60°$
10. $\sin 75°$	**(d)** $2 \sin 150° \cos 150°$	**(i)** $\cos (-35°)$
11. $\sin 300°$		**(j)** $\cot 125°$
12. $\cos 300°$		

For each item in List I give the letter of the item in List II which completes an identity.

List I				List II			
13. $\csc x$	**(a)** $\dfrac{1}{\sin x}$	**(d)** $\dfrac{1}{\cot^2 x}$	**(g)** $\dfrac{1}{\sin^2 x}$				
14. $\tan x$							
15. $\cot x$	**(b)** $\dfrac{1}{\cos x}$	**(e)** $\dfrac{1}{\cos^2 x}$	**(h)** $1 - \cos^2 x$				
16. $\sin^2 x$							
17. $\tan^2 x + 1$	**(c)** $\dfrac{\sin x}{\cos x}$	**(f)** $\dfrac{\cos x}{\sin x}$					
18. $\tan^2 x$							

Use identities to express each of the following in terms of $\sin \theta$ and $\cos \theta$ and simplify.

19. $\sec^2 \theta - \tan^2 \theta$ **20.** $\dfrac{\cot \theta}{\sec \theta}$ **21.** $\tan^2 \theta (1 + \cot^2 \theta)$

22. $\csc \theta + \cot \theta$ **23.** $\csc^2 \theta + \sec^2 \theta$ **24.** $\tan \theta - \sec \theta \csc \theta$

Show that each of the following is an identity.

25. $\dfrac{\sin 2x}{\sin x} = \dfrac{2}{\sec x}$

26. $2 \cos A - \sec A = \cos A - \dfrac{\tan A}{\csc A}$

27. $\dfrac{2 \tan B}{\sin 2B} = \sec^2 B$

28. $\tan \beta = \dfrac{1 - \cos 2\beta}{\sin 2\beta}$

29. $1 + \tan^2 \alpha = 2 \tan \alpha \csc 2\alpha$

30. $-\dfrac{\sin (A - B)}{\sin (A + B)} = \dfrac{\cot A - \cot B}{\cot A + \cot B}$

31. $\dfrac{\sin t}{1 - \cos t} = \cot \dfrac{t}{2}$

32. $2 \cos (A + B) \sin (A + B) = \sin 2A \cos 2B + \sin 2B \cos 2A$

33. $\dfrac{2 \cot x}{\tan 2x} = \csc^2 x - 2$

34. $\sin t = \dfrac{\cos t \sin 2t}{1 + \cos 2t}$

35. $\tan \theta \sin 2\theta = 2 - 2 \cos^2 \theta$

36. $\csc A \sin 2A - \sec A = \cos 2A \sec A$

37. $2 \tan x \csc 2x - \tan^2 x = 1$

38. $2 \cos^2 \theta - 1 = \dfrac{1 - \tan^2 \theta}{1 + \tan^2 \theta}$

39. $\sin^3 \theta = \sin \theta - \cos^2 \theta \sin \theta$

40. $\dfrac{\sin^2 x}{2 - 2 \cos x} = \cos^2 \dfrac{x}{2}$

41. $\cos^4 \theta = \dfrac{3}{8} + \dfrac{1}{2} \cos 2\theta + \dfrac{1}{8} \cos 4\theta$

42. $8 \sin^2 \dfrac{\gamma}{2} \cos^2 \dfrac{\gamma}{2} = 1 - \cos 2\gamma$

43. $\cos^2 \dfrac{x}{2} = \dfrac{1 + \sec x}{2 \sec x}$

44. $\tan \theta \cos^2 \theta = \dfrac{2 \tan \theta \cos^2 \theta - \tan \theta}{1 - \tan^2 \theta}$

45. $\tan 8k - \tan 8k \cdot \tan^2 4k = 2 \tan 4k$

46. $\sec^2 \alpha - 1 = \dfrac{\sec 2\alpha - 1}{\sec 2\alpha + 1}$

47. $\dfrac{\sin 3t + \sin 2t}{\sin 3t - \sin 2t} = \dfrac{\tan \dfrac{5t}{2}}{\tan \dfrac{t}{2}}$

48. $\sin 2\alpha = \dfrac{2(\sin \alpha - \sin^3 \alpha)}{\cos \alpha}$

49. $\tan 2\beta - \sec 2\beta = \dfrac{\tan \beta - 1}{\tan \beta + 1}$

50. $\dfrac{\sin^3 t - \cos^3 t}{\sin t - \cos t} = \dfrac{2 + \sin 2t}{2}$

51. $-\cot \dfrac{x}{2} = \dfrac{\sin 2x + \sin x}{\cos 2x - \cos x}$

52. $2 \cos^3 x - \cos x = \dfrac{\cos^2 x - \sin^2 x}{\sec x}$

Find all solutions for the following equations in the interval $[0, 2\pi)$.

53. $\sin^2 x = 1$

54. $2 \tan x - 1 = 0$

55. $2 \sin^2 x - 5 \sin x + 2 = 0$

56. $\tan x = \cot x$

57. $\sec^4 2x = 4$

58. $\tan^2 2x - 1 = 0$

59. $\sin \dfrac{x}{2} = \cos \dfrac{x}{2}$

60. $\sec \dfrac{x}{2} = \cos \dfrac{x}{2}$

61. $\cos 2x + \cos x = 0$

62. $\sin x \cos x = \dfrac{1}{4}$

Find all solutions for the following equations in the interval $[0°, 360°)$.

63. $2 \cos \theta = 1$

64. $(\tan \theta + 1)\left(\sec \theta - \dfrac{1}{2}\right) = 0$

65. $\sin^2 \theta + 3 \sin \theta + 2 = 0$

66. $\sin 2\theta = \cos 2\theta + 1$

67. $\dfrac{\sin \theta}{\cos \theta} = \tan^2 \theta$

68. $2 \sin 2\theta = 1$

69. Recall Snell's law from Exercises 77–80 of Section 4.5:

$$\frac{c_1}{c_2} = \frac{\sin \theta_1}{\sin \theta_2},$$

where c_1 is the speed of light in one medium, c_2 is the speed of light in a second medium, and θ_1 and θ_2 are the angles shown in the figure.

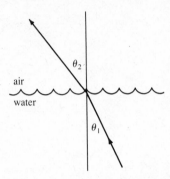

Suppose that a light is shining up through water into the air as in the figure. As θ_1 increases, θ_2 approaches 90°, at which point no light will emerge from the water. Assume the ratio c_1/c_2 in this case is .752.

(a) For what value of θ_1 does $\theta_2 = 90$°? This value of θ_1 is called the *critical angle* for water.

(b) What happens when θ_1 is greater than the critical angle?

70. The angle between the downward vertical position and another position of a rhythmically moving arm is given by

$$y = \frac{1}{3} \sin \frac{4\pi t}{3},$$

where t is time in seconds. See the figure below.

(a) Solve the equation for t.

(b) At what time(s) does the arm form an angle of .3 radians?

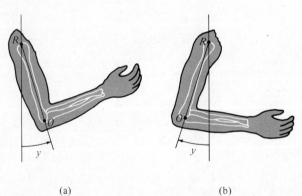

(a)   (b)

Schematic diagrams of a rhythmically moving arm. The upper arm *RO* rotates back and forth about the point *R*; the position of the arm is measured by the angle y between the actual position and the downward vertical position.

Exact values of the trigonometric functions of 15° can be found by the following method, an alternative to the use of the half-angle formulas. Start with a right triangle ABC having a 60° angle at A and a 30° angle at B. Let the hypotenuse of this triangle have length 2. Extend side BC and draw a semicircle with diameter along BC extended, center at B, and radius AB. Draw segment AE. (See the figure.) Since any angle inscribed in a semicircle is a right angle, triangle AED is a right triangle.

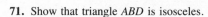

71. Show that triangle ABD is isosceles.

72. Show that angle ABD is 150°.

73. Show that angle DAB is 15°, as is angle ADB.

74. Show that DC has length $2 + \sqrt{3}$.

75. Since AC has length 1, the length of AD is given by

$$(AD)^2 = 1^2 + (2 + \sqrt{3})^2.$$

Reduce this to $\sqrt{8 + 4\sqrt{3}}$, and show that this result equals $\sqrt{6} + \sqrt{2}$.

76. Use angle ADB of triangle ADE and find cos 15°.

77. Show that AE has length $\sqrt{6} - \sqrt{2}$.   **78.** Find sin 15°.

79. Use triangle ACE and find tan 15°.   **80.** Find cot 15°.

The following exercises are taken with permission from a standard calculus book.*

81. Let α and β be two given numbers.
 (a) Prove that if sin $(\alpha + \beta)$ = sin $(\alpha - \beta)$, then either α is an odd multiple of $\pi/2$ or β is a multiple of π, or both.
 (b) Prove that if cos $(\alpha + \beta)$ = cos $(\alpha - \beta)$, then either α is a multiple of π or β is a multiple of π, or both.
 (c) Prove that if tan $(\alpha + \beta)$ = tan $(\alpha - \beta)$, then β is a multiple of π.

82. Note that for $\alpha = 18°$, we have cos 3α = sin 2α. Use the formula

$$\cos 3\alpha = 3 \sin \alpha - 4 \sin^3 \alpha$$

and show that sin 18° = $(\sqrt{5} - 1)/4$.

*Reproduced from *Calculus,* 2nd edition, by Leonard Gillman and Robert H. McDowell, by permission of W. W. Norton & Company, Inc. Copyright © 1978, 1973 by W. W. Norton & Company, Inc.

5

Applications of Trigonometry

Every triangle has three sides and three angles. This chapter shows that if any three of these six measures of a triangle are known (if at least one measure is a side), then the other three measures can be found, by a process called **solving a triangle.**

5.1 Right-Triangle Applications

A few right-triangle applications of trigonometry were discussed in Chapter 6; more applications are discussed in this section. First, recall that if θ is an acute angle of a right triangle, then $\sin \theta$, $\cos \theta$, and $\tan \theta$ can be found from the lengths of the sides of the triangle:

$$\sin \theta = \frac{\text{side opposite}}{\text{hypotenuse}} \qquad \cos \theta = \frac{\text{side adjacent}}{\text{hypotenuse}} \qquad \tan \theta = \frac{\text{side opposite}}{\text{side adjacent}}.$$

The other three values, $\cot \theta$, $\sec \theta$, and $\csc \theta$, can be found in a similar way. See Figure 1.

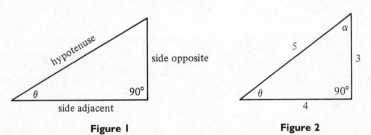

Figure 1

Figure 2

Example 1 The right triangle of Figure 2 has sides of lengths 3, 4, and 5. Find each of the following.

(a) $\sin \theta$, $\cos \theta$, and $\tan \theta$
Use the quotients given above.

$$\sin \theta = \frac{\text{side opposite}}{\text{hypotenuse}} = \frac{3}{5} \qquad \cos \theta = \frac{\text{side adjacent}}{\text{hypotenuse}} = \frac{4}{5}$$

$$\tan \theta = \frac{\text{side opposite}}{\text{side adjacent}} = \frac{3}{4}$$

(b) The degree measure of angle θ, to the nearest tenth of a degree.
Since $\sin \theta = 3/5 = .6000$, using a calculator or interpolating in Table 3 would give

$$\theta = 36.9°.$$

(c) $\sin \alpha$, $\cos \alpha$, and $\tan \alpha$
Use the appropriate values from above.

$$\sin \alpha = \frac{4}{5}, \qquad \cos \alpha = \frac{3}{5} \qquad \text{and} \qquad \tan \alpha = \frac{4}{3}$$

(d) The degree measure of angle α
The sum of the two acute angles of a right triangle is 90°, and $\theta = 36.9°$ from part (b), so

$$\alpha = 90° - 36.9° = 53.1°. \qquad ■$$

When trigonometry is used to solve triangles or to find the measures of all sides and all angles, it is convenient to use a to represent the length of the side opposite angle A, b for the length of the side opposite angle B, and so on. The letter c is used for the hypotenuse in a right triangle.

Example 2 Solve right triangle ABC, with $A = 34° \, 30'$ and $c = 12.7$. See Figure 3.
This triangle can be solved by finding the measures of the remaining sides and angles. By the definitions given above, $\sin A = a/c$, where $A = 34° \, 30'$ and $c = 12.7$. Substitute to get

$$\sin A = \frac{a}{c}$$

$$\sin 34° \, 30' = \frac{a}{12.7}$$

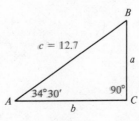

Figure 3

or, upon multiplying both sides by 12.7,

$$a = 12.7 \sin 34° \; 30' \approx 12.7(.5664) \approx 7.19.$$

The value of b could be found with the Pythagorean theorem. It is better, however, to use the information given in the problem rather than a result just calculated. If a mistake were to be made in finding a, then b would also be incorrect. Using $\cos A$ gives

$$\cos A = \frac{\text{side adjacent}}{\text{hypotenuse}} = \frac{b}{c}$$

$$\cos 34° \; 30' = \frac{b}{12.7}$$

or
$$b = 12.7 \cos 34° \; 30' \approx 12.7 \, (.8241) \approx 10.5.$$

Once b has been found, the Pythagorean theorem could be used as a check. All that is still needed for solving triangle ABC is to find B. Since $A + B = 90°$, so that $B = 90° - A$, and $A = 34° \; 30'$,

$$B = 90° - 34° \; 30' = 89° \; 60' - 34° \; 30' = 55° \; 30'.$$

Triangle ABC is now solved—the lengths of all sides and the measures of all angles are known. ∎

In Example 2 above, c was given as 12.7. Since this measure for c is given to three significant digits, the values of both a and b must be given to no more than three significant digits. The number of significant digits in an answer cannot exceed the least number of significant digits in any number used to find the answer. Also, an angle given to the nearest degree is assumed to have *two* significant digits; to the nearest ten minutes implies *three* significant digits; to the nearest minute implies *four* significant digits.

Many problems with right triangles involve the angle of elevation or the angle of depression. The **angle of elevation** from point X to point Y (above X) is the angle made by line XY and a horizontal line through X. The angle of elevation is always measured from the horizontal. See Figure 4. The **angle of depression** from point X to point Y (below X) is the angle made by line XY and a horizontal line through X.

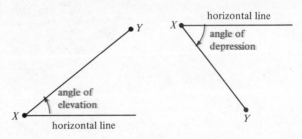

Figure 4

Example 3   Wilma Spence knows that when she stands 123 feet from the base of a flagpole, the angle of elevation to the top is 26° 40′. If her eyes are 5.30 feet above the ground, find the height of the flagpole.

The length of the side adjacent to Spence is known (see Figure 5) and the length of the side opposite her must be found. The ratio that involves these two values is the tangent.

$$\tan A = \frac{\text{side opposite}}{\text{side adjacent}}$$

$$\tan 26° \, 40′ = \frac{a}{123}$$

$$a = 123 \tan 26° \, 40′ = 61.8 \text{ feet}$$

Since Spence's eyes are 5.30 feet above the ground, the height of the flagpole is

$$61.8 + 5.30 = 67.1 \text{ feet.} \quad \blacksquare$$

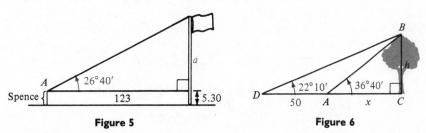

Figure 5   **Figure 6**

Example 4   Francisco needs to know the height of a tree. From a given point on the ground he finds that the angle of elevation to the top of the tree is 36° 40′. He then moves back 50 ft. From the second point, the angle of elevation to the top of the tree is 22° 10′. See Figure 6. Find the height of the tree.

The figure shows two unknowns, x, the distance from the center of the trunk of the tree to the point where the first observation was made, and h, the height of the tree. Since nothing is given about the length of the hypotenuse of either triangle ABC or triangle BCD, use a ratio that does not involve the hypotenuse, the tangent.

in triangle ABC, $\tan 36° \, 40′ = \dfrac{h}{x}$ or $h = x \tan 36° \, 40′$

in triangle BCD, $\tan 22° \, 10′ = \dfrac{h}{50 + x}$ or $h = (50 + x) \tan 22° \, 10′.$

Since each of these two expressions equals h, these expressions must be equal.

$$x \tan 36° \, 40′ = (50 + x) \tan 22° \, 10′$$

Now use algebra to solve for x.

$$x \tan 36° \, 40′ = 50 \tan 22° \, 10′ + x \tan 22° \, 10′$$

$$x \tan 36° \, 40′ - x \tan 22° \, 10′ = 50 \tan 22° \, 10′$$

$$x(\tan 36° \, 40′ - \tan 22° \, 10′) = 50 \tan 22° \, 10′$$

$$x = \frac{50 \tan 22° \, 10′}{\tan 36° \, 40′ - \tan 22° \, 10′}$$

It was shown above that $h = x \tan 36° 40'$. Substituting for x gives

$$h = \left(\frac{50 \tan 22° 10'}{\tan 36° 40' - \tan 22° 10'} \right)(\tan 36° 40').$$

From Table 3 or a calculator,

$$\tan 36° 40' = .7445$$

and

$$\tan 22° 10' = .4074,$$

so

$$\tan 36° 40' - \tan 22° 10' = .7445 - .4074 = .3371,$$

with

$$h = \left(\frac{50(.4074)}{.3371} \right)(.7445) = 45 \text{ ft.} \quad \blacksquare$$

Many applications of trigonometry involve **bearing,** an important idea in navigation and surveying. Bearing is used to give directions. There are two common systems used to express bearing. When a single angle is given, such as 164°, bearing is measured in a clockwise direction from due north. Several sample bearings using this system are shown in Figure 7.

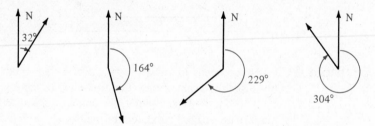

Figure 7

Example 5   Radar stations A and B are on an east-west line, 3.7 kilometers apart. Station A detects a plane at C on a bearing of 61°. Station B detects the same plane on a bearing of 331°. Find the distance from A to C.

Draw a sketch showing the given information, as in Figure 8. Angle C is a right angle, since angles CAB and CBA are complementary. The necessary distance, b, can be found by using cosine.

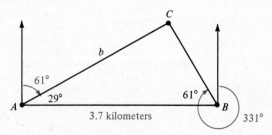

Figure 8

As the figure suggests,

$$\cos 29° = \frac{b}{3.7}$$

$$3.7 \cos 29° = b$$

$$3.7(.8746) = b$$

$$b = 3.2 \text{ kilometers} \quad \blacksquare$$

The other common system for expressing bearing starts with a north-south line and uses an acute angle to show the direction, either east or west, from this line. Figure 9 shows several sample bearings using this system. The letter N or S always comes first, followed by an acute angle, and then E or W.

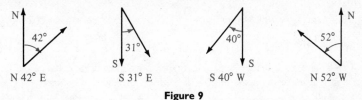

Figure 9

Example 6   The bearing from A to C is S 52° E. The bearing from A to B is N 84° E. The bearing from B to C is S 38° W. A plane flying at 250 kilometers per hour takes 2.4 hours to go from A to B. Find the distance from A to C.

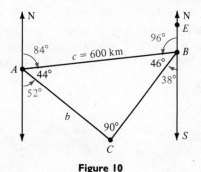

Figure 10

Figure 10 shows a sketch of the given information. Since the bearing from A to B is 84°, angle ABE is 180° − 84° = 96°, with angle ABC equal to 46°. Also, angle BAC is 180° − (84° + 52°) = 44°. Angle C is 180° − (44° + 46°) = 90°. Since a plane flying at 250 kilometers per hour takes 2.4 hours to go from A to B, the distance from A to B is 2.4(250) = 600 kilometers. To find b, the distance from A to C, use sine. (Cosine could also have been used.)

$$\sin 46° = \frac{b}{c} = \frac{b}{600}$$

$$600 \sin 46° = b$$

$$b = 430 \text{ kilometers} \quad \blacksquare$$

5.1 Exercises

Solve each of the following right triangles. Angle C is the right angle.

1. $B = 73°\ 00'$, $b = 128$ in
2. $A = 61°\ 00'$, $b = 39.2$ cm
3. $a = 76.4$ yd, $b = 39.3$ yd
4. $a = 958$ m, $b = 489$ m
5. $a = 18.9$ cm, $c = 46.3$ cm
6. $b = 219$ m, $c = 647$ m
7. $A = 53°\ 24'$, $c = 387.1$ ft
8. $A = 13°\ 47'$, $c = 1285$ m
9. $B = 39°\ 09'$, $c = .6231$ m
10. $B = 82°\ 51'$, $c = 4.825$ cm
11. $c = 7.813$ m, $b = 2.467$ m
12. $c = 44.91$ mm, $a = 32.71$ mm
13. $B = 42.432°$, $a = 157.49$ m
14. $A = 36.704°$, $c = 1461.3$ cm
15. $A = 57.209°$, $c = 186.49$ cm
16. $B = 12.099°$, $b = 7.0463$ m
17. $b = 173.921$ m, $c = 208.543$ m
18. $a = 864.003$ cm, $c = 1092.84$ cm

Solve each of the following.

19. A 39.4 meter fire-truck ladder is leaning against a wall. Find the distance the ladder goes up the wall if it makes an angle of $42°\ 30'$ with the ground.

20. A swimming pool is 50.0 feet long and 4.00 feet deep at one end. If it is 12.0 feet deep at the other end, find the total distance along the bottom.

21. A guy wire 87.4 meters long is attached to the top of a tower that is 69.4 meters high. Find the angle that the wire makes with the ground.

22. Find the length of a guy wire that makes an angle of $42°\ 10'$ with the ground if the wire is attached to the top of a tower 79.6 meters high.

23. A representation of an aerial photograph of a complex of buildings is shown on the left below.* If the sun was at an angle of $26.5°$ when the photograph was taken, how high is the building diagramed on the right? Use .48 cm as the length of the shadow.

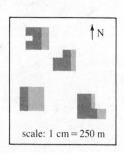
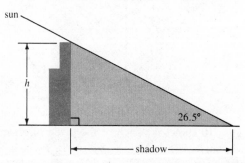

24. The figure at the top of the next page represents an aerial photograph of a cliff in a remote region of Antarctica. Compute the height of the cliff if the angle of elevation of the sun was $19.0°$ when the photograph was taken.

*Exercises 23 and 24 from *Plane Trigonometry* by Bernard J. Rice and Jerry D. Strange. Copyright © 1981 by Prindle, Weber & Schmidt. Reprinted by permission.

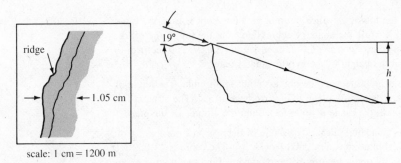

scale: 1 cm = 1200 m

Work the following problems involving angles of elevation or depression.

25. Suppose the angle of elevation of the sun is 28° 10′. Find the length of the shadow cast by a man 6.0 feet tall.

26. The shadow of a vertical tower is 58.2 meters long when the angle of elevation of the sun is 36° 20′. Find the height of the tower.

27. Find the angle of elevation of the sun if a 53.9 foot flagpole casts a shadow 74.6 feet long.

28. The angle of depression from the top of a building to a point on the ground is 34° 50′. How far is the point on the ground from the top of the building if the building is 368 meters high?

29. A television camera is to be mounted on a bank wall so as to have a good view of the head teller (see the figure). Find the angle of depression that the lens should make with the horizontal.

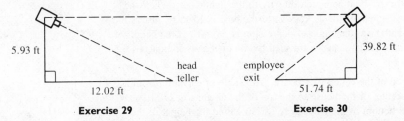

Exercise 29   Exercise 30

30. A company safety committee has recommended that a floodlight be mounted in a parking lot so as to illuminate the employee exit (see the figure). Find the angle of depression of the light.

31. Priscilla drives her Peterbilt up a straight road inclined at an angle of 4° 10′ with the horizontal. She starts at an elevation of 680 feet above sea level and drives 12,400 feet along the road. Find her final altitude.

32. The road into Death Valley is straight; it makes an angle of 4° 10′ with the horizontal. Starting at sea level, the road descends to −121 feet. Find the distance it is necessary to travel along the road to reach bottom.

33. A tower stands on top of a hill. From a point on the ground 148 m from a point directly under the tower, the angle of elevation to the *bottom* of the tower is 18° 20′. From the same point, the angle of elevation to the *top* of the tower is 34° 10′. Find the height of the tower.

34. The angle of elevation from the top of an office building in New York City to the top of the World Trade Center is 68°, while the angle of depression from the top of the office building to the bottom of the Trade Center is 63°. The office building is 290 feet from the World Trade Center. Find the height of the World Trade Center.

35. Mt. Rogers, with an altitude of 5700 feet, is the highest point in Virginia. The angle of elevation from the top of Mt. Rogers to a plane flying overhead is 33°. The straight line distance from the mountaintop to the plane is 4600 feet. Find the altitude of the plane.

36. The highest point in Texas is Guadalupe Peak. The angle of depression from the top of this peak to a small miner's cabin at elevation 2000 feet is 26°. The cabin is 14,000 feet horizontally from a point directly under the top of the mountain. Find the altitude of the top of the mountain.

Find h in each of the following.

37.

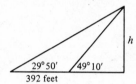

38.

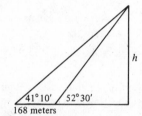

39. The angle of elevation from a point on the ground to the top of a pyramid is 35° 30'. The angle of elevation from a point 135 feet further back to the top of the pyramid is 21° 10'. Find the height of the pyramid.

40. A lighthouse keeper is watching a boat approach directly to the lighthouse. When she first begins watching the boat, the angle of depression of the boat is 15° 50'. Just as the boat turns away from the lighthouse, the angle of depression is 35° 40'. If the height of the lighthouse is 68.7 meters, find the distance traveled by the boat as it approaches the lighthouse.

41. A television antenna is on top of the center of a house. The angle of elevation from a point 28.0 meters from the center of the house to the top of the antenna is 27° 10', and the angle of elevation to the bottom of the antenna is 18° 10'. Find the height of the antenna.

42. The angle of elevation from Lone Pine to the top of Mt. Whitney is 10° 50'. If I drive 7.00 kilometers along a straight level road toward Mt. Whitney, I find the angle of elevation to be 22° 40'. Find the height of the top of Mt. Whitney above the level of the road.

In each of the following exercises, generalize problems from above by finding formulas for h in terms of k, A, and B. Assume $A < B$ in Exercise 43 and $A > B$ in Exercise 44.

43.

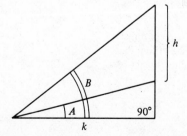

44.

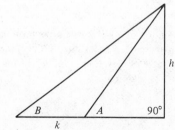

Solve each of the following problems involving bearing.

45. A ship leaves port and sails on a bearing of 28° 10′. Another ship leaves the same port at the same time and sails on a bearing of 118° 10′. If the first ship sails at 20.0 miles per hour and the second sails at 24.0 miles per hour, find the distance between the two ships after five hours.

46. Radio direction finders are set up at points A and B, which are 2.00 miles apart on an east-west line. From A it is found that the bearing of the signal from a radio transmitter is 36° 20′, while from B the bearing of the same signal is 306° 20′. Find the distance between the transmitter and B.

47. The bearing from Winston-Salem, North Carolina to Danville, Virginia is N 42° E. The bearing from Danville to Goldsboro, North Carolina is S 48° E. A small plane traveling at 60 miles per hour takes 1 hour to go from Winston-Salem to Danville and 1.8 hours to go from Danville to Goldsboro. Find the distance from Winston-Salem to Goldsboro.

48. The bearing from Atlanta to Macon is S 27° E, while the bearing from Macon to Augusta is N 63° E. A plane traveling at 60 miles per hour needs 1¼ hours to go from Atlanta to Macon and 1¾ hours to go from Macon to Augusta. Find the distance from Atlanta to Augusta.

49. The airline distance from Philadelphia to Syracuse is 260 miles, on a bearing of 335°. The distance from Philadelphia to Cincinnati is 510 miles, on a bearing of 245°. Find the bearing from Cincinnati to Syracuse.

50. A ship travels 70 kilometers on a bearing of 27°, and then turns on a bearing of 117° for 180 kilometers. Find the distance of the end of the trip from the starting point.

51. A pendulum is m cm long. Suppose it is moved from its vertical position by an angle α ($0° \leq \alpha \leq 90°$). By how much has the end of the pendulum been raised vertically?

52. A woman is standing h meters above the water on a ship at sea. If the radius of the earth is r kilometers, find the distance she can see to the horizon.

53. Atoms in metals can be arranged in patterns called **unit cells.** One such unit cell, called a **primitive cell,** is a cube with an atom at each corner. A right triangle can be formed from one edge of the cell, a face diagonal and a cube diagonal as in the figure below. If each cell edge is 3.00×10^{-8} cm and the face diagonal is 4.24×10^{-8} cm, what is the angle between the cell edge and a cube diagonal?

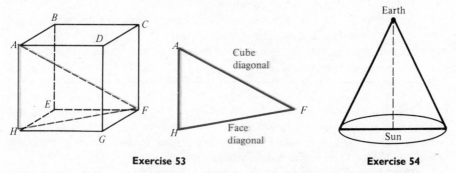

Exercise 53   **Exercise 54**

54. To determine the diameter of the sun, an astronomer might sight with a transit (a device used by surveyors for measuring angles) first to one edge of the sun and then to the other, finding that the included angle equals 1° 4′. Assuming that the distance from the earth to the sun is 92,919,800 mi, calculate the diameter of the sun. See the figure.

55. Very accurate measurements have shown that the distance between California's Owens Valley Radio Observatory and the Haystack Observatory in Massachusetts is 2441.2938 miles. Suppose the two observatories focus on a distant star and find that angles E and E' in the figure are both 89.99999°. Find the distance to the star from Haystack. (Assume the earth is flat.)

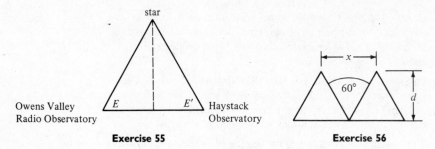

Exercise 55   Exercise 56

56. The figure shows a magnified view of the threads of a bolt. Find x if d is 2.894 mm.

Use a right triangle to find each of the following.

57. $\sin(\cos^{-1} 1/4)$ **58.** $\tan(\sin^{-1} 4/7)$ **59.** $\cos(\sin^{-1} 2/3)$ **60.** $\sin(\tan^{-1} 1/5)$

5.2 Oblique Triangles and the Law of Sines

The methods of the previous section apply only to right triangles. In the next few sections these methods are generalized to include all triangles, not just right triangles. A triangle that is not a right triangle is called an **oblique triangle.** As with right triangles, the measures of the three sides and the three angles of an oblique triangle can be found if at least one side and any other two measures are known. There are four possible cases.

Solving Oblique Triangles

1. One side and two angles are known.
2. Two sides and one angle not included between the two sides are known. This case may lead to more than one triangle.
3. Two sides and the angle included between the two sides are known.
4. Three sides are known.

The first two cases require the *law of sines,* which is discussed in this section. The last two cases require the *law of cosines,* discussed in Section 3.

To derive the law of sines, start with an oblique triangle, such as the acute triangle in Figure 11(a) or the obtuse triangle in Figure 11(b).

The following discussion applies to both triangles. First, construct the perpendicular from B to side AC. Let h be the length of this perpendicular. Then c is the

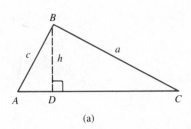
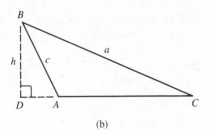

(a)   (b)

Figure 11

hypotenuse of right triangle *ADB*, and *a* is the hypotenuse of right triangle *BDC*. By results given earlier,

$$\text{in triangle } ADB, \qquad \sin A = \frac{h}{c} \quad \text{or} \quad h = c \sin A,$$

$$\text{in triangle } BDC, \qquad \sin C = \frac{h}{a} \quad \text{or} \quad h = a \sin C.$$

Since $h = c \sin A$ and $h = a \sin C$,

$$a \sin C = c \sin A,$$

or, upon dividing both sides by $\sin A \sin C$,

$$\frac{a}{\sin A} = \frac{c}{\sin C}.$$

In a similar way, by constructing the perpendiculars from other vertices,

$$\frac{a}{\sin A} = \frac{b}{\sin B} \quad \text{and} \quad \frac{b}{\sin B} = \frac{c}{\sin C},$$

proving the following theorem, called the **law of sines.**

Law of Sines

> In any triangle *ABC*, with sides *a*, *b*, and *c*,
>
> $$\frac{a}{\sin A} = \frac{b}{\sin B}, \qquad \frac{a}{\sin A} = \frac{c}{\sin C}, \quad \text{and} \quad \frac{b}{\sin B} = \frac{c}{\sin C}.$$

The three formulas of the law of sines can be written in a more compact form as

$$\frac{a}{\sin A} = \frac{b}{\sin B} = \frac{c}{\sin C} \quad \text{or} \quad \frac{\sin A}{a} = \frac{\sin B}{b} = \frac{\sin C}{c}.$$

In some cases, this second form is easier to use.

   If two angles and one side of a triangle are known, the law of sines can be used to solve the triangle.

Example 1 Solve triangle ABC if $A = 32.0°$, $B = 81.8°$, and $a = 42.9$ cm. See Figure 12.

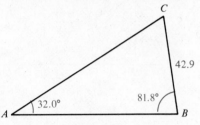

Figure 12

Start by drawing a triangle, roughly to scale, and labeling the given parts as in Figure 12. Since the values of A, B, and a are known, use the part of the law of sines that involves these variables.

$$\frac{a}{\sin A} = \frac{b}{\sin B}$$

Substituting the known values gives

$$\frac{42.9}{\sin 32.0°} = \frac{b}{\sin 81.8°}.$$

Multiply both sides of the equation by $\sin 81.8°$.

$$b = \frac{42.9 \sin 81.8°}{\sin 32.0°}$$

When using a calculator to find b, keep intermediate answers in the calculator until the final result is found. Then round to the proper number of significant digits. In this case, find $\sin 81.8°$, and then multiply that number by 42.9. Keep the result in the calculator while you find $\sin 32.0°$, and then divide. This final answer should be rounded to 3 significant figures.

$$b \approx 80.1 \text{ cm}$$

Find C from the fact that the sum of the angles of any triangle is 180°.

$$A + B + C = 180°$$
$$C = 180° - A - B$$
$$C = 180° - 32.0° - 81.8° = 66.2°$$

Now use the law of sines again to find c. (Why should the Pythagorean theorem not be used?)

$$\frac{a}{\sin A} = \frac{c}{\sin C}$$
$$\frac{42.9}{\sin 32.0°} = \frac{c}{\sin 66.2°}$$
$$c = \frac{42.9 \sin 66.2°}{\sin 32.0°} \approx \frac{42.9(.9150)}{.5299} \approx 74.1$$

The length of side c is 74.1 cm. ∎

Example 2 Shawn Johnson wishes to measure the distance across the Big Muddy River. See Figure 13. She finds that $C = 112° \, 53'$, $A = 31° \, 06'$, and $b = 347.6$ ft. Find the required distance.

Before the law of sines can be used to find a, angle B must be found.

$$B = 180° - A - C$$
$$= 180° - 31° \, 06' - 112° \, 53' = 36° \, 01'$$

Use the part of the law of sines involving A, B, and b.

$$\frac{a}{\sin A} = \frac{b}{\sin B}$$

Substitute the known values.

$$\frac{a}{\sin 31° \, 06'} = \frac{347.6}{\sin 36° \, 01'}$$

$$a = \frac{347.6 \sin 31° \, 06'}{\sin 36° \, 01'}$$

$$a = 305.3 \text{ ft} \quad \blacksquare$$

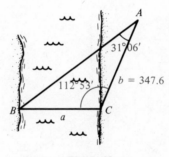

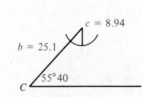

Figure 13   **Figure 14**

Example 3 Solve triangle ABC if $C = 55° \, 40'$, $c = 8.94$ m, and $b = 25.1$ m.

Let us first look for angle B. The work is easier if the unknown is in the numerator, so start with

$$\frac{\sin B}{b} = \frac{\sin C}{c}.$$

Substitute the given values.

$$\frac{\sin B}{25.1} = \frac{\sin 55° \, 40'}{8.94}$$

$$\sin B = \frac{25.1 \sin 55° \, 40'}{8.94} = 2.3184$$

By this result, $\sin B$ is greater than 1. This is impossible, since $-1 \leq \sin B \leq 1$, for any angle B. For this reason, triangle ABC does not exist. See Figure 14. $\quad \blacksquare$

When any two angles and the length of a side of a triangle are given, the law of sines can be applied directly to solve the triangle. However, if only one angle and two sides are given, the triangle may not exist, as in Example 3, or there may be more than one triangle satisfying the given conditions. For example, suppose the measure of acute angle A of triangle ABC is known, along with the length of side a and the length of side b. To show this information, draw angle A and measure off a length b along its terminal side. Now draw a side of length a opposite angle A. The chart in Figure 15 shows that there might be more than one possible outcome, giving the **ambiguous case of the law of sines.**

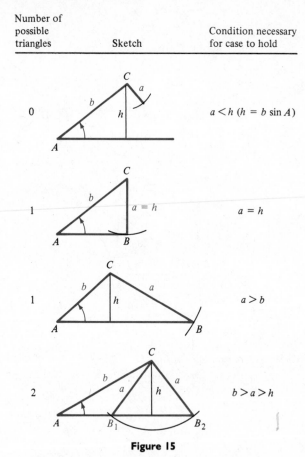

Number of possible triangles	Sketch	Condition necessary for case to hold
0		$a < h$ $(h = b \sin A)$
1		$a = h$
1		$a > b$
2		$b > a > h$

Figure 15

If angle A is obtuse, there are two possible outcomes as shown in the chart in Figure 16.

It is possible to derive formulas that show which of the various cases exist for a particular set of numerical data. However, this work is unnecessary with the law of sines. For example, if the law of sines is used, and gives $\sin B > 1$, there is no triangle at all. (Why?) A case producing two different triangles is illustrated in the next example.

Number of possible triangles	Sketch	Condition necessary for case to hold
0	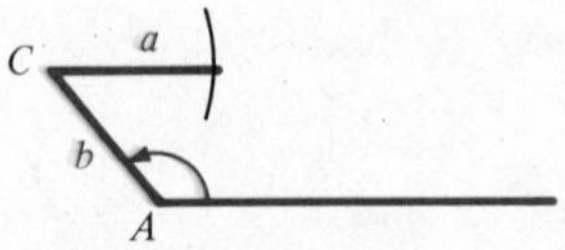	$a \leq b$
1	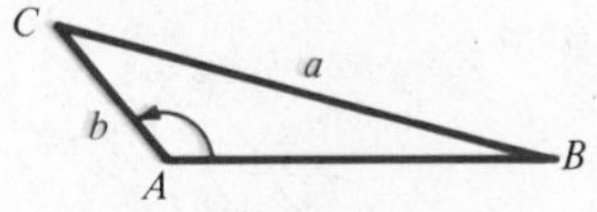	$a > b$

Figure 16

Example 4 Solve triangle ABC if $A = 55° \ 20'$, $a = 22.8$, and $b = 24.9$.

To begin, use the law of sines to find angle B.

$$\frac{a}{\sin A} = \frac{b}{\sin B}$$

$$\frac{22.8}{\sin 55° \ 20'} = \frac{24.9}{\sin B}$$

$$\sin B = \frac{24.9 \sin 55° \ 20'}{22.8}$$

$$\sin B = .8982$$

Since $\sin B < 1$, there is at least one triangle. Figure 17 shows the case if there are two triangles. Assume there are two triangles and find two possible values of B. Since $\sin B = .8982$, one value of B is

$$B = 64° \ 00'.$$

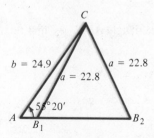

Figure 17

From the identity $\sin (180° - B) = \sin B$, another possible value of B is

$$B = 180° - 64° \ 00'$$

$$B = 116° \ 00'.$$

To keep track of these two different values of B, let

$$B_1 = 116° \ 00' \quad \text{and} \quad B_2 = 64° \ 00'.$$

Now separately solve triangles AB_1C_1 and AB_2C_2 as shown in Figure 18.

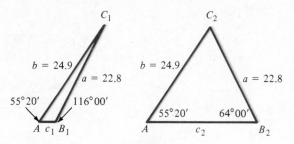

Figure 18

Since B_1 is the larger of the two values of B, find C_1 next.

$$C_1 = 180° - A - B_1$$
$$C_1 = 8° \ 40'.$$

Had this answer been negative, there would have been only one triangle. This is why the larger angle was used first. Now, use the law of sines to find c_1.

$$\frac{a}{\sin A} = \frac{c_1}{\sin C_1}$$

$$\frac{22.8}{\sin 55° \ 20'} = \frac{c_1}{\sin 8° \ 40'}$$

$$c_1 = \frac{22.8 \sin 8° \ 40'}{\sin 55° \ 20'}$$

$$c_1 = 4.18$$

Solve triangle AB_2C_2 by first finding C_2.

$$C_2 = 180° - A - B_2$$
$$C_2 = 60° \ 40'$$

By the law of sines,

$$\frac{22.8}{\sin 55° \ 20'} = \frac{c_2}{\sin 60° \ 40'}$$

$$c_2 = \frac{22.8 \sin 60° \ 40'}{\sin 55° \ 20'}$$

$$c_2 = 24.2. \quad \blacksquare$$

Area   The method used to derive the law of sines can also be used to derive a useful formula for the area of a triangle. A familiar formula for the area of a triangle is $K = (1/2)bh$, where K represents the area, b the base, and h the height. This formula cannot always be used, since in practice h is often unknown. To find a more useful formula, refer to acute triangle ABC in Figure 19(a) or obtuse triangle ABC in Figure 19(b).

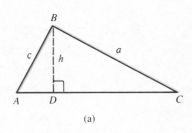

(a)   (b)

Figure 19

A perpendicular has been drawn from B to the base of the triangle. This perpendicular forms two right triangles. Using triangle ABD,

$$\sin A = \frac{h}{c},$$

or

$$h = c \sin A.$$

Substituting into the formula $K = \frac{1}{2}bh$,

$$K = \frac{1}{2}b(c \sin A)$$

$$K = \frac{1}{2}bc \sin A.$$

Any other pair of sides and the angle between them could have been used, as in the next theorem.

Area of a Triangle   The area of a triangle is given by half the product of the lengths of two sides and the sine of the angle between the two sides.

Example 5   Find the area of triangle MNP if $m = 29.7$ m, $n = 53.9$ m, and $P = 28° \, 40'$.
By the last result, the area of the triangle is

$$\frac{1}{2}(29.7)(53.9) \sin 28° \, 40' = 384 \text{ m}^2. \quad \blacksquare$$

5.2 Exercises

Solve each of the following triangles that exist.

1. $A = 46° 30'$, $B = 52° 50'$, $b = 87.3$ mm
2. $A = 59° 30'$, $B = 48° 20'$, $b = 32.9$ m
3. $A = 27.2°$, $C = 115.5°$, $c = 76.0$ ft
4. $B = 124.1°$, $C = 18.7°$, $c = 94.6$ m
5. $A = 68.41°$, $B = 54.23°$, $a = 12.75$ ft
6. $C = 74.08°$, $B = 69.38°$, $c = 45.38$ m
7. $A = 87.2°$, $b = 75.9$ yd, $C = 74.3°$
8. $B = 38° 40'$, $a = 19.7$ cm, $C = 91° 40'$
9. $B = 20° 50'$, $C = 103° 10'$, $AC = 132$ ft
10. $A = 35.3°$, $B = 52.8°$, $AC = 675$ ft
11. $A = 39.70°$, $C = 30.35°$, $b = 39.74$ m
12. $C = 71.83°$, $B = 42.57°$, $a = 2.614$ cm
13. $B = 42.88°$, $C = 102.40°$, $b = 3974$ ft
14. $A = 18.75°$, $B = 51.53°$, $c = 2798$ yd
15. $A = 39° 54'$, $a = 268.7$ m, $B = 42° 32'$
16. $C = 79° 18'$, $c = 39.81$ mm, $A = 32° 57'$

Find the missing angles in each of the following triangles.

17. $A = 29.7°$, $b = 41.5$ ft, $a = 27.2$ ft
18. $B = 48.2°$, $a = 890$ cm, $b = 697$ cm
19. $C = 41° 20'$, $b = 25.9$ m, $c = 38.4$ m
20. $B = 48° 50'$, $a = 3850$ in, $b = 4730$ in
21. $B = 74.3°$, $a = 859$ m, $b = 783$ m
22. $C = 82.2°$, $a = 10.9$ km, $c = 7.62$ km
23. $A = 142.13°$, $b = 5.432$ ft, $a = 7.297$ ft
24. $B = 113.72°$, $a = 189.6$ yd, $b = 243.8$ yd
25. $C = 129° 18'$, $a = 372.9$ cm, $c = 416.7$ cm
26. $A = 132° 07'$, $b = 7.481$ mi, $a = 8.219$ mi

Solve each of the following triangles that exist.

27. $A = 42.5°$, $a = 15.6$ ft, $b = 8.14$ ft
28. $C = 52.3°$, $a = 32.5$ yd, $c = 59.8$ yd
29. $B = 72.2°$, $b = 78.3$ m, $c = 145$ m
30. $C = 68.5°$, $c = 258$ cm, $b = 386$ cm
31. $A = 38° 40'$, $a = 9.72$ km, $b = 11.8$ km
32. $C = 29° 50'$, $a = 8.61$ m, $c = 5.21$ m
33. $B = 32° 50'$, $a = 7540$ cm, $b = 5180$ cm
34. $C = 22° 50'$, $b = 159$ mm, $c = 132$ mm
35. $A = 96.80°$, $b = 3.589$ ft, $a = 5.818$ ft
36. $C = 88.70°$, $b = 56.87$ yd, $c = 112.4$ yd
37. $B = 39.68°$, $a = 29.81$ m, $b = 23.76$ m
38. $A = 51.20°$, $c = 7986$ cm, $a = 7208$ cm

Solve each of the following exercises. Recall that bearing was discussed in Section 5.1.

39. To find the distance AB across a river, a distance $BC = 354$ meters is laid off on one side of the river. In triangle ABC, it is found that $B = 112° 10'$ and $C = 15° 20'$. Find AB.

40. To determine the distance RS across a deep canyon, Joanna lays off a distance $TR = 582$ yards. She then finds that in triangle RST, $T = 32° 50'$ and $R = 102° 20'$. Find RS.

41. Radio direction finders are placed at points A and B, which are 3.46 miles apart on an east-west line, with A west of B. From A the bearing of a certain radio transmitter is $47° 40'$, while from B the bearing is $302° 30'$. Find the distance between the transmitter and A.

42. A ship is sailing due north. Captain Odjakjian notices that the bearing of a lighthouse 12.5 kilometers distant is $38° 50'$. Later on, the captain notices that the bearing of the lighthouse has become $135° 50'$. How far did the ship travel between the two observations of the lighthouse?

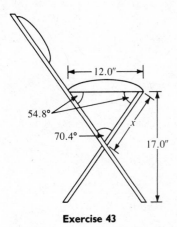

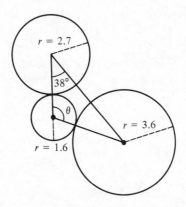

Exercise 43   **Exercise 45**

43. A folding chair is to have a seat 12.0 in deep with angles as shown in the figure. How far down from the seat should the crossing legs be joined? (Find x in the figure.)

44. Mark notices that the bearing of a tree on the opposite bank of a river is 115° 20′. Lisa is on the same bank as Mark but 428 meters away. She notices that the bearing of the tree is 45° 20′. The river is flowing north between parallel banks. What is the distance across the river?

45. Three gears are arranged as shown in the figure above. Find angle θ.

46. Three atoms with atomic radii of 2, 3, and 4.5 are arranged as in the figure below. Find the distance between the centers of atoms A and C.

47. A surveyor reported the following data about a piece of property: "The property is triangular in shape, with dimensions as shown in the figure." Use the law of sines to see if such a piece of property could exist.

48. The surveyor tries again: "A second triangular piece of property has dimensions as shown." This time it turns out that the surveyor did not consider every possible case. Use the law of sines to show why.

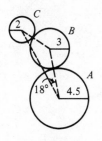

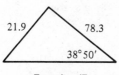

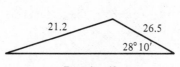

Exercise 46   **Exercise 47**   **Exercise 48**

Find the area of each of the following triangles.

49. $A = 42.5°$, $b = 13.6$ m, $c = 10.1$ m

50. $C = 72.2°$, $b = 43.8$ ft, $a = 35.1$ ft

51. $B = 124.5°$, $a = 30.4$ cm, $c = 28.4$ cm

52. $C = 142.7°$, $a = 21.9$ km, $b = 24.6$ km

53. $A = 56.80°$, $b = 32.67$ in, $c = 52.89$ in

54. $A = 34.97°$, $b = 35.29$ m, $c = 28.67$ m

55. $A = 24° 25′$, $B = 56° 20′$, $c = 78.40$ cm

56. $B = 48° 30′$, $C = 74° 20′$, $a = 462$ km

57. A painter is going to apply a special coating to a triangular metal plate on a new building. Two sides measure 16.1 m and 15.2 m. She knows that the angle between these sides is 125°. How many square meters should she plan to cover with the coating?

58. A real estate salesman wants to find the area of a triangular lot. A surveyor takes measurements for him. He finds that two sides are 52.1 m and 21.3 m, and the angle between them is 42.2°. What is the area of the lot?

Prove that each of the following statements are true for any triangle ABC, with corresponding sides a, b, and c.

59. $\dfrac{a + b}{b} = \dfrac{\sin A + \sin B}{\sin B}$

60. $\dfrac{a - b}{a + b} = \dfrac{\sin A - \sin B}{\sin A + \sin B}$

61. $\dfrac{a + b}{c} = \dfrac{\cos \dfrac{1}{2}(A - B)}{\sin \dfrac{1}{2} C}$

62. $\dfrac{a - b}{c} = \dfrac{\sin \dfrac{1}{2}(A - B)}{\cos \dfrac{1}{2} C}$

63. In any triangle having sides a, b, and c, it must be true that $a + b > c$. Use this fact and the law of sines to show that $\sin A + \sin B > \sin (A + B)$ for any two angles A and B of a triangle.

64. Show that the area of a triangle having sides a, b, and c and corresponding angles, A, B, and C is given by

$$\frac{a^2 \sin B \sin C}{2 \sin A}.$$

65. Prove the law of sines if angle A is a right angle.

66. Derive the formula for the area of a triangle, $K = (1/2) bc \sin A$, if angle A is a right angle.

5.3 The Law of Cosines

If two sides and the angle between the two sides are given, the law of sines cannot be used to solve the triangle. Also, if all three of the sides of a triangle are given, the law of sines cannot be used to find the unknown angles. Both of these cases require the law of cosines.

To derive this law, let ABC be any oblique triangle. Choose a coordinate system with vertex B at the origin and side BC along the positive x-axis. See Figure 20.

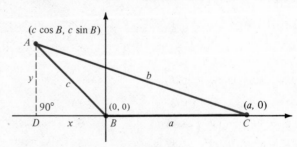

Figure 20

Let (x, y) be the coordinates of vertex A of the triangle. Verify that for angle B, whether obtuse or acute,

$$\sin B = \frac{y}{c} \quad \text{and} \quad \cos B = \frac{x}{c}.$$

(Here x is negative if B is obtuse.) From these results

$$y = c \sin B \quad \text{and} \quad x = c \cos B,$$

so that the coordinates of point A become

$$(c \cos B, c \sin B).$$

Point C has coordinates $(a, 0)$ and AC has length b. By the distance formula,

$$b = \sqrt{(c \cos B - a)^2 + (c \sin B)^2}.$$

Squaring both sides and simplifying gives

$$\begin{aligned} b^2 &= (c \cos B - a)^2 + (c \sin B)^2 \\ &= c^2 \cos^2 B - 2ac \cos B + a^2 + c^2 \sin^2 B \\ &= a^2 + c^2 (\cos^2 B + \sin^2 B) - 2ac \cos B \\ &= a^2 + c^2 (1) - 2ac \cos B = a^2 + c^2 - 2ac \cos B. \end{aligned}$$

This result is one form of the **law of cosines.** (If $B = 90°$, this result reduces to the Pythagorean theorem.) In the work above, A or C could just as easily have been placed at the origin, giving the same result but with the variables rearranged. These various forms of the law of cosines are summarized in the following theorem.

The Law of Cosines

In any triangle ABC, with sides a, b, and c,
$$a^2 = b^2 + c^2 - 2bc \cos A$$
$$b^2 = a^2 + c^2 - 2ac \cos B$$
$$c^2 = a^2 + b^2 - 2ab \cos C.$$

Example 1 Solve triangle ABC if $A = 42.3°$, $b = 12.9$, and $c = 15.4$. See Figure 21.

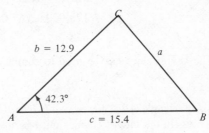

Figure 21

Find a with the law of cosines.

$$a^2 = b^2 + c^2 - 2bc \cos A$$

Substitute the given values.

$$a^2 = (12.9)^2 + (15.4)^2 - 2(12.9)(15.4)\cos 42.3°$$
$$a^2 = 166.41 + 237.16 - (397.32)(.7396)$$
$$a^2 = 403.57 - 293.86 = 109.71$$
$$a = 10.5$$

Now that a, b, c, and A are known, the law of sines can be used to find either angle B or angle C. If there is an obtuse angle in the triangle, it will be the larger of B and C. Since we cannot tell from the sine of the angle whether it is acute or obtuse, it is a good idea to find the smaller angle (which will be acute) first. In this triangle, $B < C$ because $b < c$, so use the law of sines to find B.

$$\frac{\sin 42.3°}{10.5} = \frac{\sin B}{12.9}$$
$$\sin B = \frac{12.9 \sin 42.3°}{10.5}$$
$$\sin B = .8268$$
$$B = 55.8°$$

Now find C.

$$C = 180° - A - B = 81.9°. \quad \blacksquare$$

Example 2   Solve triangle ABC if $C = 132° 40'$, $b = 259$, and $a = 423$.

Since C is given, but c is unknown, use the law of cosines in the form $c^2 = a^2 + b^2 - 2ab \cos C$. Inserting the given data gives

$$c^2 = (423)^2 + (259)^2 - 2(423)(259) \cos 132° 40'.$$

The value of $\cos 132° 40'$ can be found with a calculator or reference angles. By either method,

$$\cos 132° 40' = -\cos 47° 20' = -.6777.$$

Now continue finding c.

$$c^2 = (423)^2 + (259)^2 - 2(423)(259)(-.6777)$$
$$c^2 = 178,929 + 67,081 + 148,494 = 394,504$$
$$c = 628$$

The law of sines can be used to complete the solution. Check that $A = 29° 40'$ and $B = 17° 40'$. $\quad \blacksquare$

Example 3   Solve triangle ABC if $a = 9.47$, $b = 15.9$, and $c = 21.1$.

Again the law of cosines must be used. It is a good idea to find the largest angle first in case it is obtuse. Since c has the greatest length, angle C will be the largest angle. Start with

$$c^2 = a^2 + b^2 - 2ab \cos C,$$

or
$$\cos C = \frac{a^2 + b^2 - c^2}{2ab}.$$

Substituting the given values leads to

$$\cos C = \frac{(9.47)^2 + (15.9)^2 - (21.1)^2}{2(9.47)(15.9)}$$

$$= \frac{-102.7191}{301.146}$$

$$= -.341094,$$

and
$$C = 109.9°.$$

(Angle C is obtuse since $\cos C$ is negative.) Use the law of sines to find B. Verify that $B = 45.1°$. Since $A = 180° - B - C,$

$$A = 25.0°. \quad \blacksquare$$

As mentioned above, there are four possible cases that can arise in solving an oblique triangle. These cases are summarized as follows. (The first two cases require the law of sines, while the second two require the law of cosines. In all four cases, assume that the given information actually produces a triangle.)

Solving Triangles

Case	Abbreviation	Example
One side and two angles are known.	SAA	a, B, A known, find b $\qquad b = \frac{a \sin B}{\sin A}$
Two sides and one angle (not included between the two sides) are known (watch for the ambiguous case—there may be two triangles).	SSA	b, c, B known, find C $\qquad \sin C = \frac{c \sin B}{b}$
Three sides are known.	SSS	$a, b, c,$ known, find A $\qquad \cos A = \frac{b^2 + c^2 - a^2}{2bc}$
Two sides and the angle included between the two sides are known.	SAS	a, B, c known, find b $\qquad b^2 = a^2 + c^2 - 2ac \cos B$

Area The law of cosines can be used to find a formula for the area of a triangle when only the lengths of the three sides of the triangle are known. This formula, called Heron's area formula, is given as the next theorem. For a proof see Exercise 56.

Heron's Area Formula	If a triangle has sides of lengths a, b, and c, and if the semiperimeter s is $$s = \frac{1}{2}(a + b + c),$$ then the area of the triangle is $$K = \sqrt{s(s - a)(s - b)(s - c)}.$$

Example 4 Find the area of the triangle having sides of lengths $a = 29.7$ ft, $b = 42.3$ ft, and $c = 38.4$ ft.

To use Heron's area formula, first find s.

$$s = \frac{1}{2}(a + b + c)$$

$$s = \frac{1}{2}(29.7 + 42.3 + 38.4)$$

$$s = 55.2$$

The area is then given by

$$K = \sqrt{s(s - a)(s - b)(s - c)}$$
$$= \sqrt{55.2(55.2 - 29.7)(55.2 - 42.3)(55.2 - 38.4)}$$
$$K = \sqrt{55.2(25.5)(12.9)(16.8)}$$
$$K = 552 \text{ ft}^2. \quad \blacksquare$$

5.3 Exercises

Solve each of the following triangles. Use a calculator or interpolate as necessary.

1. $C = 28.3°$, $b = 5.71$ in, $a = 4.21$ in

2. $A = 41.4°$, $b = 2.78$ yd, $c = 3.92$ yd

3. $C = 45.6°$, $b = 8.94$ m, $a = 7.23$ m

4. $A = 67.3°$, $b = 37.9$ km, $c = 40.8$ km

5. $A = 80° \, 40'$, $b = 143$ cm, $c = 89.6$ cm

6. $C = 72° \, 40'$, $a = 327$ ft, $b = 251$ ft

7. $B = 74.80°$, $a = 8.919$ in, $c = 6.427$ in

8. $C = 59.70°$, $a = 3.725$ mi, $b = 4.698$ mi

9. $A = 112.8°$, $b = 6.28$ m, $c = 12.2$ m

10. $B = 168.2°$, $a = 15.1$ cm, $c = 19.2$ cm

11. $C = 24° \, 49'$, $a = 251.3$ m, $b = 318.7$ m

12. $B = 52° \, 28'$, $a = 7598$ in, $c = 6973$ in

Find all the angles in each of the following triangles. Round answers to the nearest tenth of a degree.

13. $a = 3.00$ ft, $b = 5.00$ ft, $c = 6.00$ ft

14. $a = 4.00$ ft, $b = 5.00$ ft, $c = 8.00$ ft

15. $a = 9.31$ cm, $b = 5.73$ cm, $c = 8.24$ cm

16. $a = 28.3$ in, $b = 47.1$ in, $c = 57.9$ in

17. $a = 42.9$ m, $b = 37.6$ m, $c = 62.7$ m

18. $a = 189$ yd, $b = 214$ yd, $c = 325$ yd

19. $AB = 1240$ ft, $AC = 876$ ft, $BC = 918$ ft

20. $AB = 298$ m, $AC = 421$ m, $BC = 324$ m

Use a calculator or interpolation to find all the angles in each of the following triangles.
Round answers to the nearest minute.

21. $a = 12.54$ in, $b = 16.83$ in, $c = 21.62$ in

22. $a = 250.8$ ft, $b = 212.7$ ft, $c = 324.1$ ft

23. $a = 7.095$ m, $b = 5.613$ m, $c = 11.53$ m

24. $a = 15,250$ m, $b = 17,890$ m, $c = 27,840$ m

Solve each of the following problems. Use the laws in this chapter as necessary.

25. Points A and B are on opposite sides of Lake Yankee. From a third point, C, the angle between the lines of sight to A and B is 46.3°. If AC is 350 m long and BC is 286 m long, find AB.

26. The sides of a parallelogram are 4.0 cm and 6.0 cm. One angle is 58° while another is 122°. Find the lengths of the diagonals of the figure.

27. Airports A and B are 450 km apart, on an east-west line. Tom flies in a northeast direction from A to airport C. From C he flies 359 km on a bearing of 128° 40′ to B. How far is C from A?

28. Two ships leave a harbor together, traveling on courses that have an angle of 135° 40′ between them. If they each travel 402 mi, how far apart are they?

29. The plans for a mountain cabin show the dimensions given in the figure below left. Find x.

30. A hill slopes at an angle of 12.47° with the horizontal. From the base of the hill, the angle of elevation of a 459.0 ft tower at the top of the hill is 35.98°. How much rope would be required to reach from the top of the tower to the bottom of the hill?

31. A crane with a counterweight is shown in the figure below center. Find the distance between points A and B.

32. A weight is supported by cables attached to both ends of a balance beam. See the figure below right. What angles are formed between the beam and the cables?

33. Two factories blow their whistles at 5 o'clock exactly. A man hears the two blasts at 3 seconds and 6 seconds after 5, respectively. The angle between his lines of sight to the two factories is 42° 10′. If sound travels 344 m per second, how far apart are the factories?

34. A parallelogram has sides of length 25.9 cm and 32.5 cm. The longer diagonal has a length of 57.8 cm. Find the angle opposite the diagonal.

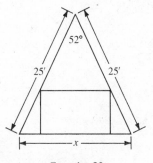

Exercise 29

Exercise 31

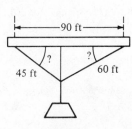

Exericse 32

35. A plane flying a straight course observes a mountain at a bearing 24° 10′ to the right of its course. At that time, the plane is 7.92 km from the mountain. After flying awhile, the bearing to the mountain becomes 32° 40′. How far is the plane from the mountain when the second bearing is found?

36. The aircraft carrier *Tallahassee* is traveling at sea on a steady course with a bearing of 30° at 32 miles per hour. Patrol planes on the carrier have enough fuel for 2.6 hours of flight when traveling at a speed of 520 miles per hour. One of the pilots takes off on a bearing of 338° and then turns and heads in a straight line, so as to be able to catch the carrier, landing on the deck at the exact instant that his fuel runs out. If the pilot left at 2 P.M., at what time did he turn to head for the carrier? See the figure below.

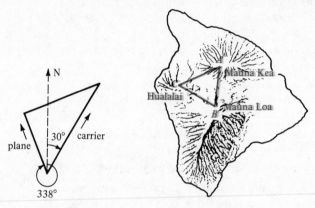

Exercise 36   **Exercises 37, 38**

To help predict eruptions from the volcano Mauna Loa on the island of Hawaii, scientists keep track of the volcano's movement by using a "super triangle" with vertices on the three volcanos shown on the map above. (For example, in a recent year, Mauna Loa moved six inches north and northwest—a result of increasing internal pressure.) The data in the following exercises have been rounded.

37. $AB = 22.47928$ mi, $AC = 28.14276$ mi, $A = 58.56989°$; find BC

38. $AB = 22.47928$ mi, $BC = 25.24983$ mi, $A = 58.56989°$; find B.

The diagram on the following page is an engineering drawing used in the construction of Michigan's Mackinac Straits Bridge.*

39. Find the angles of the triangle formed by the Mackinac West Base, Green Island, and St. Ignace West Base.

40. Find the angles of the triangle formed by A_2, St. Ignace West Base, and St. Ignace East Base.

*Reproduced with permission from *Mackinac Bridge*, by G. Edwin Pidcock, *Civil Engineering*, May 1956, p. 43.

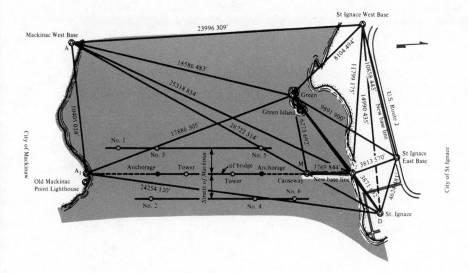

Find the area of each of the following triangles.

41. $a = 15$ in, $b = 19$ in, $c = 24$ in

42. $a = 27$ m, $b = 40$ m, $c = 34$ m

43. $a = 154$ cm, $b = 179$ cm, $c = 183$ cm

44. $a = 25.4$ yd, $b = 38.2$ yd, $c = 19.8$ yd

45. $a = 76.3$ ft, $b = 109$ ft, $c = 98.8$ ft

46. $a = 15.89$ in, $b = 21.74$ in, $c = 10.92$ in

47. $a = 74.14$ ft, $b = 89.99$ ft, $c = 51.82$ ft

48. $a = 1.096$ km, $b = 1.142$ km, $c = 1.253$ km

49. Sam wants to paint a triangular region 75 by 68 by 85 m. A can of paint covers 75 m^2 of area. How many cans (rounded to the larger number of cans) will he need?

50. How many cans would be needed if the region were 8.2 by 9.4 by 3.8 m?

Use the information that $\cos A = (b^2 + c^2 - a^2)/(2bc)$ and $s = (1/2)(a + b + c)$ to show that each of the following is true.

51. $1 + \cos A = \dfrac{(b + c + a)(b + c - a)}{2bc}$

52. $1 - \cos A = \dfrac{(a - b + c)(a + b - c)}{2bc}$

53. $\cos \dfrac{A}{2} = \sqrt{\dfrac{s(s - a)}{bc}}$ $\left(\text{Recall: } \cos \dfrac{A}{2} = \sqrt{\dfrac{1 + \cos A}{2}}\right)$

54. $\sin \dfrac{A}{2} = \sqrt{\dfrac{(s - b)(s - c)}{bc}}$ $\left(\text{Recall: } \sin \dfrac{A}{2} = \sqrt{\dfrac{1 - \cos A}{2}}\right)$

55. The area of a triangle having sides b and c and angle A is given by $\dfrac{1}{2} bc \sin A$. Show that this result can be written as

$$\sqrt{\frac{1}{2} bc(1 + \cos A) \cdot \frac{1}{2} bc(1 - \cos A)}.$$

56. Use the results of Exercises 51–55 to prove Heron's area formula.

57. Let a and b be the equal sides of an isosceles triangle. Prove that $c^2 = 2a^2(1 - \cos C)$.

58. Let point D on side AB of triangle ABC be such that CD bisects angle C. Show that $AD/DB = b/a$.

5.4 Vectors and Applications

As shown in earlier sections, the measure of all six parts of a triangle can be found, given at least one side and any other two measures. In this section, applications of this work to *vectors* are discussed.

Many quantities in mathematics involve magnitudes, such as 45 lb or 60 mph. These quantities are called **scalars.** Other quantities, called **vector quantities,** involve both magnitude and direction. Typical vector quantities are velocity, acceleration, and force.

A vector quantity is often represented with a directed line segment, which is called a **vector.** The length of the vector represents the magnitude of the vector quantity. The direction of the vector, indicated with an arrowhead, represents the direction of the quantity. For example, the vector in Figure 22 represents a force of 10 lb applied at an angle of 30° from the horizontal.

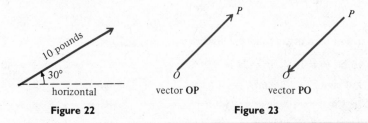

Figure 22   **Figure 23**

The symbol for a vector is often printed in boldface type. To write vectors by hand, it is customary to use an arrow over the letter or letters. Thus **OP** and $\overrightarrow{OP}$ both represent vector OP. Vectors may be named with either one lowercase or uppercase letter, or two uppercase letters. When two letters are used, the first indicates the *initial point* and the second indicates the *terminal point* of the vector. Knowing these points gives the direction of the vector. For example, vectors **OP** and **PO** in Figure 23 are not the same vectors. They have the same magnitude, but opposite directions. The magnitude of vector **OP** is written $|\mathbf{OP}|$.

Two vectors are *equal* if and only if they both have the same directions and the same magnitudes. In Figure 24, vectors **A** and **B** are equal, as are vectors **C** and **D.**

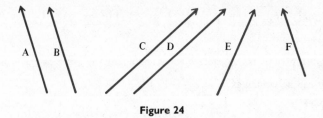

Figure 24

As Figure 24 shows, equal vectors need not coincide, but they must be parallel. Vectors **A** and **E** are unequal because they do not have the same direction, while **A** ≠ **F** because they have different magnitudes, as indicated by their different lengths.

To find the **sum** of two vectors **A** and **B**, written **A** + **B**, place the initial point of vector **B** at the terminal point of vector **A**, as shown in Figure 25. The vector with the same initial point as **A** and the same terminal point as **B** is the sum **A** + **B**. The sum of two vectors is also a vector.

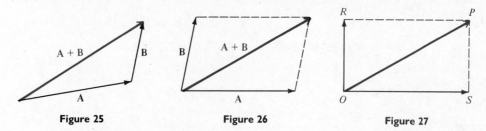

Figure 25   Figure 26   Figure 27

Another way to find the sum of two vectors is to use the **parallelogram rule**. Place vectors **A** and **B** so that their initial points coincide. Then complete a parallelogram which has **A** and **B** as two sides. The diagonal of the parallelogram with the same initial point as **A** and **B** is the same vector sum **A** + **B** found by the definition. See Figure 26.

Parallelograms can be used to show that vector **B** + **A** is the same as vector **A** + **B**, or that

$$A + B = B + A,$$

so that vector addition is **commutative**.

The vector sum **A** + **B** is the **resultant** of vectors **A** and **B**. Each of the vectors **A** and **B** is a **component** of vector **A** + **B**. In many practical applications, such as surveying, it is necessary to break a vector into its **vertical** and **horizontal components**. These components are two vectors, one vertical and one horizontal, whose resultant is the original vector. As shown in Figure 27, vector **OR** is the vertical component and vector **OS** is the horizontal component of **OP**.

For every vector **v** there is a vector −**v** which has the same magnitude as **v** but opposite direction. Vector −**v** is called the **opposite** of **v**. See Figure 28. The sum of **v** and −**v** has magnitude 0 and is called the **zero vector**. As with real numbers, to *subtract* vector **B** from vector **A**, find the vector sum **A** + (−**B**). See Figure 29.

The **scalar product** of a real number (or scalar) k and a vector **u** is the vector $k \cdot \mathbf{u}$ which has magnitude $|k|$ times the magnitude of **u**. As suggested by Figure 30, the vector $k \cdot \mathbf{u}$ has the same direction as **u** if $k > 0$, and opposite direction if $k < 0$.

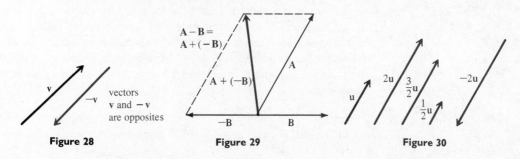

Figure 28   Figure 29   Figure 30

Example 1   Vector **w** has magnitude 25.0 and is inclined at an angle of 41.7° from the horizontal. Find the magnitudes of the horizontal and vertical components of the vector.

In Figure 31, the vertical component is labeled **v** and the horizontal component is labeled **u**. Vectors **u**, **v**, and **w** form a right triangle in which $\sin 41.7° = |v|/|w| = |v|/25.0$. From this,

$$|v| = 25.0 \sin 41.7° = 25.0(.6652) = 16.6.$$

In the same way, $\cos 41.7° = |u|/25.0$, and

$$|u| = 25.0 \cos 41.7° = 18.7. \quad\blacksquare$$

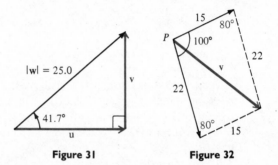

Figure 31   **Figure 32**

Example 2   Two forces of 15 and 22 newtons (a newton is a unit of force used in physics) act on a point in the plane. If the angle between the forces is 100°, find the magnitude of the resultant force.

As shown in Figure 32, a parallelogram that has the forces as adjacent sides can be formed. The angles of the parallelogram adjacent to angle P each measure 80°, since adjacent angles of a parallelogram are supplementary. Opposite sides of the parallelogram are equal in length. The resultant force divides the parallelogram into two triangles. Use the law of cosines with either triangle to get

$$|v|^2 = 15^2 + 22^2 - 2(15)(22) \cos 80°$$
$$= 225 + 484 - 115$$
$$|v|^2 = 594$$
$$|v| = 24. \quad\blacksquare$$

Let vector **u** be placed in a plane so that the initial point of the vector is at the origin, $(0, 0)$, and the endpoint is at the point (a, b). A vector with initial point at the origin is called a **position vector** or (sometimes) a **radius vector.** A position vector having endpoint at the point (a, b) is called the **vector (a, b).** For simplicity, the vector (a, b) is written as $\langle a, b \rangle$. The numbers a and b are called the ***x*-component** and ***y*-component,** respectively. Figure 33 shows the vector $\mathbf{u} = \langle a, b \rangle$. As the figure suggests,

---

the length of vector $\mathbf{u} = \langle a, b \rangle$ is given by

$$|u| = \sqrt{a^2 + b^2}.$$

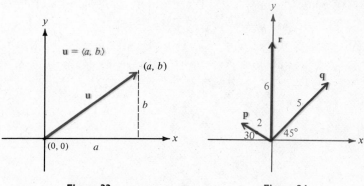

Figure 33   Figure 34

Example 3   Write each of the position vectors of Figure 34 in the form $\langle a, b \rangle$.

Vector **p** has length 2 and makes an angle of 30° with the negative x-axis. From knowledge of 30°–60° right triangles, $\mathbf{p} = \langle -\sqrt{3}, 1 \rangle$. Vector **q** has length 5 and makes an angle of 45° with the positive x-axis, so $\mathbf{q} = \langle 5\sqrt{2}/2, 5\sqrt{2}/2 \rangle$. Finally, $\mathbf{r} = \langle 0, 6 \rangle$.   ■

Let vector **OM** in Figure 35 be given by $\langle a, b \rangle$, while **ON** is given by $\langle c, d \rangle$. Let **OP** be given by $\langle a + c, b + d \rangle$. With facts from geometry, points O, N, M, and P can be shown to form the vertices of a parallelogram. Since a diagonal of this parallelogram gives the resultant of **OM** and **ON**, vector **OP** is given by **OP = OM + ON**, with the resultant of $\langle a, b \rangle$ and $\langle c, d \rangle$ given by $\langle a + c, b + d \rangle$. In the same way, $k \cdot \langle a, b \rangle = \langle ka, kb \rangle$ for any real number k. In summary,

Vector Operations

for any real numbers a, b, c, d, and k,

$$\langle a, b \rangle + \langle c, d \rangle = \langle a + c, b + d \rangle$$

$$k \cdot \langle a, b \rangle = \langle ka, kb \rangle.$$

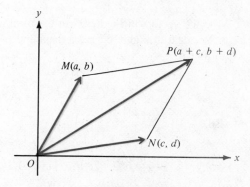

Figure 35

Example 4   Let $\mathbf{u} = \langle -2, 1 \rangle$ and $\mathbf{v} = \langle 4, 3 \rangle$. Find each of the following. See Figure 36.

(a) $\mathbf{u} + \mathbf{v} = \langle -2, 1 \rangle + \langle 4, 3 \rangle = \langle -2 + 4, 1 + 3 \rangle = \langle 2, 4 \rangle$

(b) $-2\mathbf{u} = -2 \cdot \langle -2, 1 \rangle = \langle -2(-2), -2(1) \rangle = \langle 4, -2 \rangle$

(c) $4\mathbf{u} + 3\mathbf{v} = 4 \cdot \langle -2, 1 \rangle + 3 \cdot \langle 4, 3 \rangle = \langle -8, 4 \rangle + \langle 12, 9 \rangle$
$$= \langle -8 + 12, 4 + 9 \rangle = \langle 4, 13 \rangle \quad \blacksquare$$

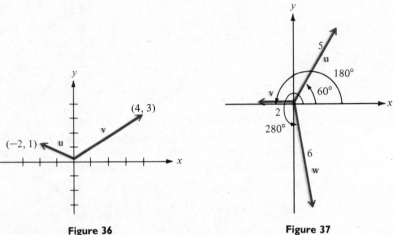

Figure 36   Figure 37

The angle between the positive x-axis and a vector, measured in a counterclockwise direction, is called the **direction angle** for the vector. In Figure 37, $\mathbf{u}$ has direction angle $60°$, while $\mathbf{v}$ has direction angle $180°$, and $\mathbf{w}$ has direction angle $280°$. Using the definition of direction angle and earlier results,

---

if a vector $\mathbf{u}$ has direction angle θ and magnitude r, then
$$\mathbf{u} = \langle r \cos \theta, r \sin \theta \rangle.$$

Example 5   Write the vectors in Figure 37 in the form $\langle a, b \rangle$.

Vector $\mathbf{u}$ in Figure 37 has a magnitude of 5 and direction angle $60°$. By the result just above,

$$\mathbf{u} = \langle 5 \cos 60°, 5 \sin 60° \rangle = \left\langle 5 \cdot \frac{1}{2}, 5 \cdot \frac{\sqrt{3}}{2} \right\rangle = \left\langle \frac{5}{2}, \frac{5\sqrt{3}}{2} \right\rangle.$$

Also,

$$\mathbf{v} = \langle 2 \cos 180°, 2 \sin 180° \rangle = \langle 2(-1), 2(0) \rangle = \langle -2, 0 \rangle.$$

Finally, $\mathbf{w} = \langle 6 \cos 280°, 6 \sin 280° \rangle$

or $\mathbf{w} \approx \langle 1.0419, -5.9088 \rangle. \quad \blacksquare$

The next theorem summarizes some basic properties of vectors.

Basic Results for Vectors

Let a vector have direction angle θ and magnitude r. Then the x-component of the vector is

$$x = r \cos \theta,$$

and the y-component is

$$y = r \sin \theta.$$

Also,

$$x^2 + y^2 = r^2 \qquad \text{and} \qquad \tan \theta = \frac{y}{x}, \quad x \neq 0.$$

Example 6   Figure 38 shows vector $\mathbf{u} = \langle 3, -2 \rangle$. Find the magnitude and direction angle for $\mathbf{u}$.
The magnitude is

$$\sqrt{3^2 + (-2)^2} = \sqrt{13}.$$

To find the direction angle θ, start with

$$\tan \theta = \frac{y}{x} = \frac{-2}{3} = -\frac{2}{3}.$$

Vector $\mathbf{u}$ has positive x-component and negative y-component, placing the vector in quadrant IV. Use the table or a calculator to show that a quadrant IV angle satisfying $\tan \theta = -2/3$ is

$$\theta = -33° \; 40',$$

or $$-33° \; 40' + 360° = 326° \; 20'.$$

The direction angle is $326° \; 20'$. ∎

As shown in Figure 39, vector $\mathbf{u}$ can be thought of as the resultant of two vectors; one on the x-axis, having magnitude given by the absolute value of the x-component; and one on the y-axis, having magnitude given by the absolute value of the y-component.

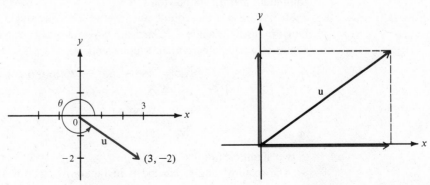

Figure 38   **Figure 39**

This idea applies to any vector $\mathbf{u} = \langle a, b \rangle$.

$$\mathbf{u} = \langle a, b \rangle = \langle a, 0 \rangle + \langle 0, b \rangle$$
$$= a \langle 1, 0 \rangle + b \langle 0, 1 \rangle.$$

The vector $\langle 1, 0 \rangle$ is called the **unit vector i,** while $\langle 0, 1 \rangle$ is the unit vector **j.** ("Unit vector" refers to the fact that the magnitude is 1.) With these unit vectors, any vector $\mathbf{u} = \langle a, b \rangle$ may be written as

$$\mathbf{u} = a\mathbf{i} + b\mathbf{j}.$$

The vector sum $a\mathbf{i} + b\mathbf{j}$ is a **linear combination** of vectors **i** and **j.** As an example, vector $\mathbf{u} = \langle 3, -2 \rangle$ in Figure 40 can be written with unit vectors as

$$\mathbf{u} = 3\mathbf{i} - 2\mathbf{j}.$$

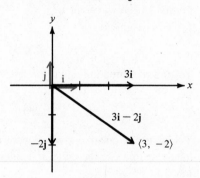

Figure 40

Applications of Vectors Vectors have many applications, especially in physics. Some of these applications use the idea of an *equilibrant:* if the resultant of two forces is **u,** then $-\mathbf{u}$ is the equilibrant of the two forces. The **equilibrant** is the force necessary to counterbalance the joint action of two forces.

Example 7 Find the magnitude of the equilibrant of forces of 48 and 60 newtons acting on a point A, if the angle between the forces is 50°. Then find the angle between the equilibrant and the 48 newton force.

In Figure 41, the equilibrant is $-\mathbf{v}$. The magnitude of **v,** and also of $-\mathbf{v}$, is found by using triangle ABC and the law of cosines. Angle B is 130°, since adjacent angles of a parallelogram are supplementary.

$$|\mathbf{v}|^2 = 48^2 + 60^2 - 2(48)(60) \cos 130°$$
$$= 2304 + 3600 - 5760 \,(-.6428)$$
$$|\mathbf{v}|^2 = 9606.5,$$

or $\qquad\qquad |\mathbf{v}| = 98,$

to two significant digits.

The required angle, labeled α in Figure 41, can be found by subtracting angle

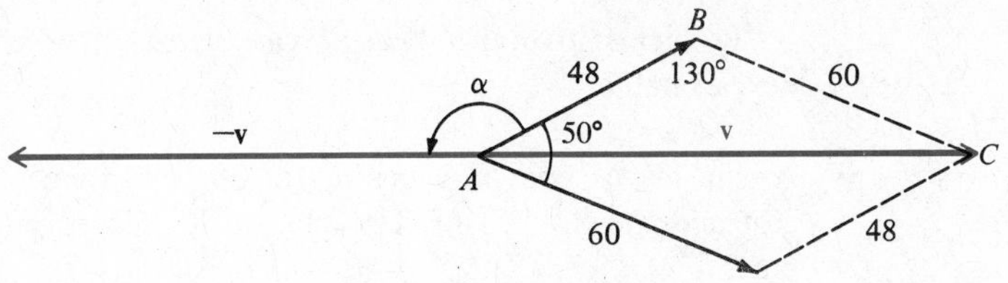

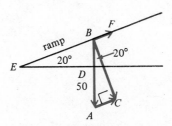

Figure 41

CAB from 180°. Use the law of sines to find angle *CAB*.

$$\frac{98}{\sin 130°} = \frac{60}{\sin CAB}$$

$$\sin CAB = .4690$$

From a calculator or table,

$$\text{angle } CAB = 28°.$$

Finally, $\alpha = 180° - 28° = 152°.$ ■

Example 8 Find the force required to pull a 50-lb weight up a ramp inclined at 20° to the horizontal.

Figure 42

In Figure 42, the vertical 50-lb force represents the force due to gravity. The component **BC** represents the force with which the body pushes against the ramp, while the component **BF** represents a force that would pull the body up the ramp. Since vectors **BF** and **AC** are equal, $|\mathbf{AC}|$ gives the required force.

Vectors **BF** and **AC** are parallel, so angle *EBD* equals angle *A*. Since angle *BDE* and angle *C* are right angles, triangles *ABC* and *DEB* have two corresponding angles equal and so are similar triangles. Therefore, angle *ABC* equals angle *E*, which is 20°. From right triangle *ABC*,

$$\sin 20° = \frac{|\mathbf{AC}|}{50}$$

$$|\mathbf{AC}| = 50 \sin 20°$$

$$|\mathbf{AC}| = 17.$$

To the nearest pound, a 17-lb force will be required to pull the weight up the ramp. ■

Problems involving bearing can also be worked with vectors, as shown in the next example.

Example 9 A ship leaves port on a bearing of 28° and travels 8.2 mi. The ship then turns due east and travels 4.3 mi. How far is the ship from port? What is its bearing from port?

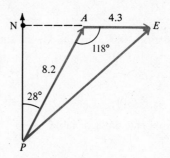

Figure 43

In Figure 43, vectors **PA** and **AE** represent the ship's lines of travel. The magnitude and bearing of the resultant **PE** must be found. Triangle *PNA* is a right triangle, so angle *NAP* = 90° − 28° = 62°. Then angle *PAE* = 180° − 62° = 118°. Use the law of cosines to find |**PE**|, the magnitude of vector **PE.**

$$|\mathbf{PE}|^2 = 8.2^2 + 4.3^2 - 2(8.2)(4.3) \cos 118°$$
$$= 67.24 + 18.49 - 70.52(-.4695)$$
$$|\mathbf{PE}|^2 = 118.84$$

Therefore, |**PE**| = 10.9.

To find the bearing of the ship from port, first find angle *APE*. Use the law of sines, along with the value of |**PE**| before rounding.

$$\frac{\sin APE}{4.3} = \frac{\sin 118°}{10.9}$$

$$\sin APE = \frac{4.3 \sin 118°}{10.9}$$

angle *APE* = 20.4°

After rounding to two significant digits, angle *APE* is 20°, and the ship is 11 mi from port on a bearing of 28° + 20° = 48°. ■

In air navigation, the air speed of a plane is its speed relative to the air, while the ground speed is its speed relative to the ground. Because of the wind, these two speeds are usually different. The ground speed of the plane is represented by the vector sum of the air speed and wind speed vectors. See Figure 44.

course and ground speed
(actual direction of the plane)

N

← wind direction and speed

← heading and air speed

drift angle

Figure 44

N

A

x
121°

O

192

α

C
15.9

B

Figure 45

Example 10   A plane with an air speed of 192 mph is headed on a bearing of 121°. A north wind is blowing (from north to south) at 15.9 mph. Find the ground speed and the actual bearing of the plane.

In Figure 45 the ground speed is represented by $|\mathbf{x}|$. We must find angle α to find the bearing, which will be $121° + \alpha$. From Figure 45, angle BCO equals angle AOC, which equals 121°. Find $|\mathbf{x}|$ by the law of cosines:

$$|\mathbf{x}|^2 = 192^2 + (15.9)^2 - 2(192)(15.9) \cos 121°$$
$$|\mathbf{x}|^2 = 36,864 + 252.81 - 6105.6(-.5150) = 40,261.$$

Therefore, $|\mathbf{x}| = 200.7$,

or 201 mph. Now find α by using the law of sines. As before, use the value of $|\mathbf{x}|$ before rounding.

$$\frac{\sin \alpha}{15.9} = \frac{\sin 121°}{200.7}$$
$$\sin \alpha = .0679$$
$$\alpha = 3.89°$$

After rounding, α is 3.9°. The ground speed is about 201 mph, on a bearing of 124.9°, or 125° to three significant digits. ■

5.4 Exercises

In Exercises 1–4 refer to the vectors below. Name all pairs of vectors

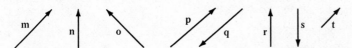

m   n   o   p   q   r   s   t

1. which appear to be equal.   **2.** which are opposites.

3. where the first is a scalar multiple of the second, with the scalar positive.

4. where the first is a scalar multiple of the second, with the scalar negative.

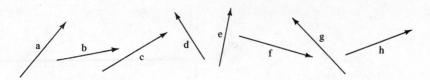

In Exercises 5–20 refer to the vectors pictured above. Draw a sketch to represent each of the following vectors.

5. $-\mathbf{b}$	**6.** $-\mathbf{g}$	**7.** $3\mathbf{a}$	**8.** $2\mathbf{h}$
9. $\mathbf{a} + \mathbf{c}$	**10.** $\mathbf{a} + \mathbf{b}$	**11.** $\mathbf{h} + \mathbf{g}$	**12.** $\mathbf{e} + \mathbf{f}$
13. $\mathbf{a} + \mathbf{h}$	**14.** $\mathbf{b} + \mathbf{d}$	**15.** $\mathbf{h} + \mathbf{d}$	**16.** $\mathbf{a} + \mathbf{f}$
17. $\mathbf{a} - \mathbf{c}$	**18.** $\mathbf{d} - \mathbf{e}$	**19.** $\mathbf{a} + (\mathbf{b} + \mathbf{c})$	**20.** $(\mathbf{a} + \mathbf{b}) + \mathbf{c}$

Let $\mathbf{u} = \langle -2, 5 \rangle$, $\mathbf{v} = \langle 3, -2 \rangle$, and $\mathbf{w} = \langle -4, 6 \rangle$. Find each of the following vectors.

21. $\mathbf{u} + \mathbf{v}$ **22.** $\mathbf{u} - \mathbf{w}$ **23.** $6\mathbf{v} - 2\mathbf{u}$ **24.** $-3\mathbf{u} + 4\mathbf{w}$

In each of the following exercises, $\mathbf{v}$ has the given direction angle and magnitude. Find the x- and y-components of $\mathbf{v}$.

25. $\theta = 45°$, $|\mathbf{v}| = 20$ **26.** $\theta = 75°$, $|\mathbf{v}| = 100$

27. $\theta = 60° \ 10'$, $|\mathbf{v}| = 28.6$ **28.** $\theta = 35° \ 50'$, $|\mathbf{v}| = 47.8$

29. $\theta = 128° \ 30'$, $|\mathbf{v}| = 198$ **30.** $\theta = 146° \ 10'$, $|\mathbf{v}| = 238$

31. $\theta = 251° \ 10'$, $|\mathbf{v}| = 69.1$ **32.** $\theta = 302° \ 40'$, $|\mathbf{v}| = 7890$

Find the magnitude and direction angle for each of the following vectors.

33. $\langle 1, 1 \rangle$ **34.** $\langle -4, 4\sqrt{3} \rangle$ **35.** $\langle 8\sqrt{2}, -8\sqrt{2} \rangle$ **36.** $\langle \sqrt{3}, -1 \rangle$

37. $\langle 15, -8 \rangle$ **38.** $\langle -7, 24 \rangle$ **39.** $\langle -6, 0 \rangle$ **40.** $\langle 0, -12 \rangle$

Write each of the following vectors in the form $a\mathbf{i} + b\mathbf{j}$.

41. $\langle -5, 8 \rangle$ **42.** $\langle 6, -3 \rangle$ **43.** $\langle 2, 0 \rangle$ **44.** $\langle 0, -4 \rangle$

45. direction angle 45°, magnitude 8 **46.** direction angle 210°, magnitude 3

47. direction angle 115°, magnitude .6 **48.** direction angle 208°, magnitude .9

In each of the following, two forces act at a point in the plane. The angle between the two forces is given. Find the magnitude of the resultant force.

49. Forces of 250 and 450 newtons, forming an angle of 85°

50. Forces of 19 and 32 newtons, forming an angle of 118°

51. Forces of 17.9 and 25.8 lb, forming an angle of 105° 30′

52. Forces of 75.6 and 98.2 lb, forming an angle of 82° 50′

53. Forces of 116 and 139 lb, forming an angle of 140° 50′

54. Forces of 37.8 and 53.7 lb, forming an angle of 68° 30′

Solve each of the following problems.

55. Two forces of 692 and 423 newtons act on a point. The resultant force is 786 newtons. Find the angle between the forces.

56. Two forces of 128 and 253 lb act on a point. The equilibrant force is 320 lb. Find the angle between the forces.

57. Find the force required to push a 100-lb box up a ramp inclined 10° with the horizontal.

58. Find the force required to keep a 3000-lb car parked on a hill which makes an angle of 15° with the horizontal.

59. A force of 25 lb is required to push an 80-lb lawn mower up a hill. What angle does the hill make with the horizontal?

60. A force of 500 lb is required to pull a boat up a ramp inclined at 18° with the horizontal. How much does the boat weigh?

61. Anna and Kerry are little dogs. Anna pulls on a rope attached to their doggie dish with a force of 3.89 lb. Kerry pulls on another rope with a force of 4.72 lb. The angle between the forces is 142.8°. Find the direction and magnitude of the equilibrant.

62. Two people are carrying a box. One person exerts a force of 150 lb at an angle of 62.4° with the horizontal. The other person exerts a force of 114 lb at an angle of 54.9°. Find the weight of the box.

63. A crate is supported by two ropes. One rope makes an angle of 46° 20′ with the horizontal and has a tension of 89.6 lb on it. The other rope is horizontal. Find the weight of the crate and the tension in the horizontal rope.

64. Three forces acting at a point are in equilibrium. The forces are 980 lb, 760 lb, and 1220 lb. Find the angles between the directions of the forces. (*Hint:* Arrange the forces to form the sides of a triangle.)

65. A force of 176 lb makes an angle of 78° 50′ with a second force. The resultant of the two forces makes an angle of 41° 10′ with the first force. Find the magnitude of the second force and of the resultant.

66. A force of 28.7 lb makes an angle of 42° 10′ with a second force. The resultant of the two forces makes an angle of 32° 40′ with the first force. Find the magnitude of the second force and of the resultant.

67. A plane flies 650.0 mph on a bearing of 175.3°. A 25.00 mph wind, from a direction of 266.6°, blows against the plane. Find the resulting bearing of the plane.

68. A pilot wants to fly on a bearing of 74.9°. By flying due east, he finds that a 42.0 mph wind blowing from the south puts him on course. Find the air speed and the ground speed.

69. Starting at point *A*, a ship sails 18.5 km on a bearing of 189°, then turns and sails 47.8 km on a bearing of 317°. Find the distance of the ship from point *A*.

70. The distance between points *A* and *B* is 1.7 mi. In between *A* and *B* is a dark forest. To avoid the forest, John walks from *A* a distance of 1.1 mi on a bearing of 325°, and then turns and walks 1.4 mi to *B*. Find the bearing of *B* from *A*.

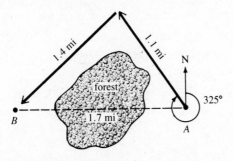

71. The airline route from San Francisco to Honolulu is on a bearing of 233°. A jet flying at 450 mi per hour on that bearing runs into a wind blowing at 39.0 mi per hour from a direction 114°. Find the resulting bearing and ground speed of the plane.

72. A pilot is flying at 168 mph. She wants her flight path to be on a bearing of 57° 40′. A wind is blowing from the south at 27.1 mph. Find the bearing the pilot should fly, and find the plane's ground speed.

73. What bearing and air speed are required for a plane to fly 400 mi due north in 2.5 hours, if the wind is blowing from a direction 328° at 11 mi per hour?

74. Paula and Steve are pulling their daughter Jessie on a sled. Steve pulls with a force of 18 lb at an angle of 10°. Paula pulls with a force of 12 lb at an angle of 15°. Find the resultant. See the figure.

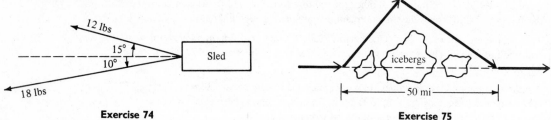

Exercise 74   **Exercise 75**

75. A ship sailing the North Atlantic has been warned to change course to avoid a group of icebergs. The captain turns and sails on a bearing of 62° for a while, then changes course again to a bearing of 115° until the ship reaches its original course. (See the figure.) How much farther did the ship have to travel to avoid the icebergs?

76. A car going around a banked curve is subject to the forces shown in the figure. If the radius of the curve is 100 ft, what value of θ would allow an automobile to travel around the curve at a speed of 40 ft per sec without depending on friction?

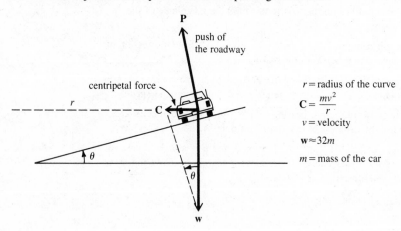

r = radius of the curve

$$C = \frac{mv^2}{r}$$

v = velocity

$w \approx 32m$

m = mass of the car

Let $\mathbf{u} = \langle a_1, b_1 \rangle$, $\mathbf{v} = \langle a_2, b_2 \rangle$, $\mathbf{w} = \langle a_3, b_3 \rangle$, and $\mathbf{0} = \langle 0, 0 \rangle$. Let k and m be any real numbers. Prove each of the following statements.

77. $\mathbf{u} + \mathbf{v} = \mathbf{v} + \mathbf{u}$

78. $\mathbf{u} + (\mathbf{v} + \mathbf{w}) = (\mathbf{u} + \mathbf{v}) + \mathbf{w}$

79. $-1(\mathbf{u}) = -\mathbf{u}$

80. $k(\mathbf{u} + \mathbf{v}) = k\mathbf{u} + k\mathbf{v}$

81. $\mathbf{u} + \mathbf{0} = \mathbf{u}$

82. $\mathbf{u} + (-\mathbf{u}) = \mathbf{0}$

Let $\mathbf{u} = a_1\mathbf{i} + b_1\mathbf{j}$ and $\mathbf{v} = a_2\mathbf{i} + b_2\mathbf{j}$. The *dot product,* or *inner product,* of $\mathbf{u}$ and $\mathbf{v}$, written $\mathbf{u} \cdot \mathbf{v}$, is defined as

$$\mathbf{u} \cdot \mathbf{v} = a_1a_2 + b_1b_2.$$

Find $\mathbf{u} \cdot \mathbf{v}$ for each of the following pairs of vectors.

83. $\mathbf{u} = 6\mathbf{i} - 2\mathbf{j}, \mathbf{v} = -3\mathbf{i} + 2\mathbf{j}$ **84.** $\mathbf{u} = 3\mathbf{i} + 2\mathbf{j}, \mathbf{v} = -3\mathbf{i} + 7\mathbf{j}$

85. $\mathbf{u} = \langle -6, 8 \rangle$ and $\mathbf{v} = \langle 3, -4 \rangle$ **86.** $\mathbf{u} = \langle 0, -2 \rangle$ and $\mathbf{v} = \langle -2, 6 \rangle$

87. Let α be the angle between the vectors $\mathbf{u}$ and $\mathbf{v}$, where $0° \le \alpha \le 180°$. Show that $\mathbf{u} \cdot \mathbf{v}$ = $|\mathbf{u}| \cdot |\mathbf{v}| \cdot \cos \alpha$.

Use the result of Exercise 87 to find the angle between the following pairs of vectors.

88. $\mathbf{u} = \langle -2, 5 \rangle$ and $\mathbf{v} = \langle 3, -4 \rangle$ **89.** $\mathbf{u} = \langle 1, 8 \rangle$ and $\mathbf{v} = \langle 2, -5 \rangle$

90. $\mathbf{u} = \langle -6, -2 \rangle$ and $\mathbf{v} = \langle 3, -1 \rangle$

Prove each of the following properties of the dot product. Assume that $\mathbf{u}$, $\mathbf{v}$, and $\mathbf{w}$ are vectors, and k is a nonzero real number.

91. $\mathbf{u} \cdot \mathbf{v} = \mathbf{v} \cdot \mathbf{u}$ **92.** $\mathbf{u} \cdot \mathbf{u} = |\mathbf{u}|^2$

93. $\mathbf{u} \cdot (\mathbf{v} + \mathbf{w}) = \mathbf{u} \cdot \mathbf{v} + \mathbf{u} \cdot \mathbf{w}$ **94.** $(k \cdot \mathbf{u}) \cdot \mathbf{v} = k \cdot (\mathbf{u} \cdot \mathbf{v})$

95. If $\mathbf{u}$ and $\mathbf{v}$ are not $\mathbf{0}$, and if $\mathbf{u} \cdot \mathbf{v} = 0$, then $\mathbf{u}$ and $\mathbf{v}$ are perpendicular.

96. If $\mathbf{u}$ and $\mathbf{v}$ are perpendicular, then $\mathbf{u} \cdot \mathbf{v} = 0$.

Chapter 5 Summary

Key Words

solving a triangle	law of sines	resultant
angle of elevation	law of cosines	position vector
angle of depression	Heron's area formula	unit vector
oblique triangle	vector	equilibrant

Review Exercises

Find the indicated parts in each of the following right triangles. Assume that the right angle is at C.

1. $A = 47° \, 20'$, $b = 39.6$ cm; find B and c **2.** $A = 15° \, 20'$, $c = 301$ m; find B

3. $b = 68.6$ m, $c = 122.8$ m; find A and B **4.** $A = 42° \, 10'$, $a = 689$ cm; find b and c

5. $B = 88° \, 20'$, $b = 402$ ft; find a and c **6.** $A = 51.74°$, $b = 29.62$ ft; find a and c

7. When the angle of elevation of the sun is $15° \, 50'$, the shadow of a tower is 84.2 feet long. Find the height of the tower.

8. From the top of a cliff, the angle of depression to a river below is $32° \, 10'$. The river is 850 feet from a point directly below the top of the cliff. How high is the cliff?

9. From a point at the base of a mountain, the angle of elevation to the top is $21° \, 10'$. From a point 2000 feet back, the angle of elevation is $18° \, 00'$. Find the height of the mountain.

10. The figure is an illustration of one of the first steam engines.* Steam pressure forced the beam to pivot up and down at point X. The beam, in turn, moved the shaft. Gear A then revolved in a path around a gear B of equal diameter, causing gear B to rotate and turn the attached wheel. Suppose that the beam moved from Y to Z, sweeping out an angle of 40°. If YZ equals the diameter of the path of gear A around gear B, and if $XY = 203$ cm, find the diameter of gear B.

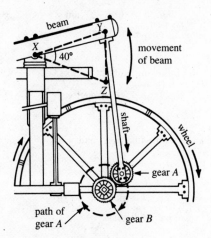

Find the indicated parts of each of the following triangles.

11. $A = 100° 10'$, $B = 25° 00'$, $a = 165$ m; find b

12. $A = 82° 50'$, $C = 62° 10'$, $b = 12.8$ cm; find a

13. $B = 39° 50'$, $b = 268$ m, $a = 430$ m; find A

14. $C = 79° 20'$, $c = 97.4$ mm, $a = 75.3$ mm; find A

15. $A = 25° 10'$, $a = 6.92$ feet, $b = 4.82$ feet; find B

16. $C = 74° 10'$, $c = 96.3$ m, $B = 39° 30'$; find b

Solve each of the following problems.

17. The angle of elevation from the top of a cliff to the top of a second cliff 290 ft away is 68°, while the angle of depression from the top of the first cliff to the bottom of the second cliff is 63°. Find the height of the second cliff.

18. A lot has the shape of the quadrilateral in the figure. What is its area?

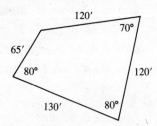

19. A tree leans at an angle of 8° from the vertical. From a point 7 m from the bottom of the tree, the angle of elevation to the top of the tree is 68°. How tall is the tree?

7 m

20. A hill makes an angle of 14.3° with the horizontal. From the base of the hill, the angle of elevation to the top of a tree on top of the hill is 27.2°. The distance along the hill from the base to the tree is 212 ft. Find the height of the tree.

Find the indicated parts in each of the following triangles.

21. $A = 129°\ 40'$, $a = 127$ feet, $b = 69.8$ feet; find B

22. $C = 51°\ 20'$, $c = 68.3$ m, $b = 58.2$ m; find B

23. $a = 86.1$ inches, $b = 253$ inches, $c = 241$ inches; find A

24. $a = 14.8$ m, $b = 19.7$ m, $c = 31.8$ m; find B

25. $A = 46°\ 10'$, $b = 18.4$ m, $c = 19.2$ m; find a

26. $B = 120.7°$, $a = 127$ ft, $c = 69.8$ ft; find b

27. Solve the triangle having $A = 25°\ 10'$, $a = 6.92$ yd, $b = 4.82$ yd.

28. Solve the triangle having $A = 61.7°$, $a = 78.9$ m, $b = 86.4$ m.

Find the area of each of the following triangles.

29. $b = 841$ m, $c = 716$ m, $A = 149°\ 30'$

30. $a = 94.6$ yards, $b = 123$ yards, $c = 109$ yards

31. $a = 27.6$ cm, $b = 19.8$ cm, $C = 42°\ 30'$

32. Raoul plans to paint a triangular wall in his A-frame cabin. Two sides measure 7 m each and the third side measures 6 m. How much paint will he need if a can of paint covers 7.5 m²?

In Exercises 33–35, use the vectors shown here. Find each of the following.

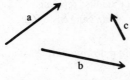

33. a + b   **34. a − b**   **35. a + 3c**

Find the horizontal and vertical components of each of the following vectors.

36. $\alpha = 45°$, magnitude 50   **37.** $\alpha = 75°$, magnitude 69.2   **38.** $\alpha = 154° \ 20'$, magnitude 964

Suppose vector **v** has the given direction angle and magnitude. Find the x- and y-components of **v**.

39. $\theta = 45°$, $|\mathbf{v}| = 2\sqrt{2}$ **40.** $\theta = 120°$, $|\mathbf{v}| = 5\sqrt{3}$

41. $\theta = 210°$, $|\mathbf{v}| = 8$ **42.** $\theta = 302°$, $|\mathbf{v}| = 25$

Find the magnitude and direction angles for each of the following vectors.

43. $\langle -6, 2 \rangle$ **44.** $\langle -6\sqrt{2}, 6\sqrt{2} \rangle$ **45.** $\langle 0, -2 \rangle$ **46.** $\langle \pi, 0 \rangle$

Write each of the following vectors in the form $a\mathbf{i} + b\mathbf{j}$.

47. $\langle 2, -1 \rangle$ **48.** $\langle -6, 3 \rangle$

49. direction angle 30°, magnitude 20   **50.** direction angle 162°, magnitude 5

Let **u** and **v** be as given. Find $\mathbf{u} \cdot \mathbf{v}$.

51. $\mathbf{u} = \langle 2, 6 \rangle$, $\mathbf{v} = \langle -3, 2 \rangle$ **52.** $\mathbf{u} = \langle 4, -5 \rangle$, $\mathbf{v} = \langle -2, 1 \rangle$

53. $\mathbf{u} = 5\mathbf{i}$, $\mathbf{v} = 2\mathbf{i} + 3\mathbf{j}$ **54.** $\mathbf{u} = -\mathbf{i} + 8\mathbf{j}$, $\mathbf{v} = 4\mathbf{i} + 3\mathbf{j}$

Find the angle between each of the following pairs of vectors.

55. $\mathbf{u} = 6\mathbf{i} + 2\mathbf{j}$, $\mathbf{v} = 3\mathbf{i} - 2\mathbf{j}$ **56.** $\mathbf{u} = 4\mathbf{i} - 3\mathbf{j}$, $\mathbf{v} = 2\mathbf{i} + \mathbf{j}$

57. $\mathbf{u} = \langle 2\sqrt{3}, 2 \rangle$, $\mathbf{v} = \langle 5, 5\sqrt{3} \rangle$ **58.** $\mathbf{u} = \langle 2\sqrt{2}, 2\sqrt{2} \rangle$, $\mathbf{v} = \langle -1, 1 \rangle$

Given two forces and the angle between them, find the magnitude of the resultant force.

59. Forces of 15 and 23 lb, forming an angle of 87°

60. Forces of 142 and 215 newtons, forming an angle of 112°

61. Forces of 85.2 and 69.4 newtons, forming an angle of 58° 20′

62. Forces of 475 and 586 lb, forming an angle of 78° 20′

Solve each of the following problems.

63. A force of 150 pounds acts at a right angle to a force of 225 pounds. Find the magnitude of the equilibrant and the angle it makes with the 150-pound force.

64. Forces of 320 and 294 grams act on an object. The angle between the forces is 62.5°. Find the magnitude of the resultant.

65. A box of chickens is supported above the ground to keep the foxes out. The box hangs from two ropes. One makes an angle of 52° 40′ with the horizontal. The tension in this rope is 89.6 lbs. The second rope makes an angle of 82° 30′ with the first rope, and has a tension of 61.7 lbs. The box weighs 10.0 lbs. Find the weight of the chickens. (*Hint:* Add the y-components of each tension vector.)

66. A force of 186 pounds just keeps a 2800 pound Toyota from rolling down a hill. What angle does the hill make with the horizontal?

67. A plane has a still-air speed of 520 miles per hour. The pilot wishes to fly on a bearing of 310°. A wind of 37.0 miles per hour is coming from a bearing of 212°. What direction should the pilot fly, and what will be her actual speed?

68. A boat travels 15 kilometers per hour in still water. The boat is traveling across a large river, on a bearing of 130°. The current in the river, coming from the west, is at a speed of 7.0 kilometers per hour. Find the resulting speed of the boat and its resulting direction of travel.

The following identities involve all six parts of a triangle, *ABC,* and are thus useful for checking answers. Prove each of these results.

69. Newton's formula $$\dfrac{a + b}{c} = \dfrac{\cos \frac{1}{2}(A - B)}{\sin \frac{1}{2}C}$$

70. *Mollweide's formula* $$\dfrac{a - b}{c} = \dfrac{\sin \frac{1}{2}(A - B)}{\cos \frac{1}{2}C}$$

The law of sines can be used to prove the identity for sin $(A + B)$. Let *ABC* be any triangle, and go through the following steps.

71. Show that $c = a \cos B + b \cos A$.

72. Multiply the terms of the result in Exercise 71 by the corresponding terms in the law of sines:

$$\frac{\sin C}{c} = \frac{\sin A}{a} = \frac{\sin B}{b}$$

73. Since $A + B + C = \pi$, we have $C = \pi - (A + B)$. Use the fact that $\sin (\pi - s) = \sin s$ to show that $\sin C = \sin (A + B)$.

74. Finally, obtain the identity for sin $(A + B)$.

(This proof is only valid for values A and B that might be angles of a triangle. Adjustments would have to be made to generalize this result for other angles. For more details, see *Mathematics Magazine,* vol. 35 (1962), p. 229.)

75. Let $v_1 = \langle a_1, b_1 \rangle$ and $v_2 = \langle a_2, b_2 \rangle$ be two nonzero vectors.
(a) Show that if $v_1 = kv_2$ for some scalar k, then $a_1b_2 = a_2b_1$.
(b) Show that if $a_1b_2 = a_2b_1$, then there exists a scalar k such that $v_1 = kv_2$.

76. Let **P** and **Q** be nonzero vectors. Prove that
(a) if **P** and **Q** are perpendicular, then $|P + Q|^2 = |P|^2 + |Q|^2$
(b) if $|P + Q|^2 = |P|^2 + |Q|^2$, then **P** and **Q** are perpendicular.

77. Let A, B, C be the interior angles of a triangle in which no angle is a right angle. By calculating tan $(A + B + C)$ and observing that $A + B + C = 180°$, show that the equation

$$\tan A + \tan B + \tan C = \tan A \tan B \tan C$$

must be satisfied.

6

Complex Numbers

So far, this text has dealt only with real numbers. However, the set of real numbers does not include all the numbers needed in trigonometry and algebra. For example, there is no real number solution of the equation $x^2 + 1 = 0$. In this chapter is discussed a set of numbers having the real numbers as a subset, that is, the set of complex numbers.

6.1 Complex Numbers

Using real numbers alone, it is not possible to find a number whose square is -1. Finding such a number requires that the real number system be extended. To achieve this extension of the real number system, a new number i is defined as follows.

Definition of i

$$i = \sqrt{-1} \quad \text{or} \quad i^2 = -1$$

Numbers of the form $a + bi$, where a and b are real numbers, are called **complex numbers.** Each real number is a complex number, since a real number a may be thought of as the complex number $a + 0i$. A complex number of the form $0 + bi$, where b is nonzero, is called **an imaginary number** (sometimes a *pure* imaginary number). Both the set of real numbers and the set of imaginary numbers are subsets of the set of complex numbers. (See Figure 1, which is an extension of Figure 1 in Section 1.1.) A complex number that is written in the form $a + bi$ or $a + ib$ is in **standard form.** (The form $a + ib$ is used to simplify certain symbols such as $i\sqrt{5}$, since $\sqrt{5}i$ could be too easily mistaken for $\sqrt{5i}$.)

Example 1 The following statements identify different kinds of complex numbers.

(a) -8 and $\sqrt{7}$ and π are real numbers and complex numbers.

(b) $3i$ and $-11i$ and $i\sqrt{14}$ are imaginary numbers and complex numbers.

(c) $1 - 2i$ and $8 - 8i\sqrt{3}$ are complex numbers.   ■

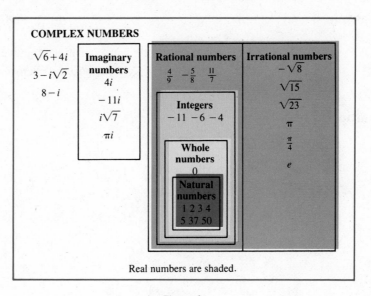

Figure 1

Example 2 The list below shows several numbers, along with the standard form of the number.

Number	Standard form
$6i$	$0 + 6i$
-9	$-9 + 0i$
0	$0 + 0i$
$9 - i$	$9 - i$
$i - 1$	$-1 + i$ ■

Equality for complex numbers is defined as follows.

Equality

For real numbers a, b, c, and d,
$$a + bi = c + di \quad \text{if and only if} \quad a = c \text{ and } b = d.$$

Example 3 Solve $2 + mi = k + 3i$ for real numbers m and k.

By the definition of equality, $2 + mi = k + 3i$ if and only if $2 = k$ and $m = 3$.   ■

Simplify an expression of the form $\sqrt{-a}$, where a is a positive real number, with the following definition.

$\sqrt{-a}$ | If $a > 0$, then $\sqrt{-a} = i\sqrt{a}$.

Example 4 (a) $\sqrt{-16} = i\sqrt{16} = 4i$ (b) $\sqrt{-70} = i\sqrt{70}$ ∎

Products or quotients with square roots of negative numbers may be simplified using the fact that $\sqrt{-a} = i\sqrt{a}$ for positive numbers a. The next example shows how to do this.

Example 5 (a) $\sqrt{-7} \cdot \sqrt{-7} = i\sqrt{7} \cdot i\sqrt{7}$
$$= i^2 \cdot (\sqrt{7})^2$$
$$= (-1) \cdot 7 = -7$$
(b) $\sqrt{-6} \cdot \sqrt{-10} = i\sqrt{6} \cdot i\sqrt{10}$
$$= i^2 \cdot \sqrt{6 \cdot 10}$$
$$= -1 \cdot 2\sqrt{15} = -2\sqrt{15}$$
(c) $\dfrac{\sqrt{-20}}{\sqrt{-2}} = \dfrac{i\sqrt{20}}{i\sqrt{2}} = \sqrt{10}$
(d) $\dfrac{\sqrt{-48}}{\sqrt{24}} = \dfrac{i\sqrt{48}}{\sqrt{24}} = i\sqrt{2}$ ∎

When working with negative radicands, use the definition $\sqrt{-a} = i\sqrt{a}$ before using any of the other rules for radicals. In particular, the rule $\sqrt{c} \cdot \sqrt{d} = \sqrt{cd}$ is valid only when c and d are *not* both negative. For example,

$$\sqrt{(-4)(-9)} = \sqrt{36} = 6,$$

while
$$\sqrt{-4} \cdot \sqrt{-9} = 2i(3i) = 6i^2 = -6,$$

so
$$\sqrt{(-4)(-9)} \neq \sqrt{-4} \cdot \sqrt{-9}.$$

Operations on Complex Numbers Complex numbers may be added, subtracted, multiplied, and divided, as shown by the following definitions and examples.

The *sum* of two complex numbers $a + bi$ and $c + di$ is defined as follows.

Sum | $$(a + bi) + (c + di) = (a + c) + (b + d)i$$

Example 6 (a) $(3 - 4i) + (-2 + 6i) = [3 + (-2)] + [-4 + 6]i$
$$= 1 + 2i$$
(b) $(-9 + 7i) + (3 - 15i) = -6 - 8i$ ∎

Since $(a + bi) + (0 + 0i) = a + bi$ for all complex numbers $a + bi$, the number $0 + 0i$ is called the **addition identity** for complex numbers. The sum of $a + bi$ and $-a - bi$ is $0 + 0i$, so the number $-a - bi$ is called the **negative** or **addition inverse** of $a + bi$.

Using this definition of addition inverse, *subtraction* of the complex numbers $a + bi$ and $c + di$ is defined as follows:

$$(a + bi) - (c + di) = (a + bi) + (-c - di).$$

This definition is often written as

Subtraction

$$(a + bi) - (c + di) = (a - c) + (b - d)i.$$

Example 7 Subtract as indicated.

(a) $(-4 + 3i) - (6 - 7i) = (-4 - 6) + [3 - (-7)]i$
$$= -10 + 10i$$

(b) $(12 - 5i) - (8 - 3i) = 4 - 2i$ ∎

The product of two complex numbers can be found by multiplying as if the numbers were binomials and using the fact that $i^2 = -1$, as follows.

$$(a + bi)(c + di) = ac + adi + bic + bidi$$
$$= ac + adi + bci + bdi^2$$
$$= ac + (ad + bc)i + bd(-1)$$
$$(a + bi)(c + di) = (ac - bd) + (ad + bc)i$$

Based on this, the *product* of the complex numbers $a + bi$ and $c + di$ is defined in the following way.

Product

$$(a + bi)(c + di) = (ac - bd) + (ad + bc)i$$

This definition is hard to remember. To find a given product, it is usually better to multiply as with binomials. The next example shows this.

Example 8 Find each of the following products.

(a) $(2 - 3i)(3 + 4i) = 2(3) - 3i(3) + 2(4i) - 3i(4i)$
$$= 6 - 9i + 8i - 12i^2$$
$$= 6 - i - 12(-1) = 18 - i$$

(b) $(5 - 4i)(7 - 2i) = 5(7) - 4i(7) + 5(-2i) - 4i(-2i)$
$$= 35 - 28i - 10i + 8i^2$$
$$= 35 - 38i + 8(-1)$$
$$= 27 - 38i$$

(c) $(6 + 5i)(6 - 5i) = 6 \cdot 6 - 6 \cdot 5i + 6 \cdot 5i - (5i)^2$

$$= 36 - 25i^2$$
$$= 36 - 25(-1)$$
$$= 36 + 25$$
$$= 61$$

(d) i^{15}

Since $i^2 = -1$, the value of a power of i is found by writing the given power as a product involving i^2. For example, $i^3 = i^2 \cdot i = (-1) \cdot i = -i$. Also, $i^4 = i^2 \cdot i^2 = (-1)(-1) = 1$. Using i^4 to rewrite i^{15} gives

$$i^{15} = i^{12} \cdot i^3 = (i^4)^3 \cdot i^3 = (1)^3 (-i) = -i. \quad \blacksquare$$

Methods similar to those of part (d) of this last example give the following list of **powers of i:**

Powers of i

$i^1 = i$	$i^5 = i$	$i^9 = i$
$i^2 = -1$	$i^6 = -1$	$i^{10} = -1$
$i^3 = -i$	$i^7 = -i$	$i^{11} = -i$
$i^4 = 1$	$i^8 = 1$	$i^{12} = 1,$

and so on.

Example 8(c) showed that $(6 + 5i)(6 - 5i) = 61$. The numbers $6 + 5i$ and $6 - 5i$ differ only in their middle signs; these numbers are **conjugates** of each other. Conjugates are useful because the product of a complex number and its conjugate is always a real number (see Exercise 100).

Example 9   The following list shows several pairs of conjugates, along with the products of the conjugates.

Number	Conjugate	Product
$3 - i$	$3 + i$	$(3 - i)(3 + i) = 10$
$2 + 7i$	$2 - 7i$	$(2 + 7i)(2 - 7i) = 53$
$-6i$	$6i$	$(-6i)(6i) = 36$ $\quad \blacksquare$

The fact that the product of a complex number and its conjugate is always a real number is used to find the *quotient* of two complex numbers, as shown in the next example.

Example 10   **(a)** Find $\dfrac{3 + 2i}{5 - i}$.

Multiply numerator and denominator by the conjugate of $5 - i$.

$$\frac{3 + 2i}{5 - i} = \frac{(3 + 2i)(5 + i)}{(5 - i)(5 + i)}$$

$$= \frac{15 + 3i + 10i + 2i^2}{25 - i^2}$$

$$= \frac{13 + 13i}{26} = \frac{1}{2} + \frac{1}{2}i$$

To check this answer, show that

$$(5 - i)\left(\frac{1}{2} + \frac{1}{2}i\right) = 3 + 2i.$$

(b) $\dfrac{3}{i} = \dfrac{3(-i)}{i(-i)}$ $-i$ is the conjugate of i

$$= \frac{-3i}{-i^2}$$

$$= \frac{-3i}{1}$$ $-i^2 = -(-1) = 1$

$$= -3i$$ $0 - 3i$ in standard form ∎

6.1 Exercises

Identify each number as real, imaginary, or complex.

1. $-9i$ **2.** 6 **3.** π **4.** $-\sqrt{7}$

5. $i\sqrt{6}$ **6.** $-3i$ **7.** $2 + 5i$ **8.** $-7 - 6i$

Write each of the following in standard form.

9. $\sqrt{-100}$ **10.** $\sqrt{-169}$ **11.** $-\sqrt{-400}$ **12.** $-\sqrt{-225}$

13. $-\sqrt{-39}$ **14.** $-\sqrt{-95}$ **15.** $5 + \sqrt{-4}$ **16.** $-7 + \sqrt{-100}$

17. $-6 - \sqrt{-196}$ **18.** $13 + \sqrt{-16}$ **19.** $9 - \sqrt{-50}$ **20.** $-11 - \sqrt{-24}$

21. $\sqrt{-5} \cdot \sqrt{-5}$ **22.** $\sqrt{-20} \cdot \sqrt{-20}$ **23.** $\sqrt{-8} \cdot \sqrt{-2}$ **24.** $\sqrt{-27} \cdot \sqrt{-3}$

25. $\dfrac{\sqrt{-40}}{\sqrt{-10}}$ **26.** $\dfrac{\sqrt{-190}}{\sqrt{-19}}$ **27.** $\dfrac{\sqrt{-6} \cdot \sqrt{-2}}{\sqrt{3}}$ **28.** $\dfrac{\sqrt{-12} \cdot \sqrt{-6}}{\sqrt{8}}$

Add or subtract. Write each result in standard form.

29. $(3 + 2i) + (4 - 3i)$ **30.** $(4 - i) + (2 + 5i)$

31. $(-2 + 3i) - (-4 + 3i)$ **32.** $(-3 + 5i) - (-4 + 3i)$

33. $(2 - 5i) - (3 + 4i) - (-2 + i)$ **34.** $(-4 - i) - (2 + 3i) + (-4 + 5i)$

35. $-i - 2 - (3 - 4i) - (5 - 2i)$ **36.** $3 - (4 - i) - 4i + (-2 + 5i)$

Multiply. Write each result in standard form.

37. $(2 + i)(3 - 2i)$ **38.** $(-2 + 3i)(4 - 2i)$ **39.** $(2 + 4i)(-1 + 3i)$

40. $(1 + 3i)(2 - 5i)$ **41.** $(-3 + 2i)^2$ **42.** $(2 + i)^2$

43. $(3 + i)(-3 - i)$ **44.** $(-5 - i)(5 + i)$ **45.** $(2 + 3i)(2 - 3i)$

46. $(6 - 4i)(6 + 4i)$ **47.** $(\sqrt{6} + i)(\sqrt{6} - i)$ **48.** $(\sqrt{2} - 4i)(\sqrt{2} - 4i)$

49. $i(3 - 4i)(3 + 4i)$ **50.** $i(2 + 7i)(2 - 7i)$ **51.** $3i(2 - i)^2$

52. $-5i(4 - 3i)^2$

Divide. Write each result in standard form.

53. $\dfrac{1 + i}{1 - i}$ **54.** $\dfrac{2 - i}{2 + i}$ **55.** $\dfrac{4 - 3i}{4 + 3i}$ **56.** $\dfrac{5 - 2i}{6 - i}$

57. $\dfrac{3 - 4i}{2 - 5i}$ **58.** $\dfrac{1 - 3i}{1 + i}$ **59.** $\dfrac{-3 + 4i}{2 - i}$ **60.** $\dfrac{5 + 6i}{5 - 6i}$

61. $\dfrac{2}{i}$ **62.** $\dfrac{-7}{3i}$ **63.** $\dfrac{1 - \sqrt{-5}}{3 + \sqrt{-4}}$ **64.** $\dfrac{2 + \sqrt{-3}}{1 - \sqrt{-9}}$

Find each of the following powers of i.

65. i^5 **66.** i^8 **67.** i^9 **68.** i^{11}

69. i^{12} **70.** i^{25} **71.** i^{43} **72.** $1/i^9$

73. $1/i^{12}$ **74.** i^{-6} **75.** i^{-15} **76.** i^{-49}

Perform the indicated operations and write your answers in standard form.

77. $\dfrac{2 + i}{3 - i} \cdot \dfrac{5 + 2i}{1 + i}$ **78.** $\dfrac{1 - i}{2 + i} \cdot \dfrac{4 + 3i}{1 + i}$ **79.** $\dfrac{6 + 2i}{5 - i} \cdot \dfrac{1 - 3i}{2 + 6i}$ **80.** $\dfrac{5 - 3i}{1 + 2i} \cdot \dfrac{2 - 4i}{1 + i}$

81. $\dfrac{5 - i}{3 + i} + \dfrac{2 + 7i}{3 + i}$ **82.** $\dfrac{4 - 3i}{2 + 5i} + \dfrac{8 - i}{2 + 5i}$ **83.** $\dfrac{6 + 2i}{1 + 3i} + \dfrac{2 - i}{1 - 3i}$ **84.** $\dfrac{4 - i}{3 + 4i} - \dfrac{3 + 2i}{3 - 4i}$

85. $\dfrac{6 + 3i}{1 - i} - \dfrac{2 - i}{4 + i}$ **86.** $\dfrac{2 - 3i}{2 + i} + \dfrac{6 + i}{3 + 5i}$

Use the definition of equality for complex numbers to solve the following equations for real numbers a and b.

87. $a + bi = 23 + 5i$ **88.** $a + bi = -2 + 4i$ **89.** $a + bi = 18 - 3i$ **90.** $2 + bi = a - 4i$

91. $a + 3i = 5 + 3bi + 2a$ **92.** $4a - 2bi + 7 = 3i + 3a + 5$

93. $i(2b + 6) - 3 = 4(bi + a)$ **94.** $3i + 2(a - 1) = 4 + 2i(b + 3)$

95. Let $z = 6 - 5i$ and find $4i - 3z$.   **96.** Let $z = 1 - 7i$ and find $2z - 9i$.

Let $z = a + bi$ for real numbers a and b, and let $\bar{z} = a - bi$, the conjugate of z. For example, if $z = 8 - 9i$, then $\bar{z} = 8 + 9i$. Prove each of the following properties of conjugates.

97. $\bar{\bar{z}} = z$ **98.** $\bar{z} = z$ if and only if $b = 0$

99. $\overline{-z} = -\bar{z}$ **100.** $z \cdot \bar{z}$ is a real number

Prove that the complex numbers $z_1 = a + bi$, $z_2 = c + di$, and $z_3 = e + fi$ satisfy each of the following properties.

101. commutative property for addition:
$z_1 + z_2 = z_2 + z_1$

102. commutative property for multiplication:
$z_1 z_2 = z_2 z_1$

103. associative property for addition:
$(z_1 + z_2) + z_3 = z_1 + (z_2 + z_3)$

104. associative property for multiplication:
$(z_1 z_2) z_3 = z_1 (z_2 z_3)$

105. distributive property: $z_1 (z_2 + z_3) = z_1 z_2 + z_1 z_3$

106. closure property of addition: $z_1 + z_2$ is a complex number

107. closure property of multiplication: $z_1 z_2$ is a complex number

Evaluate $8z - z^2$ by replacing z with the indicated complex number.

108. $2 + i$ **109.** $4 - 3i$ **110.** $-6i$

Find all complex numbers $a + bi$ such that the square $(a + bi)^2$ is

111. real

112. imaginary

113. Show that $\dfrac{\sqrt{2}}{2} + \dfrac{\sqrt{2}}{2} i$ is a square root of i.

114. show that $\dfrac{\sqrt{3}}{2} - \dfrac{1}{2} i$ is a cube root of $-i$.

6.2 **Trigonometric Form of Complex Numbers**

This section shows how trigonometry and vectors can be used with the complex numbers introduced in Section 6.1. Graphing complex numbers such as $2 - 3i$ requires a modification of the familiar coordinate system. One way to graph complex numbers is to call the horizontal axis the **real axis** and the vertical axis the **imaginary axis.** Then the complex number $2 - 3i$ can be graphed as shown in Figure 2.

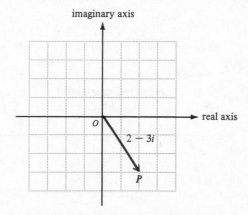

Figure 2

Each nonzero complex number graphed in this way determines a unique directed line segment, the segment from the origin to the point representing the complex number. Recall from Chapter 5 that such directed line segments (like **OP** of Figure 2) are called vectors.

With the definitions of the previous section, the sum of the two complex numbers $4 + i$ and $1 + 3i$ is

$$(4 + i) + (1 + 3i) = 5 + 4i.$$

Graphically, the sum of two complex numbers is represented by the vector which is the resultant of the vectors corresponding to the two numbers. The vectors representing the complex numbers $4 + i$ and $1 + 3i$, and the resultant vector which represents their sum, $5 + 4i$, are shown in Figure 3.

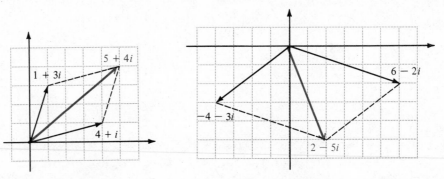

Figure 3   **Figure 4**

Example 1   Find the resultant of $6 - 2i$ and $-4 - 3i$. Graph both complex numbers and their resultant.

The resultant is found by adding the two numbers.

$$(6 - 2i) + (-4 - 3i) = 2 - 5i$$

The graphs are shown in Figure 4.   ■

Figure 5 shows the complex number $x + yi$ that corresponds to a vector **OP** with direction angle θ and magnitude r. The following familiar relationships among r, θ, x, and y can be verified with the aid of Figure 5.

$$x = r \cos \theta \qquad r = \sqrt{x^2 + y^2}$$

$$y = r \sin \theta \qquad \tan \theta = \frac{y}{x}, \quad x \neq 0$$

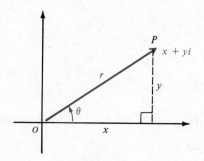

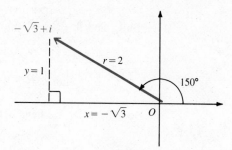

Figure 5   Figure 6

Substituting $x = r \cos \theta$ and $y = r \sin \theta$ from the results above into $x + yi$ gives

$$x + yi = r \cos \theta + (r \sin \theta)i$$
$$= r(\cos \theta + i \sin \theta).$$

Trigonometric or Polar Form of a Complex Number	The expression $$r(\cos \theta + i \sin \theta)$$ is called the **trigonometric form** or **polar form** of the complex number $x + yi$.*

The number r is called the **modulus** or **absolute value** of $x + yi$, while θ is the **argument** of $x + yi$.

Example 2   Write the following complex numbers in trigonometric form.

(a) $-\sqrt{3} + i$

First find r and θ. Since $x = -\sqrt{3}$ and $y = 1$,

$$r = \sqrt{x^2 + y^2} = \sqrt{3 + 1} = 2,$$

Find θ from

$$\tan \theta = \frac{y}{x} = \frac{1}{-\sqrt{3}} = -\frac{\sqrt{3}}{3}.$$

A negative value of x and a positive value of y place the complex number in quadrant II. (See Figure 6.) In quadrant II,

$$\tan 150° = -\frac{\sqrt{3}}{3},$$

so that the trigonometric form of $-\sqrt{3} + i$ is

$$x + yi = r(\cos \theta + i \sin \theta)$$
$$-\sqrt{3} + i = 2(\cos 150° + i \sin 150°).$$

*The expression $\cos \theta + i \sin \theta$ is sometimes abbreviated as cis θ. With this notation $r(\cos \theta + i \sin \theta)$ is written as r cis θ.

(b) $-2 - 2i$

First find r and θ.

$$r = \sqrt{x^2 + y^2} = \sqrt{4 + 4} = 2\sqrt{2}$$

$$\tan \theta = \frac{y}{x} = \frac{-2}{-2} = 1$$

Since both x and y are negative, the complex number is in quadrant III, where

$$\tan 225° = 1.$$

These results make the trigonometric form of $-2 - 2i$

$$2\sqrt{2}(\cos 225° + i \sin 225°). \quad \blacksquare$$

Example 3   Express $2(\cos 300° + i \sin 300°)$ in standard form.

Recall that $a + bi$ is the standard form for a complex number. Since $\cos 300° = 1/2$ and $\sin 300° = -\sqrt{3}/2$,

$$2(\cos 300° + i \sin 300°) = 2\left(\frac{1}{2} - i\frac{\sqrt{3}}{2}\right) = 1 - i\sqrt{3}. \quad \blacksquare$$

Products of Complex Numbers   The product of the two complex numbers $1 + i\sqrt{3}$ and $-2\sqrt{3} + 2i$ can be found by the methods shown earlier:

$$(1 + i\sqrt{3})(-2\sqrt{3} + 2i) = -2\sqrt{3} + 2i - 2i(3) + 2i^2\sqrt{3}$$
$$= -2\sqrt{3} + 2i - 6i - 2\sqrt{3} \qquad \text{since } i^2 = -1$$
$$(1 + i\sqrt{3})(-2\sqrt{3} + 2i) = -4\sqrt{3} - 4i.$$

This same product also can be found by first converting the complex numbers $1 + i\sqrt{3}$ and $-2\sqrt{3} + 2i$ to trigonometric form. Using the method explained above,

$$1 + i\sqrt{3} = 2(\cos 60° + i \sin 60°)$$

and $$-2\sqrt{3} + 2i = 4(\cos 150° + i \sin 150°).$$

If the trigonometric forms are now multiplied together and if the trigonometric identities for the cosine and the sine of the sum of two angles are used, the result is

$$[2(\cos 60° + i \sin 60°)][4(\cos 150° + i \sin 150°)]$$
$$= 2 \cdot 4(\cos 60° \cdot \cos 150° + i \sin 60 \cdot \cos 150°$$
$$\qquad + i \cos 60° \cdot \sin 150° + i^2 \sin 60° \cdot \sin 150°)$$
$$= 8[(\cos 60° \cdot \cos 150° - \sin 60° \cdot \sin 150°)$$
$$\qquad + i(\sin 60° \cdot \cos 150° + \cos 60° \cdot \sin 150°)]$$
$$= 8[\cos (60° + 150°) + i \sin (60° + 150°)]$$
$$= 8(\cos 210° + i \sin 210°).$$

The modulus of the product, 8, is equal to the product of the moduli of the factors, $2 \cdot 4$, while the argument of the product, 210°, is the sum of the arguments of the factors, $60° + 150°$.

Generalizing, the product of the two complex numbers, $r_1(\cos \theta_1 + i \sin \theta_1)$ and $r_2(\cos \theta_2 + i \sin \theta_2)$ is

$$[r_1(\cos \theta_1 + i \sin \theta_1)] \cdot [r_2(\cos \theta_2 + i \sin \theta_2)]$$
$$= r_1r_2(\cos \theta_1 \cos \theta_2 + i \sin \theta_1 \cos \theta_2 + i \cos \theta_1 \sin \theta_2 + i^2 \sin \theta_1 \sin \theta_2)$$
$$= r_1r_2[(\cos \theta_1 \cos \theta_2 - \sin \theta_1 \sin \theta_2) + i(\sin \theta_1 \cos \theta_2 + \cos \theta_1 \sin \theta_2)]$$
$$= r_1r_2[\cos (\theta_1 + \theta_2) + i \sin (\theta_1 + \theta_2)].$$

This work proves the following product theorem.

Product Theorem

> If $r_1 (\cos \theta_1 + i \sin \theta_1)$ and $r_2(\cos \theta_2 + i \sin \theta_2)$ are any two complex numbers, then
>
> $$[r_1(\cos \boldsymbol{\theta_1} + i \sin \boldsymbol{\theta_1})] \cdot [r_2(\cos \boldsymbol{\theta_2} + i \sin \boldsymbol{\theta_2})]$$
> $$= r_1r_2[\cos (\boldsymbol{\theta_1} + \boldsymbol{\theta_2}) + i \sin (\boldsymbol{\theta_1} + \boldsymbol{\theta_2})].$$

Example 4   Find the product of $3(\cos 45° + i \sin 45°)$ and $2(\cos 135° + i \sin 135°)$.
Using the product theorem,

$$[3(\cos 45° + i \sin 45°)][2(\cos 135° + i \sin 135°)]$$
$$= 3 \cdot 2 \, [\cos (\boldsymbol{45°} + \boldsymbol{135°}) + i \sin (\boldsymbol{45°} + \boldsymbol{135°})\,]$$
$$= 6(\cos 180° + i \sin 180°),$$

which can be expressed as $6(-1 + i \cdot 0) = 6(-1) = -6$. The two complex numbers in this example are complex factors of -6. ∎

Quotients of Complex Numbers   In standard form, the quotient of the complex numbers $1 + i\sqrt{3}$ and $-2\sqrt{3} + 2i$ is

$$\frac{1 + i\sqrt{3}}{-2\sqrt{3} + 2i} = \frac{(1 + i\sqrt{3})(-2\sqrt{3} - 2i)}{(-2\sqrt{3} + 2i)(-2\sqrt{3} - 2i)}$$

$$= \frac{-2\sqrt{3} - 2i - 6i - 2i^2\sqrt{3}}{12 - 4i^2}$$

$$\frac{1 + i\sqrt{3}}{-2\sqrt{3} + 2i} = \frac{-8i}{16} = -\frac{1}{2}i.$$

Writing $1 + i\sqrt{3}$, $-2\sqrt{3} + 2i$, and $-\frac{1}{2}i$ in trigonometric form gives

$$1 + i\sqrt{3} = 2(\cos 60° + i \sin 60°)$$
$$-2\sqrt{3} + 2i = 4(\cos 150° + i \sin 150°)$$
$$-\frac{1}{2}i = \frac{1}{2}\,[\cos (-90°) + i \sin (-90°)].$$

The modulus of the quotient, 1/2, is the quotient of the two moduli, 2 and 4. The argument of the quotient, $-90°$, is the difference of the two arguments, $60° - 150°$ $= -90°$. It would be easier to find the quotient of these two complex numbers in trigonometric form than in standard form. Generalizing from this example leads to the following theorem. (For the proof of this result, see Exercise 90 below.)

Quotient Theorem

> If $r_1(\cos \theta_1 + i \sin \theta_1)$ and $r_2(\cos \theta_2 + i \sin \theta_2)$ are complex numbers, where $r_2(\cos \theta_2 + i \sin \theta_2) \neq 0$, then
>
> $$\frac{r_1(\cos \theta_1 + i \sin \theta_1)}{r_2(\cos \theta_2 + i \sin \theta_2)} = \frac{r_1}{r_2}[\cos (\theta_1 - \theta_2) + i \sin (\theta_1 - \theta_2)].$$

Example 5   Find the quotient of $10[\cos(-60°) + i \sin(-60°)]$ and $5(\cos 150° + i \sin 150°)$. Write the result in standard form.

By the quotient theorem,

$$\frac{10[\cos(-60°) + i \sin(-60°)]}{5(\cos 150° + i \sin 150°)}$$

$$= \frac{10}{5}[\cos(-60° - 150°) + i \sin (-60° - 150°)]$$

$$= \frac{10}{5}[\cos(-210°) + i \sin(-210°)].$$

Since angles of $-210°$ and $150°$ are coterminal, replace $-210°$ with $150°$ to get

$$\frac{10[\cos(-60°) + i \sin(-60°)]}{5(\cos 150° + i \sin 150°)} = 2(\cos 150° + i \sin 150°).$$

Because $\cos 150° = -\sqrt{3}/2$ and $\sin 150° = 1/2$,

$$2(\cos 150° + i \sin 150°) = 2\left(\frac{-\sqrt{3}}{2} + i \cdot \frac{1}{2}\right)$$

$$= -\sqrt{3} + i.$$

The quotient in standard form is $-\sqrt{3} + i$. ■

6.2 Exercises

Graph each of the following complex numbers.

1. $-2 + 3i$ 2. $-4 + 5i$ 3. $8 - 5i$ 4. $6 - 5i$
5. $2 - 2i\sqrt{3}$ 6. $4\sqrt{2} + 4i\sqrt{2}$ 7. $-4i$ 8. $3i$
9. -8 10. 2

Find the resultant of each of the following pairs of complex numbers.

11. $2 - 3i, -1 + 4i$ 12. $-4 - 5i, 2 + i$ 13. $-5 + 6i, 3 - 4i$ 14. $8 - 5i, -6 + 3i$

15. $-2, 4i$ **16.** $5, -4i$ **17.** $2 + 6i, -2i$ **18.** $4 - 2i, 5$

19. $7 + 6i, 3i$ **20.** $-5 - 8i, -1$

Write the following complex numbers in standard form.

21. $2(\cos 45° + i \sin 45°)$ **22.** $4(\cos 60° + i \sin 60°)$ **23.** $10(\cos 90° + i \sin 90°)$

24. $8(\cos 270° + i \sin 270°)$ **25.** $4(\cos 240° + i \sin 240°)$ **26.** $2(\cos 330° + i \sin 330°)$

27. $(\cos 30° + i \sin 30°)$ **28.** $3(\cos 150° + i \sin 150°)$ **29.** $5(\cos 300° + i \sin 300°)$

30. $6(\cos 135° + i \sin 135°)$ **31.** $\sqrt{2}(\cos 180° + i \sin 180°)$ **32.** $\sqrt{3}(\cos 315° + i \sin 315°)$

Using a calculator or Table 3, complete the following chart to the nearest ten minutes.

	Standard form	Trigonometric form
33.	$2 + 3i$	
34.		$(\cos 35° + i \sin 35°)$
35.		$3(\cos 250° \ 10' + i \sin 250° \ 10')$
36.	$-4 + i$	
37.	$-1.8794 + .6840i$	
38.		$2(\cos 310° \ 20' + i \sin 310° \ 20')$
39.	$3 + 5i$	
40.		$(\cos 110° \ 30' + i \sin 110° \ 30')$

Write each of the following complex numbers in trigonometric form.

41. $3 - 3i$ **42.** $-2 + 2i\sqrt{3}$ **43.** $-3 - 3i\sqrt{3}$ **44.** $1 + i\sqrt{3}$

45. $\sqrt{3} - i$ **46.** $4\sqrt{3} + 4i$ **47.** $-5 - 5i$ **48.** $-\sqrt{2} + i\sqrt{2}$

49. $2 + 2i$ **50.** $-\sqrt{3} + i$ **51.** -4 **52.** $5i$

Find each of the following products. Write the result in standard form.

53. $[3(\cos 60° + i \sin 60°)][2(\cos 90° + i \sin 90°)]$

54. $[4(\cos 30° + i \sin 30°)][5(\cos 120° + i \sin 120°)]$

55. $[2(\cos 45° + i \sin 45°)][2(\cos 225° + i \sin 225°)]$

56. $[8(\cos 300° + i \sin 300°)][5(\cos 120° + i \sin 120°)]$

57. $[4(\cos 60° + i \sin 60°)][6(\cos 330° + i \sin 330°)]$

58. $[8(\cos 210° + i \sin 210°)][2(\cos 330° + i \sin 330°)]$

59. $[5(\cos 90° + i \sin 90°)][3(\cos 45° + i \sin 45°)]$

60. $[6(\cos 120° + i \sin 120°)][5(\cos (-30°) + i \sin (-30°))]$

61. $[\sqrt{3}(\cos 45° + i \sin 45°)][\sqrt{3}(\cos 225° + i \sin 225°)]$

62. $[\sqrt{2}(\cos 300° + i \sin 300°)][\sqrt{2}(\cos 270° + i \sin 270°)]$

Find the following quotients. Write the results in standard form.

63. $\dfrac{4(\cos 120° + i \sin 120°)}{2(\cos 150° + i \sin 150°)}$

64. $\dfrac{10(\cos 225° + i \sin 225°)}{5(\cos 45° + i \sin 45°)}$

65. $\dfrac{16(\cos 300° + i \sin 300°)}{8(\cos 60° + i \sin 60°)}$

66. $\dfrac{24(\cos 150° + i \sin 150°)}{2(\cos 30° + i \sin 30°)}$

67. $\dfrac{3(\cos 305° + i \sin 305°)}{9(\cos 65° + i \sin 65°)}$

68. $\dfrac{12(\cos 293° + i \sin 293°)}{6(\cos 23° + i \sin 23°)}$

69. $\dfrac{8}{\sqrt{3} + i}$

70. $\dfrac{2i}{-1 - i\sqrt{3}}$

71. $\dfrac{-i}{1 + i}$

72. $\dfrac{1}{2 - 2i}$

73. $\dfrac{2\sqrt{6} - 2i\sqrt{2}}{\sqrt{2} - i\sqrt{6}}$

74. $\dfrac{4 + 4i}{2 - 2i}$

In applied work in trigonometry, it is often necessary to find the resultant of more than two vectors graphically. Find the resultant of the indicated vectors in each of the following cases. Give the resultant in standard form.

75.

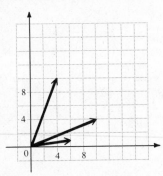

76.

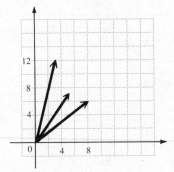

77. The alternating current in an electric inductor is

$$I = \frac{E}{Z}$$

amperes, where E is the voltage and $Z = R + X_L i$ is the impedance. If $E = 8(\cos 20° + i \sin 20°)$, $R = 6$, and $X_L = 3$, find the current. Give the answer in standard form.

78. The current I in a circuit with voltage E, resistance R, capacitive reactance X_c, and inductive reactance X_L is

$$I = \frac{E}{R + (X_L - X_c)i}.$$

Find I if $E = 12(\cos 25° + i \sin 25°)$, $R = 3$, $X_L = 4$, and $X_c = 6$. Give the answer in standard form.

Use your calculator to work each of the following problems. Give answers in standard form.

79. $[3.7(\cos 27° 15' + i \sin 27° 15')][4.1(\cos 53° 42' + i \sin 53° 42')]$

80. $[2.81(\cos 54° 12' + i \sin 54° 12')][5.8(\cos 82° 53' + i \sin 82° 53')]$

81. $\dfrac{45.3(\cos 127° 25' + i \sin 127° 25')}{12.8(\cos 43° 32' + i \sin 43° 32')}$

82. $\dfrac{2.94(\cos 1.5032 + i \sin 1.5032)}{10.5(\cos 4.6528 + i \sin 4.6528)}$

Find and graph all complex numbers c satisfying the following conditions.

83. $|c| = 1$

84. $|c| > 1$

85. The real part of c is 1.

86. The imaginary part of c is 1.

87. The real and imaginary part of c are equal.

88. The real part of c equals c itself.

89. Prove the product theorem.

90. Prove the quotient theorem.

6.3 De Moivre's Theorem and *n*th Roots

The product theorem in the previous section can be used to find the square of a complex number. For example, the square of $r(\cos \theta + i \sin \theta)$ is found by multiplying the number by itself.

$$[r(\cos \theta + i \sin \theta)]^2 = [r(\cos \theta + i \sin \theta)][r(\cos \theta + i \sin \theta)]$$
$$= r \cdot r[\cos (\theta + \theta) + i \sin (\theta + \theta)]$$
$$[r(\cos \theta + i \sin \theta)]^2 = r^2(\cos 2\theta + i \sin 2\theta)$$

In the same way,

$$[r(\cos \theta + i \sin \theta)]^3 = r^3(\cos 3\theta + i \sin 3\theta).$$

These results suggest the plausibility of the following theorem, for positive integer values of n. Although this theorem is stated and can be proved for all n, we will use it only for positive integer values of n and their reciprocals.

De Moivre's Theorem

> If $r(\cos \theta + i \sin \theta)$ is a complex number expressed in trigonometric form, and if n is any real number, then
>
> $$[r(\cos \theta + i \sin \theta)]^n = r^n(\cos n\theta + i \sin n\theta).$$

For positive integer values of n, De Moivre's theorem can be proved by the method of mathematical induction.

Example 1 Find $(1 + i\sqrt{3})^8$.

To use De Moivre's theorem, first convert $1 + i\sqrt{3}$ into trigonometric form.

$$1 + i\sqrt{3} = 2 (\cos 60° + i \sin 60°)$$

Now apply De Moivre's theorem.

$$(1 + i\sqrt{3})^8 = [2(\cos 60° + i \sin 60°)]^8$$
$$= 2^8[\cos (8 \cdot 60°) + i \sin (8 \cdot 60°)]$$
$$= 256(\cos 480° + i \sin 480°)$$
$$= 256(\cos 120° + i \sin 120°)$$
$$(1 + i\sqrt{3})^8 = 256\left(-\frac{1}{2} + i \frac{\sqrt{3}}{2}\right) = -128 + 128i\sqrt{3} \quad \blacksquare$$

De Moivre's theorem is also used to find the *n*th roots of complex numbers. By definition,

nth Root

> for a positive integer *n*, the complex number $a + bi$ is an **nth root** of the complex number $x + yi$ if
>
> $$(a + bi)^n = x + yi.$$

To find the cube roots of the complex number $8(\cos 135° + i \sin 135°)$, for example, look for a complex number, say $r(\cos \alpha + i \sin \alpha)$, that will satisfy

$$[r(\cos \alpha + i \sin \alpha)]^3 = 8(\cos 135° + i \sin 135°).$$

By De Moivre's theorem, this equation becomes

$$r^3(\cos 3\alpha + i \sin 3\alpha) = 8(\cos 135° + i \sin 135°).$$

One way to satisfy this equation is to set $r^3 = 8$ and also $\cos 3\alpha + i \sin 3\alpha = \cos 135° + i \sin 135°$. The first of these conditions implies that $r = 2$, and the second implies that

$$\cos 3\alpha = \cos 135° \quad \text{and} \quad \sin 3\alpha = \sin 135°.$$

These equations can be satisfied only if

$$3\alpha = 135° + 360° \cdot k, \quad k \text{ any integer,}$$

or

$$\alpha = \frac{135° + 360° \cdot k}{3}, \quad k \text{ any integer.}$$

If $k = 0$,

$$\alpha = \frac{135° + 0°}{3} = 45°.$$

For $k = 1$,

$$\alpha = \frac{135° + 360°}{3} = \frac{495°}{3} = 165°.$$

When $k = 2$,

$$\alpha = \frac{135° + 720°}{3} = \frac{855°}{3} = 285°.$$

In the same way, $\alpha = 405°$ when $k = 3$. However, $\sin 405° = \sin 45°$ and $\cos 405° = \cos 45°$, so all of the cube roots (three of them) can be found by letting $k = 0, 1,$ and 2.

When $k = 0$, the root is

$$2(\cos 45° + i \sin 45°).$$

When $k = 1$, the root is

$$2(\cos 165° + i \sin 165°).$$

When $k = 2$, the root is

$$2(\cos 285° + i \sin 285°).$$

In summary, the complex numbers 2(cos 45° + *i* sin 45°), 2(cos 165° + *i* sin 165°), and 2(cos 285° + *i* sin 285°) are the three cube roots of 8(cos 135° + *i* sin 135°). Generalizing this result leads to the following theorem.

nth Root Theorem

> If *n* is any positive integer and *r* is a positive real number, then the complex number *r*(cos θ + *i* sin θ) has exactly *n* distinct *n*th roots, given by
>
> $$r^{1/n}(\cos \alpha + i \sin \alpha),$$
>
> where
>
> $$\alpha = \frac{\theta + 360° \cdot k}{n}, \qquad k = 0, 1, 2, \ldots, n - 1.$$

Alternatively, write the formula for α as

$$\alpha = \frac{\theta}{n} + \left(\frac{360°}{n}\right) \cdot k, \qquad k = 0, 1, 2, \ldots, n - 1.$$

Example 2   Find all fourth roots of $-8 + 8i\sqrt{3}$.
First write $-8 + 8i\sqrt{3}$ in trigonometric form as

$$-8 + 8i\sqrt{3} = 16(\cos 120° + i \sin 120°).$$

Here *r* = 16 and θ = 120°. The fourth roots of this number have modulus $16^{1/4} = 2$ and arguments given as follows.

If *k* = 0, $\dfrac{120° + 360° \cdot \mathbf{0}}{4} = 30°$,

if *k* = 1, $\dfrac{120° + 360° \cdot \mathbf{1}}{4} = 120°$,

if *k* = 2, $\dfrac{120° + 360° \cdot \mathbf{2}}{4} = 210°$,

if *k* = 3, $\dfrac{120° + 360° \cdot \mathbf{3}}{4} = 300°$.

Using these angles, the fourth roots are

2(cos 30° + *i* sin 30°),   2(cos 120° + *i* sin 120°),

2(cos 210° + *i* sin 210°),   and   2(cos 300° + *i* sin 300°).

These four roots can be written in standard form as $\sqrt{3} + i$, $-1 + i\sqrt{3}$, $-\sqrt{3} - i$, and $1 - i\sqrt{3}$. The graphs of these roots are all on a circle that has center at the origin and radius 2, as shown in Figure 7 on the next page. Notice that the roots are equally spaced about the circle and are 90° apart. ∎

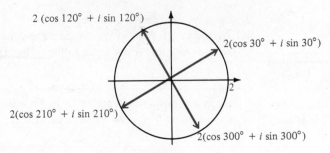

Figure 7

Example 3 Find all complex number solutions of $x^5 - 1 = 0$.
Write the equation as

$$x^5 - 1 = 0$$
$$x^5 = 1$$
$$x = \sqrt[5]{1}.$$

While there is only one real number solution, 1, there are five complex number solutions. To find these solutions, first write 1 in trigonometric form as

$$1 = 1 + 0i = 1(\cos 0° + i \sin 0°).$$

The modulus of the fifth roots is $1^{1/5} = 1$, and the arguments are given by

$$\frac{0° + 360° \cdot k}{5}, \qquad k = 0, 1, 2, 3, \text{ or } 4.$$

By using these arguments, the fifth roots are

$1(\cos 0° + i \sin 0°)$, $1(\cos 72° + i \sin 72°)$, $1(\cos 144° + i \sin 144°)$,
$1(\cos 216° + i \sin 216°)$,   and $1(\cos 288° + i \sin 288°)$.

The first of these roots equals 1; the others cannot easily be expressed in standard form. The five fifth roots all lie on a unit circle and are equally spaced around it every 72°, as shown in Figure 8. ∎

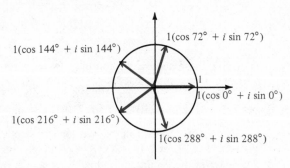

Figure 8

6.3 Exercises

Find the following powers. Write the result in standard form.

1. $[3(\cos 30° + i \sin 30°)]^3$

2. $[2(\cos 135° + i \sin 135°)]^4$

3. $\left(\cos \dfrac{\pi}{4} + i \sin \dfrac{\pi}{4}\right)^8$

4. $\left[2\left(\cos \dfrac{\pi}{3} + i \sin \dfrac{\pi}{3}\right)\right]^3$

5. $[3(\cos 100° + i \sin 100°)]^3$

6. $[3(\cos 40° + i \sin 40°)]^3$

7. $(\sqrt{3} + i)^5$

8. $(2\sqrt{2} - 2i\sqrt{2})^6$

9. $(2 - 2i\sqrt{3})^4$

10. $\left(\dfrac{\sqrt{2}}{2} - \dfrac{\sqrt{2}}{2}i\right)^8$

11. $(-2 - 2i)^5$

12. $(-1 + i)^7$

13. $(-.4283 + .5172i)^4$

14. $(1.87615 - 1.42213i)^3$

15. $\left[1.86\left(\cos \dfrac{5\pi}{9} + i \sin \dfrac{5\pi}{9}\right)\right]^{15}$

16. $\left[24.3\left(\cos \dfrac{7\pi}{12} + i \sin \dfrac{7\pi}{12}\right)\right]^3$

In Exercises 17–46, leave answers in trigonometric form.

Find and graph all cube roots of each of the following complex numbers.

17. $(\cos 0° + i \sin 0°)$

18. $(\cos 90° + i \sin 90°)$

19. $8(\cos 60° + i \sin 60°)$

20. $27(\cos 300° + i \sin 300°)$

21. $-8i$

22. $27i$

23. -64

24. 27

25. $1 + i\sqrt{3}$

26. $2 - 2i\sqrt{3}$

27. $-2\sqrt{3} + 2i$

28. $\sqrt{3} - i$

Find and graph all the following roots of 1. (Roots of 1 are sometimes called *roots of unity*.)

29. Second

30. Fourth

31. Sixth

32. Eighth

Find and graph all the following roots of *i*.

33. Second

34. Fourth

Find all solutions of each of the following equations.

35. $x^3 - 1 = 0$

36. $x^3 + 1 = 0$

37. $x^3 + i = 0$

38. $x^4 + i = 0$

39. $x^3 - 8 = 0$

40. $x^3 + 27 = 0$

41. $x^4 + 1 = 0$

42. $x^4 + 16 = 0$

43. $x^4 - i = 0$

44. $x^5 - i = 0$

45. $x^3 - (4 + 4i\sqrt{3}) = 0$ **46.** $x^4 - (8 + 8i\sqrt{3}) = 0$

Use a calculator to find all solutions of each of the following equations. Give answers in standard form.

47. $x^3 + 4 - 5i = 0$

48. $x^5 + 2 + 3i = 0$

49. $x^2 + (3.72 + 8.24i) = 0$

50. $x^4 - (5.13 - 4.27i) = 0$

Let $z = a + bi$. Solve the following equations for z. Give answers in trigonometric form.

51. $z^2 = 1 + i$

52. $z^2 = -\sqrt{2} + i\sqrt{2}$

53. $z^2 = 3 - 3i$

54. $z^2 = -\sqrt{3} - 1$

55. Let α be an *n*th root of 1 (that is, $\alpha^n = 1$). Show that, if $\alpha \neq 1$, then
$\alpha^{n-1} + \alpha^{n-2} + \cdots + \alpha + 1 = 0$. (*Hint:* Divide $\alpha^n - 1$ by $\alpha - 1$.)

56. Prove the theorem on *n*th roots.

6.4 Polar Equations

Throughout this text the Cartesian coordinate system has been used to graph equations. The **polar coordinate system** is another coordinate system that is useful for graphing. The system is based on a point, called the **pole,** and a ray, called the **polar axis.** The polar axis is usually drawn in the direction of the positive x-axis, as shown in Figure 9.

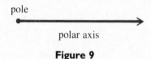

Figure 9

In Figure 10 the pole has been placed at the origin of a Cartesian coordinate system, so that the polar axis coincides with the positive x-axis. Point P has coordinates (x, y) in the Cartesian coordinate system. Point P can also be located by giving the directed angle θ from the positive x-axis to OP and the directed distance r from the pole to point P. The ordered pair (r, θ) gives the **polar coordinates** of point P.

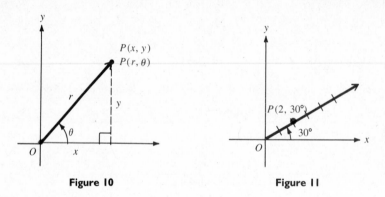

Figure 10   **Figure 11**

Example 1   Plot each point, given its polar coordinates.

(a) $P(2, 30°)$

In this case, $r = 2$ and $\theta = 30°$, so the point P is located 2 units from the origin in the positive direction on a ray $30°$ from the polar axis, as shown in Figure 11.

(b) $Q(-4, 120°)$

Since r is negative, Q is 4 units in the negative direction from the pole on an extension of the $120°$ ray. See Figure 12.

(c) $R(5, -45°)$

Point R is shown in Figure 13. Since θ is negative, the angle is measured in the clockwise direction.   ■

Parts (b) and (c) show one important difference between Cartesian coordinates and polar coordinates: While a given point in the plane can have only one pair of

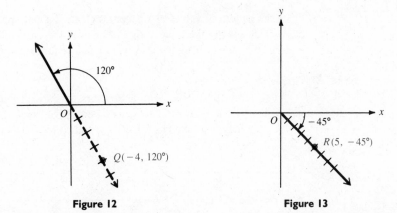

Figure 12   **Figure 13**

Cartesian coordinates, this same point can have infinitely many pairs of polar coordinates. For example, point Q in Figure 12 could also be located as $(4, 300°)$, and point R in Figure 13 is also located as $(-5, 135°)$.

Example 2   Give three other pairs of polar coordinates for the point $(3, 140°)$.

Three pairs that could be used for the point are $(3, -220°)$, $(-3, 320°)$, and $(-3, -40°)$. ■

An equation like $r = 3 \sin \theta$, where r and θ are the variables, is a **polar equation.** (Equations in x and y are called **rectangular** or **Cartesian equations.**) The simplest equation for many useful curves is often a polar equation.

Graphing a polar equation is much the same as graphing a Cartesian equation: Find some representative ordered pairs, (r, θ), satisfying the equation, and then sketch the graph. For example, to graph $r = 1 + \cos \theta$, first find and graph some ordered pairs (as in the table) and then connect the points in order—from $(2, 0°)$ to $(1.9, 30°)$ to $(1.7, 45°)$ and so on. The graph, shown in Figure 14, is called a **cardioid** because of its heart shape.

θ	0°	30°	45°	60°	90°	120°	135°	150°	180°	270°	315°
$\cos \theta$	1	.9	.7	.5	0	$-.5$	$-.7$	$-.9$	-1	0	.7
$r = 1 + \cos \theta$	2	1.9	1.7	1.5	1	.5	.3	.1	0	1	1.7

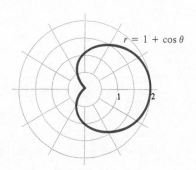

Figure 14

Once the pattern of values of r becomes clear, it is not necessary to find more ordered pairs. That is why the table above stops with the ordered pair $(1.7, 315°)$. From the pattern, the pair $(1.9, 330°)$ also would satisfy the equation.

Example 3 Graph $r^2 = \cos 2\theta$.

First complete a table of ordered pairs as shown, and then sketch the graph, as in Figure 15. The point $(-1, 0°)$, with r negative, may be plotted as $(1, 180°)$. Also, $(-.7, 30°)$ may be plotted as $(.7, 210°)$, and so on. This curve is called a **lemniscate**.

θ	0°	30°	45°	135°	150°	180°
2θ	0°	60°	90°	270°	300°	360°
$\cos 2\theta$	1	.5	0	0	.5	1
$r = \pm\sqrt{\cos 2\theta}$	± 1	$\pm.7$	0	0	$\pm.7$	± 1

Values of θ for $45° < \theta < 135°$ are not included in the table because the corresponding values of $\cos 2\theta$ are negative (quadrants II and III) and so do not have real square roots. Values of θ larger than $180°$ give 2θ larger than $360°$, and would repeat the points already found. ■

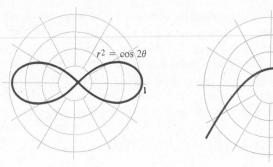

Figure 15   Figure 16

Example 4 Graph $r = \dfrac{4}{1 + \sin \theta}$.

Again complete a table of ordered pairs, which leads to the graph shown in Figure 16.

θ	0°	30°	45°	60°	90°	120°	135°	150°	180°	210°	225°
$\sin \theta$	0	.5	.7	.9	1	.9	.7	.5	0	$-.5$	$-.7$
$r = \dfrac{4}{1 + \sin \theta}$	4	2.7	2.3	2.1	2.0	2.1	2.3	2.7	4.0	8.0	13.3

With the points given in the table, the pattern of the graph should start to be clear. If it is not, continue to find additional points. ■

Example 5   Graph $r = 3 \cos 2\theta$.

Because of the 2θ, the graph requires a large number of points. A few points are given below. You should complete the table similarly through the first 360°.

θ	0°	15°	30°	45°	60°	75°	90°
2θ	0°	30°	60°	90°	120°	150°	180°
$\cos 2\theta$	1	.9	.5	0	$-.5$	$-.9$	-1
$r = 3 \cos 2\theta$	3	2.7	1.5	0	-1.5	-2.7	-3

Plotting these points in order gives the graph, called a **four-leaved rose.** Notice in Figure 17 how the graph is developed with a continuous curve, beginning with the upper half of the right horizontal leaf and ending with the lower half of that leaf. As the graph is traced, the curve goes through the pole four times.   ■

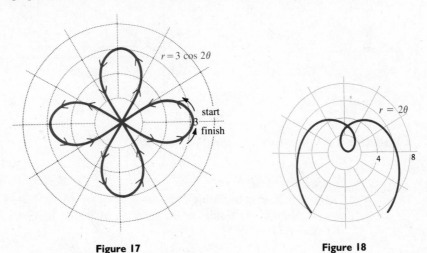

Figure 17   Figure 18

Generalizing from Example 5, the graphs of $r = \sin n\theta$ and $r = \cos n\theta$ are roses, with n petals if n is odd, and $2n$ petals if n is even.

Example 6   Graph $r = 2\theta$ (θ measured in radians).

Some ordered pairs are shown below. In Examples 3 to 5, it was not necessary to use negative values of θ because the trigonometric functions are periodic. Here negative values must be considered too. The radian measures have been rounded for simplicity.

θ (degrees)	-180	-90	-45	0	30	60	90	180	270	360
θ (radians)	-3.1	-1.6	$-.8$	0	.5	1	1.6	3.1	4.7	6.3
$r = 2\theta$	-6.2	-3.2	-1.6	0	1	2	3.2	6.2	9.4	12.6

Figure 18 shows this graph, called a **spiral of Archimedes.**   ■

Sometimes an equation given in polar form is easier to graph in Cartesian form. To convert a polar equation to a Cartesian equation use the following relationships from Section 6.2, which are derived from triangle POQ in Figure 19.

Converting Between Polar and Rectangular Coordinates

$$x = r \cos \theta \qquad r = \sqrt{x^2 + y^2}$$

$$y = r \sin \theta \qquad \tan \theta = \frac{y}{x}, \quad x \neq 0$$

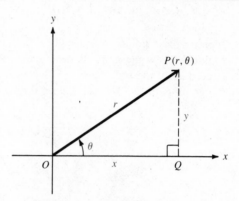

Figure 19

Example 7 Convert the equation in Example 4,

$$r = \frac{4}{1 + \sin \theta},$$

to Cartesian coordinates.

Multiply both sides of the equation by the denominator on the right, to clear the fraction.

$$r = \frac{4}{1 + \sin \theta}$$

$$r + r \sin \theta = 4$$

Now substitute $\sqrt{x^2 + y^2}$ for r and y for $r \sin \theta$.

$$\sqrt{x^2 + y^2} + y = 4$$

$$\sqrt{x^2 + y^2} = 4 - y$$

Square both sides to eliminate the radical.

$$x^2 + y^2 = (4 - y)^2$$

$$x^2 + y^2 = 16 - 8y + y^2$$

$$x^2 = -8y + 16$$

$$x^2 = -8(y - 2)$$

The final equation represents a parabola and can be graphed using rectangular coordinates. ∎

Example 8　Convert the equation $3x + 2y = 4$ to a polar equation.

Use $x = r \cos \theta$ and $y = r \sin \theta$ to get

$$3x + 2y = 4$$
$$3r \cos \theta + 2r \sin \theta = 4.$$

Now solve for r. First factor out r on the left.

$$r(3 \cos \theta + 2 \sin \theta) = 4$$
$$r = \frac{4}{3 \cos \theta + 2 \sin \theta}$$

The polar equation of the line $3x + 2y = 4$ is

$$r = 4/(3 \cos \theta + 2 \sin \theta). \quad \blacksquare$$

Example 9　Note the use of polar coordinates in the following example, taken with permission from *Calculus and Analytic Geometry*, fifth edition, by George Thomas and Ross Finney (Addison-Wesley, 1979).

Karl von Frisch has advanced the following theory about how bees communicate information about newly discovered sources of food. A scout returning to the hive from a flower bed gives away samples of the food and then, if the bed is more than about a hundred yards away, performs a dance to show where the flowers are. The bee runs straight ahead for a centimeter or so, waggling from side to side, and circles back to the starting place. The bee then repeats the straight run, circling back in the opposite direction. (See Figure 20.) The dance continues this way in regular alternation. Exceptionally excited bees have been observed to dance for more than three and a half hours.

If the dance is performed inside, it is performed on the vertical wall of a honeycomb, with gravity substituting for the sun's position. A vertical straight run means that the food is in the direction of the sun. A run 30° to the right of vertical means that the food is 30° to the right of the sun, and so on. Distance (more accurately, the amount of energy required to reach the food) is communicated by the duration of the straight-run portions of the dance. Straight runs lasting three seconds each are typical for distances of about a half mile from the hive. Straight runs that last five seconds each mean about two miles. $\quad \blacksquare$

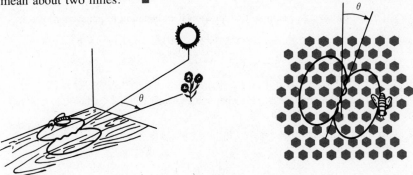

The waggle dance of a scout bee.

Figure 20

6.4 Exercises

Plot each point, given its polar coordinates.

1. $(1, 45°)$ **2.** $(3, 120°)$ **3.** $(-2, 135°)$ **4.** $(-4, 27°)$

5. $(5, -60°)$ **6.** $(2, -45°)$ **7.** $(-3, -210°)$ **8.** $(-1, -120°)$

9. $(3, 300°)$ **10.** $(4, 270°)$ **11.** $(-5, -420°)$ **12.** $(2, -435°)$

Graph each of the following equations for $0° \leq \theta \leq 180°$, unless other domains are specified.

13. $r = 2 + 2 \cos \theta$ **14.** $r = 2(4 + 3 \cos \theta)$

15. $r = 3 + \cos \theta$ (limaçon) **16.** $r = 2 - \cos \theta$ (limaçon)

17. $r = \sin 2\theta$ (four-leaved rose) **18.** $r = 3 \cos 5\theta$ (five-leaved rose); $0° \leq \theta < 360°$
(*Hint:* Use $0° \leq \theta < 360°$ every 15°.)

19. $r^2 = 4 \cos 2\theta$ (lemniscate) **20.** $r^2 = 4 \sin 2\theta$ (lemniscate); $0° \leq \theta < 360°$

21. $r = 4(1 - \cos \theta)$ (cardioid) **22.** $r = 3(2 - \cos \theta)$ (cardioid) **23.** $r = 2 \sin \theta \tan \theta$ (cissoid)

24. $r = \dfrac{\cos 2\theta}{\cos \theta}$ **25.** $r = \dfrac{3}{2 + \sin \theta}$ **26.** $r = \sin \theta \cos^2 \theta$

Graph each of the following for $-\pi \leq \theta \leq \pi$, measuring θ in radians.

27. $r = 5\theta$ (spiral of Archimedes) **28.** $r = \theta$ (spiral of Archimedes)

29. $r\theta = \pi$ (hyperbolic spiral) **30.** $r^2 = \theta$ (parabolic spiral)

31. $\ln r = \theta$ (logarithmic spiral) **32.** $\log r = \theta$ (logarithmic spiral)

For each of the following equations, find an equivalent equation in Cartesian coordinates, and sketch the graph.

33. $r = 2 \sin \theta$ **34.** $r = 2 \cos \theta$ **35.** $r = \dfrac{2}{1 - \cos \theta}$

36. $r = \dfrac{3}{1 - \sin \theta}$ **37.** $r + 2 \cos \theta = -2 \sin \theta$ **38.** $r = \dfrac{3}{4 \cos \theta - \sin \theta}$

39. $r = 2 \sec \theta$ **40.** $r = -5 \csc \theta$ **41.** $r(\cos \theta + \sin \theta) = 2$

42. $r(2 \cos \theta + \sin \theta) = 2$ **43.** $r \sin \theta + 2 = 0$ **44.** $r \sec \theta = 5$

For each of the following equations, find an equivalent equation in polar coordinates.

45. $x + y = 4$ **46.** $2x - y = 5$ **47.** $x^2 + y^2 = 16$ **48.** $x^2 + y^2 = 9$

49. $y = 2$ **50.** $x = 4$ **51.** $y^2 = 25x$ **52.** $x^2 = 4y$

53. $x^2 + 9y^2 = 36$ **54.** $16x^2 + y^2 = 16$

55. Discuss the symmetry of $r = f(\theta)$ when r is replaced with $-r$; when θ is replaced with $-\theta$; and when θ is replaced with $\pi - \theta$.

56. Show that the distance between (r_1, θ_1) and (r_2, θ_2) is $\sqrt{r_1^2 + r_2^2 - 2r_1r_2 \cos (\theta_1 - \theta_2)}$.

Chapter 6 Summary

Key Words   complex number
imaginary number
conjugate
polar coordinates
trigonometric form
De Moivre's theorem
nth root

Review Exercises

Write in standard form.

1. $(1 - i) - (3 + 4i) + 2i$

2. $(2 - 5i) + (9 - 10i) - 3$

3. $(6 - 5i) + (2 + 7i) - (3 - 2i)$

4. $(4 - 2i) - (6 + 5i) - (3 - i)$

5. $(3 + 5i)(8 - i)$

6. $(4 - i)(5 + 2i)$

7. $(2 + 6i)^2$

8. $(6 - 3i)^2$

9. $(1 - i)^3$

10. $(2 + i)^3$

11. i^{17}

12. i^{52}

13. $\dfrac{6 + 2i}{3 - i}$

14. $\dfrac{2 - 5i}{1 + i}$

15. $\dfrac{2 + i}{1 - 5i} \cdot \dfrac{1 + i}{3 - i}$

16. $\dfrac{4 + 3i}{1 - i} \cdot \dfrac{2 - 3i}{2 + i}$

17. $\dfrac{8 - i}{2 + i} + \dfrac{3 + 2i}{4i}$

18. $\dfrac{6 + 3i}{1 + i} + \dfrac{1 - i}{2 + 2i}$

19. $\sqrt{-12}$

20. $\sqrt{-18}$

Find the resultant of each of the following pairs of complex numbers.

21. $7 + 3i$ and $-2 + i$

22. $2 - 4i$ and $-1 - 2i$

Graph each of the following complex numbers.

23. $5i$

24. $-4 + 2i$

25. $3 - 3i\sqrt{3}$

26. $-5 + i\sqrt{3}$

Complete the following chart.

	Standard form	Trigonometric form
27.	$-2 + 2i$	
28.		$3 (\cos 90° + i \sin 90°)$
29.		$2 (\cos 225° + i \sin 225°)$
30.	$-4 + 4i\sqrt{3}$	
31.	$1 - i$	
32.		$4 (\cos 240° + i \sin 240°)$
33.	$-4i$	
34.		$2 (\cos 180° + i \sin 180°)$

Perform the indicated operations. Write answers in standard form.

35. $5 (\cos 90° + i \sin 90°) \cdot 6 (\cos 180° + i \sin 180°)$

36. $3 (\cos 135° + i \sin 135°) \cdot 2 (\cos 105° + i \sin 105°)$

37. $\dfrac{2 (\cos 60° + i \sin 60°)}{8 (\cos 300° + i \sin 300°)}$

38. $\dfrac{4 (\cos 270° + i \sin 270°)}{2 (\cos 90° + i \sin 90°)}$

39. $(\sqrt{3} + i)^3$

40. $(2 - 2i)^5$

41. $(\cos 100° + i \sin 100°)^6$

42. $(\cos 20° + i \sin 20°)^3$

In Exercises 43–46 give answers in trigonometric form.

43. Find the fifth roots of $-2 + 2i$.

44. Find the cube roots of $1 - i$.

45. Find the sixth roots of 1.

46. Find the fourth roots of $\sqrt{3} + i$.

Find all solutions for the following equations.

47. $x^3 + 125 = 0$

48. $x^4 + 16 = 0$

Graph the following.

49. $r = \dfrac{3}{1 + \cos \theta}$

50. $r = \dfrac{4}{2 \sin \theta - \cos \theta}$

51. $r = \sin \theta + \cos \theta$

52. $r = 2$

Find an equivalent equation in polar coordinates.

53. $x = -3$

54. $y = x$

55. $y = x^2$

56. $x = y^2$

Let $R(z)$ and $I(z)$ denote respectively the real and imaginary parts of a complex number z. Show that the following are true.

57. $z + \bar{z} = 2R(z)$

58. $z - \bar{z} = 2iI(z)$

59. $|R(z)| \leq |z|$

60. $|z_1 + z_2|^2 = |z_1|^2 + |z_2|^2 + 2R(z_1\bar{z}_2)$

61. $|z_1 + z_2| \leq |z_1| + |z_2|$

62. Find all pairs (x_0, y_0) that are both polar and rectangular coordinates for the same point.

63. A regular pentagon in inscribed in a circle with center at the origin and radius 3. One vertex is on the positive x-axis. Write the polar coordinates of all the vertices.

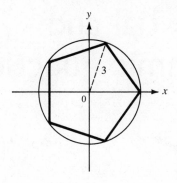

7

Exponential and Logarithmic Functions

Many applications of mathematics, particularly to growth and decay of populations, involve the closely interrelated exponential and logarithmic functions introduced in this chapter. As shown later, these two types of functions are inverses of each other.

7.1 Exponential Functions

For any positive real number a, the symbol a^m can be defined for any rational value of m. In this section, the definition of a^m is extended to include all *real*, and not just rational, values of the exponent m. For example, what is meant by $2^{\sqrt{3}}$? The exponent, $\sqrt{3}$, can be approximated more and more closely by the numbers 1.7, 1.73, 1.732, and so on, making it reasonable that $2^{\sqrt{3}}$ should be approximated more and more closely by the numbers $2^{1.7}$, $2^{1.73}$, $2^{1.732}$, and so on. (Recall, for example, that $2^{1.7} = 2^{17/10}$, which means $\sqrt[10]{2^{17}}$.) In fact, this is exactly how the number $2^{\sqrt{3}}$ is defined in a more advanced course. To show that this assumption is reasonable, Figure 1 gives the graphs of the function $f(x) = 2^x$ with three different domains.

We shall assume that the meaning given to real exponents is such that all previous rules and theorems for exponents are valid for real number exponents as well as rational ones. In addition to the properties of exponents from algebra, the following additional properties will prove useful. First, any given real value of x leads to exactly one value of 2^x. For example,

$$2^2 = 4, \quad 2^3 = 8, \quad \text{and} \quad 2^{1/2} = \sqrt{2} \approx 1.4142.$$

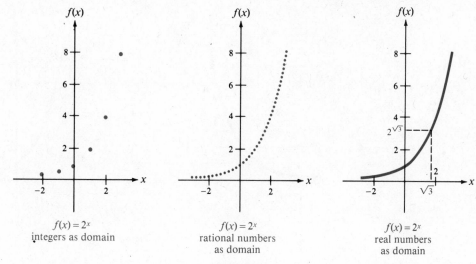

$f(x) = 2^x$
integers as domain

$f(x) = 2^x$
rational numbers
as domain

$f(x) = 2^x$
real numbers
as domain

Figure I

Furthermore, if	$3^x = 3^4$,	then $x = 4$.
And if	$p^2 = 3^2$,	then $p = 3$.

Also $\qquad 4^2 < 4^3 \qquad$ but $\qquad \left(\dfrac{1}{2}\right)^2 > \left(\dfrac{1}{2}\right)^3,$

so that when $a > 1$, increasing the exponent on a leads to a *larger* number, but if $0 < a < 1$, increasing the exponent on a leads to a *smaller* number.

These properties are generalized in the next theorem. No proof of these properties is given, since the proof requires more advanced mathematics than that of this course.

Theorem

For any real number $a > 0$, $a \neq 1$, and any real number x

(a) a^x **is a unique real number.**

(b) $a^b = a^c$ **if and only if** $b = c$.

(c) If $a > 1$ **and** $m < n$, **then** $a^m < a^n$.

(d) If $0 < a < 1$ **and** $m < n$, **then** $a^m > a^n$.

Part (a) of the theorem requires $a > 0$ so that a^x is always defined. For example, $(-6)^x$ is not a real number if $x = 1/2$. If $a > 0$, then a^x will always be positive, since a is positive. For part (b) to hold, a must not equal 1 since $1^4 = 1^5$, even though $4 \neq 5$. Figure 2(a) on the next page illustrates part (c): the base, 2, of the exponential 2^x is greater than 1, and as the x-values increase, so do the values of 2^x. Part (d) of the theorem is shown in Figure 2(b) where the base, 1/2, is between 0 and 1, and as the x-values increase, $(1/2)^x$ decreases.

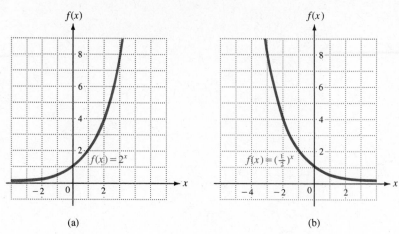

Figure 2

With all these assumptions and properties of real exponents, a function, f, can now be defined, where $f(x) = a^x$, with domain the set of all real numbers (and not just the rationals).

Exponential Function	The function f defined by $$f(x) = a^x, \qquad a > 0 \text{ and } a \neq 1,$$ is the **exponential function with base a.**

(If $a = 1$, the function is the constant function given by $f(x) = 1$.)

Example 1 If $f(x) = 2^x$, find each of the following.

(a) $f(-1)$

Replace x with -1.

$$f(-1) = 2^{-1} = \frac{1}{2}$$

(b) $f(3) = 2^3 = 8$

(c) $f(5/2) = 2^{5/2} = (2^5)^{1/2} = 32^{1/2} = \sqrt{32} = 4\sqrt{2}$ ∎

Example 2 Graph the exponential functions defined as follows.

(a) $f(x) = 2^x$.

The base of this exponential function is 2. Some ordered pairs that satisfy the equation are $(-2, 1/4)$, $(-1, 1/2)$, $(0, 1)$, $(1, 2)$, $(2, 4)$, and $(3, 8)$. Plotting these points and then drawing a smooth curve through them gives the graph in Figure 2(a). As the graph suggests, the domain of the function is the set of all real numbers, and the range is the set of all positive numbers. The function is increasing on its entire domain, making it a one-to-one function. The x-axis is a horizontal asymptote.

(b) $f(x) = (1/2)^x$

Again, plot some ordered pairs and draw a smooth curve through them. For example, $(-3, 8)$, $(-2, 4)$, $(-1, 2)$, $(0, 1)$, and $(1, 1/2)$ are on the graph shown in Figure 2(b). Like the function in part (a), this function also has the set of real numbers as domain and the set of positive real numbers as range. This graph is decreasing on the entire domain. ■

Starting with $f(x) = 2^x$ and replacing x with $-x$, gives $f(-x) = 2^{-x} = (2^{-1})^x = (1/2)^x$. For this reason, the graphs of $f(x) = 2^x$ and $f(x) = (1/2)^x$ are symmetric with respect to the y-axis. This is also suggested by the graphs in Figures 2(a) and (b).

The graph of $f(x) = 2^x$ is typical of graphs of $f(x) = a^x$ where $a > 1$. For larger values of a, the graphs rise more steeply, but the general shape is similar to the graph in Figure 2(a). When $0 < a < 1$ the graph decreases like the graph of $f(x) = (1/2)^x$ in Figure 2(b). In Figure 3, the graphs of several typical exponential functions illustrate these facts.

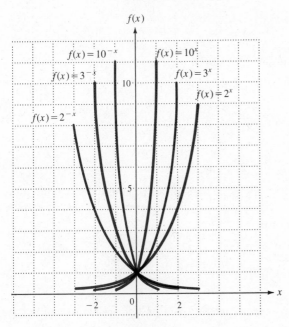

Figure 3

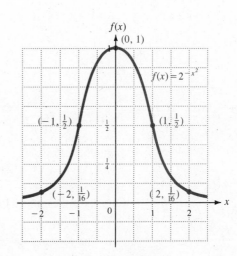

Figure 4

Example 3   Graph $f(x) = 2^{-x^2}$.

Since $f(x) = 2^{-x^2} = 1/(2^{x^2})$, we have $0 < y \le 1$ for all values of x. Plotting some typical points, such as $(-2, 1/16)$, $(-1, 1/2)$, $(0, 1)$, $(1, 1/2)$, and $(2, 1/16)$, and drawing a smooth curve through them gives the graph of Figure 4. This graph is symmetric with respect to the y-axis and has the x-axis as a horizontal asymptote. ■

The results of the last theorem can be used to solve equations with variable exponents, as shown in the next example.

Example 4   Solve $\left(\dfrac{1}{3}\right)^{x} = 81$.

First, write $1/3$ as 3^{-1}, so that $(1/3)^{x} = (3^{-1})^{x} = 3^{-x}$. Since $81 = 3^{4}$,

$$\left(\dfrac{1}{3}\right)^{x} = 81$$

becomes $$3^{-x} = 3^{4}.$$

By the second property above,

$$-x = 4, \quad \text{or} \quad x = -4.$$

The solution set of the given equation is $\{-4\}$. (Section 7.4 shows a method for solving equations of this type where both sides cannot be written as powers of the same number.)   ■

Example 5   Find b if $81 = b^{4/3}$.

Since $(b^{4/3})^{3/4} = b^{1} = b$, raise both sides of the equation to the $3/4$ power.

$$81 = b^{4/3}$$
$$81^{3/4} = (b^{4/3})^{3/4}$$
$$(\sqrt[4]{81})^{3} = b$$
$$3^{3} = b$$
$$27 = b$$

Remember that the process of raising both sides of an equation to the same power may result in false "solutions." For this reason, it is necessary to check all proposed solutions. Replacing b with 27 gives

$$27^{4/3} = (\sqrt[3]{27})^{4} = 3^{4} = 81,$$

which checks; the solution set is $\{27\}$.   ■

The formula for compound interest (interest paid on both principal and interest) defines an exponential function.

Compound Interest

> If P dollars is deposited in an account paying an annual rate of interest i compounded (paid) m times per year, the account will contain
>
> $$A = P\left(1 + \dfrac{i}{m}\right)^{nm}$$
>
> dollars after n years.

Example 6   Suppose $1000 is deposited in an account paying 8% per year compounded quarterly (four times a year). Find the total amount in the account after 10 years if no withdrawals are made. Find the amount of interest earned.

Use the compound interest formula from above with $P = 1000$, $i = .08$, $m = 4$, and $n = 10$.

$$A = P\left(1 + \frac{i}{m}\right)^{nm}$$

$$A = 1000\left(1 + \frac{.08}{4}\right)^{10(4)}$$

$$= 1000(1 + .02)^{40} = 1000(1.02)^{40}$$

The number $(1.02)^{40}$ can be found in financial tables or by using a calculator with a y^x key. To five decimal places, $(1.02)^{40} = 2.20804$. The amount on deposit after 10 years is

$$A = 1000(1.02)^{40} = 1000(2.20804) = 2208.04,$$

or $2208.04. The amount of interest earned is

$$\$2208.04 - \$1000.00 = \$1208.04. \quad \blacksquare$$

The Number e   Perhaps the single most useful base for an exponential function is the number e, an irrational number that occurs often in practical applications. To see one way the number e is used, begin with the formula for compound interest. Suppose now that a lucky investment produces annual interest of 100%, so that $i = 1.00$, or $i = 1$. Suppose also that only $1 can be deposited at this rate, and for only one year. Then $P = 1$ and $n = 1$. Substituting into the formula for compound interest gives

$$P\left(1 + \frac{i}{m}\right)^{nm} = 1\left(1 + \frac{1}{m}\right)^{1(m)}$$

$$= \left(1 + \frac{1}{m}\right)^{m}.$$

As interest is compounded more and more often, the value of this expression will increase. If interest is compounded annually, making $m = 1$, the total amount on deposit is

$$\left(1 + \frac{1}{m}\right)^{m} = \left(1 + \frac{1}{1}\right)^{1}$$

$$= 2^{1} = 2,$$

so an investment of $1 becomes $2 in one year.

A calculator with a y^x key gives the results in the table on the following page. These results have been rounded to five decimal places.

m	$\left(1 + \dfrac{1}{m}\right)^{m}$
1	2
2	2.25
5	2.48832
10	2.59374
25	2.66584
50	2.69159
100	2.70481
500	2.71557
1000	2.71692
10,000	2.71815
1,000,000	2.71828

The table suggests that as m increases, the value of $(1 + 1/m)^{m}$ gets closer and closer to some fixed number. It turns out that this is indeed the case. This fixed number is called e.

Value of e

To nine decimal places,

$$e \approx 2.718281828.$$

Table 1 in this book gives various powers of e. Also, some calculators will give values of e^{x}. In Figure 5, the functions defined by $f(x) = 2^{x}$, $f(x) = e^{x}$, and $f(x) = 3^{x}$ are graphed for comparison.

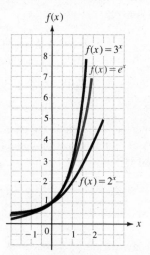

Figure 5

It can be shown that in many situations involving growth or decay of a population, the amount or number present at time t can be closely approximated by an exponential function with base e. The next example illustrates exponential growth.

Example 7 Suppose the population of a midwestern city is approximated by

$$P(t) = 10,000e^{.04t},$$

where t represents time measured in years. Find the population of the city at time **(a)** $t = 0$ **(b)** $t = 5$.

(a) The population at time $t = 0$ is

$$\begin{aligned} P(0) &= 10,000e^{(.04)0} \\ &= 10,000e^0 \\ &= 10,000(1) \\ &= 10,000, \end{aligned}$$

written $P_0 = 10,000$.

(b) The population of the city at year $t = 5$ is

$$\begin{aligned} P(5) &= 10,000e^{(.04)5} \\ &= 10,000e^{.2}. \end{aligned}$$

The number $e^{.2}$ can be found in Table 1 or by using a suitable calculator. By either of these methods, $e^{.2} = 1.22140$ (to five decimal places), so that

$$P(5) = 10,000(1.22140) = 12,214.$$

In five years the population of the city will be about 12,200. ■

7.1 Exercises

Which of the following define exponential functions?

1. $f(x) = 5^x - 1$ **2.** $f(x) = 2x^5$ **3.** $f(x) = 4x^3 - 1$ **4.** $f(x) = 2 \cdot 3^{5x-1}$

Let $f(x) = (2/3)^x$. Find each of the following values.

5. $f(2)$ **6.** $f(-1)$ **7.** $f\left(\dfrac{1}{2}\right)$ **8.** $f(0)$

9. Graph each of the functions defined as follows. Compare the graphs to that of $f(x) = 2^x$.
   (a) $f(x) = 2^x + 1$ **(b)** $f(x) = 2^x - 4$ **(c)** $f(x) = 2^{x+1}$ **(d)** $f(x) = 2^{x-4}$

10. Graph each of the functions defined as follows. Compare the graphs to that of $f(x) = 3^{-x}$.
   (a) $f(x) = 3^{-x} - 2$ **(b)** $f(x) = 3^{-x} + 4$ **(c)** $f(x) = 3^{-x-2}$ **(d)** $f(x) = 3^{-x+4}$

Graph each of the functions defined as follows.

11. $f(x) = 3^x$ **12.** $f(x) = 4^x$ **13.** $f(x) = (3/2)^x$ **14.** $f(x) = 3^{-x}$

15. $f(x) = 10^{-x}$ **16.** $f(x) = 10^x$ **17.** $f(x) = 2^{x+1}$ **18.** $f(x) = 2^{1-x}$

19. $f(x) = 2^{|x|}$ **20.** $f(x) = 2^{-|x|}$ **21.** $f(x) = 2^x + 2^{-x}$ **22.** $f(x) = (1/2)^x + (1/2)^{-x}$

Use a calculator to help graph each of the functions defined as follows.

23. $f(x) = x \cdot 2^x$ **24.** $f(x) = x^2 \cdot 2^x$ **25.** $f(x) = \dfrac{e^x - e^{-x}}{2}$ **26.** $f(x) = \dfrac{e^x + e^{-x}}{2}$

27. $f(x) = (1 - x)e^x$ **28.** $f(x) = x \cdot e^{-x}$

Solve each of the following equations.

29. $4^x = 2$ **30.** $125^r = 5$ **31.** $\left(\dfrac{1}{2}\right)^k = 4$ **32.** $\left(\dfrac{2}{3}\right)^x = \dfrac{9}{4}$

33. $2^{3-y} = 8$ **34.** $5^{2p+1} = 25$ **35.** $\dfrac{1}{27} = b^{-3}$ **36.** $\dfrac{1}{81} = k^{-4}$

37. $4 = r^{2/3}$ **38.** $z^{5/2} = 32$ **39.** $27^{4z} = 9^{z+1}$ **40.** $32^t = 16^{1-t}$

41. $125^{-x} = 25^{3x}$ **42.** $216^{3-a} = 36^a$ **43.** $\left(\dfrac{1}{8}\right)^{-2p} = 2^{p+3}$ **44.** $3^{-h} = \left(\dfrac{1}{27}\right)^{1-2h}$

45. $\left(\dfrac{1}{2}\right)^{-x} = \left(\dfrac{1}{4}\right)^{x+1}$ **46.** $\left(\dfrac{2}{3}\right)^{k-1} = \left(\dfrac{81}{16}\right)^{k+1}$ **47.** $4^{|x|} = 64$ **48.** $3^{-|x|} = \dfrac{1}{27}$

49. $\left(\dfrac{1}{5}\right)^{|x-2|} = \dfrac{1}{125}$ **50.** $\left(\dfrac{2}{5}\right)^{|3x-2|} = \dfrac{4}{25}$ **51.** $e^{-5x} = (e^2)^x$ **52.** $e^{3(1+x)} = e^{-8x}$

53. For $a > 1$, how does the graph of $y = a^x$ change as a increases? What if $0 < a < 1$?

Use the formula for compound interest,

$$A = P\left(1 + \dfrac{i}{m}\right)^{nm},$$

to find each of the following amounts.

54. \$4292 at 6% compounded annually for 10 years

55. \$965.43 at 9% compounded annually for 15 years

56. \$10,765 at 11% compounded semiannually for 7 years

57. \$1593.24 at 10½% compounded quarterly for 14 years

58. \$68,922 at 10% compounded daily (365 days) for 4 years

59. \$2964.58 at 11¼% compounded daily for 9 years (Ignore leap years.)

60. Suppose \$10,000 is left at interest for 3 years at 12%. Find the final amount on deposit if the interest is compounded   **(a)** annually   **(b)** quarterly   **(c)** daily (365 days).

61. Find the final amount on deposit if \$5800 is left at interest for 6 years at 13% and interest is compounded   **(a)** annually   **(b)** semiannually   **(c)** daily (365 days).

62. Suppose the population of a city is given by $P(t)$, where

$$P(t) = 1,000,000e^{.02t},$$

where t represents time measured in years from some initial year. Find each of the following: **(a)** $P(0)$ **(b)** $P(2)$ **(c)** $P(4)$ **(d)** $P(10)$.   **(e)** Graph $y = P(t)$.

63. Suppose the quantity in grams of a radioactive substance present at time t is

$$Q(t) = 500e^{-.05t}.$$

Let t be time measured in days from some initial day. Find the quantity present at each

of the following times: **(a)** $t = 0$ **(b)** $t = 4$ **(c)** $t = 8$ **(d)** $t = 20$.
(e) Graph $y = Q(t)$.

64. Experiments have shown that the sales of a product, under relatively stable market conditions, but in the absence of promotional activities such as advertising, tend to decline at a constant yearly rate. This rate of sales decline varies considerably from product to product, but seems to remain the same for any particular product. The sales decline can be expressed by a function of the form

$$S(t) = S_0 e^{-at},$$

where $S(t)$ is the rate of sales at time t measured in years, S_0 is the rate of sales at time $t = 0$, and a is the sales decay constant. **(a)** Suppose the sales decay constant for a particular product is $a = 0.10$. Let $S_0 = 50,000$ and find $S(1)$ and $S(3)$. **(b)** Find $S(2)$ and $S(10)$ if $S_0 = 80,000$ and $a = 0.05$.

65. *Escherichia coli* is a strain of bacteria that occurs naturally in many different organisms. Under certain conditions, the number of these bacteria present in a colony is

$$E(t) = E_0 \cdot 2^{t/30},$$

where $E(t)$ is the number of bacteria present t minutes after the beginning of an experiment, and E_0 is the number present when $t = 0$. Let $E_0 = 2,400,000$ and find the number of bacteria at the following times: **(a)** $t = 5$ **(b)** $t = 10$ **(c)** $t = 60$ **(d)** $t = 120$.

66. The higher a student's grade-point average, the fewer applications that the student need send to medical schools (other things being equal). Using information given in a guidebook for prospective medical students, we constructed the function defined by $y = 540e^{-1.3x}$ for the number of applications a student should send out. Here y is the number of applications for a student whose grade-point average is x. The domain of x is the interval $[2.0, 4.0]$. Find the number of applications that should be sent out by students having a grade-point average of **(a)** 2.0 **(b)** 3.0 **(c)** 3.5 **(d)** 3.9.

The pressure of the atmosphere, $p(h)$, in pounds per square inch, is given by

$$p(h) = p_0 e^{-kh},$$

where h is the height above sea level and p_0 and k are constants. The pressure at sea level is 15 pounds per square inch and the pressure is 9 pounds per square inch at a height of 12,000 feet.

67. Find the pressure at an altitude of 6000 feet.

68. What would be the pressure encountered by a spaceship at an altitude of 150,000 feet?

69. In our definition of exponential function, we ruled out negative values of a. However, in a textbook on mathematical economics, the author obtained a "graph" of $y = (-2)^x$ by plotting the following points.

x	-4	-3	-2	-1	0	1	2
y	1/16	$-1/8$	1/4	$-1/2$	1	-2	4

The graph, which occupies a half page in the book, oscillates very neatly from positive to negative values of y. Comment on this approach. (This example shows the dangers of relying solely on point plotting when drawing graphs.)

70. When defining an exponential function, why did we require $a > 0$?

Any points where the graphs of functions f and g cross give solutions of the equation $f(x) = g(x)$. Use this idea to estimate the *number* of solutions of the following equations.

71. $x = 2^x$ **72.** $2^{-x} = -x$ **73.** $3^{-x} = 1 - 2x$ **74.** $3x + 2 = 4^x$

Let $f(x) = a^x$ define an exponential function of base a.

75. Is f odd, even, or neither?

76. Prove that $f(m + n) = f(m) \cdot f(n)$ for any real numbers m and n.

Find examples of a function f satisfying the following conditions.

77. $f(2x) = [f(x)]^2$ **78.** $f(x + 1) = 2 \cdot f(x)$

In calculus, it is shown that

$$e^x = 1 + x + \frac{x^2}{2 \cdot 1} + \frac{x^3}{3 \cdot 2 \cdot 1} + \frac{x^4}{4 \cdot 3 \cdot 2 \cdot 1} + \frac{x^5}{5 \cdot 4 \cdot 3 \cdot 2 \cdot 1} + \cdots .$$

79. Use the terms shown here and replace x with 1 to approximate $e^1 = e$ to three decimal places. Then check your results in Table 1 or with a calculator.

80. Use the terms shown here and replace x with $-.05$ to approximate $e^{-.05}$ to four decimal places. Check your results in Table 1 or with a calculator.

81. Let $f(x) = e^x$. Show that $\dfrac{f(x + h) - f(x)}{h} = \dfrac{e^x(e^h - 1)}{h}$.

82. Let $f(x) = 1 + e^x$. Show that $\dfrac{1}{f(x)} + \dfrac{1}{f(-x)} = 1$.

Simplify each expression.

83. $\dfrac{(e^{-x} - e^x)(-e^{-x} + e^x) - (e^{-x} + e^x)(-e^{-x} - e^x)}{(e^{-x} - e^x)^2}$

84. $\dfrac{(e^x + e^{2x})(2e^{2x} + 3e^{3x}) - (e^{2x} + e^{3x})(e^x + 2e^{2x})}{(e^x + e^{2x})^2}$

7.2   Logarithmic Functions

Exponential functions defined by $f(x) = a^x$ for all positive values of a, where $a \neq 1$, were discussed in the previous section. As mentioned there, exponential functions are one-to-one, and so have inverse functions. In this section we discuss the inverses of exponential functions. The equation of the inverse comes from exchanging x and y. Doing this with $y = a^x$ gives

$$x = a^y$$

as the inverse of the exponential function defined by $f(x) = a^x$. To solve $x = a^y$ for *y,* use the following definition.

<table>
<tr><td>**Definition of
Logarithm**</td><td>For all real numbers *y,* and all positive numbers *a* and *x,* where $a \neq 1,$

$$y = \log_a x \quad \text{if and only if} \quad x = a^y.$$</td></tr>
</table>

Log is an abbreviation for *logarithm.* Read $\log_a x$ as "the logarithm of *x* to the base *a.*"

This key definition should be memorized. It is important to remember the location of the base and exponent in each part of the definition.

$$\text{logarithmic form:} \quad \overset{\text{exponent}}{\underset{\text{base}}{y = \log_a x}}$$

$$\text{exponential form:} \quad \overset{\text{exponent}}{\underset{\text{base}}{a^y = x}}$$

By the definition, a logarithm is an exponent: the exponent on the base *a* that will yield the number *x.*

Example 1   The chart below shows several pairs of equivalent statements. The same statement is written in both exponential and logarithmic forms.

Exponential form	Logarithmic form
$2^3 = 8$	$\log_2 8 = 3$
$(1/2)^{-4} = 16$	$\log_{1/2} 16 = -4$
$10^5 = 100,000$	$\log_{10} 100,000 = 5$
$3^{-4} = 1/81$	$\log_3 (1/81) = -4$
$5^1 = 5$	$\log_5 5 = 1$
$(3/4)^0 = 1$	$\log_{3/4} 1 = 0$ ∎

The definition of logarithm can be used to define the logarithmic function with base *a.*

<table>
<tr><td>**Logarithmic
Function**</td><td>If $a > 0,$ $a \neq 1,$ and $x > 0,$ then the function *f* defined by

$$f(x) = \log_a x$$

is the **logarithmic function with base** *a.*</td></tr>
</table>

Exponential and logarithmic functions are inverses of each other. Since the domain of an exponential function is the set of all real numbers, the range of a logarithmic function will also be the set of all real numbers. In the same way, both the range of an exponential function and the domain of a logarithmic function are the set of all positive real numbers, so logarithms can be found for positive numbers only.

Example 2 Graph the logarithmic functions defined as follows.

(a) $f(x) = \log_2 x$

One way to graph a logarithmic function is to begin with its inverse function. Here, the inverse has equation $y = 2^x$. The graph of the equation $y = 2^x$ is shown with a dashed curve in Figure 6(a). To get the graph of $y = \log_2 x$, reflect the graph of $y = 2^x$ about the 45° line $y = x$. The graph of the equation $y = \log_2 x$ is shown as a solid curve. As the graph shows, the function defined by $f(x) = \log_2 x$ is increasing for all its domain, is one-to-one, and has the y-axis as a vertical asymptote.

(b) $f(x) = \log_{1/2} x$

The graph of the equation of the inverse, $y = (1/2)^x$, is shown with a dashed curve in Figure 6(b). The graph of $y = \log_{1/2} x$, shown as a solid curve, is found by reflecting the graph of $y = (1/2)^x$ about the line $y = x$. The function defined by $f(x) = \log_{1/2} x$ is decreasing for all its domain, is one-to-one, and has the y-axis for a vertical asymptote. ■

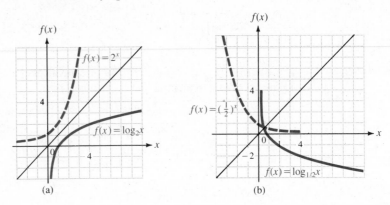

(a)   (b)

Figure 6

Example 3 Graph $f(x) = \log_2 (x - 1)$.

The graph of $f(x) = \log_2 (x - 1)$ will be the graph of $f(x) = \log_2 x$, translated one unit to the right. The asymptote is $x = 1$. The domain of the function defined by $f(x) = \log_2 (x - 1)$ is $(1, +\infty)$, since logarithms can be found only for positive numbers. See Figure 7. ■

Example 4 Graph $f(x) = \log_3 |x|$.

Use the definition of logarithm to write the equation $y = \log_3 |x|$ as $|x| = 3^y$. Then choose values for y and find the corresponding x-values. Some ordered pairs for this equation are shown below; the graph is given in Figure 8.

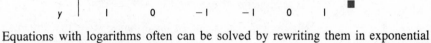

Figure 7   **Figure 8**

x	-3	-1	$-1/3$	$1/3$	1	3
y	1	0	-1	-1	0	1

Equations with logarithms often can be solved by rewriting them in exponential form as shown in the next example.

Example 5   Solve each of the following equations.

(a) $\log_x \dfrac{8}{27} = 3$

First, write the expression in exponential form; then solve.

$$x^3 = \frac{8}{27}$$

$$x^3 = \left(\frac{2}{3}\right)^3$$

$$x = \frac{2}{3}$$

The solution set is $\{2/3\}$.

(b) $\log_4 x = 5/2$

In exponential form, the given statement becomes

$$4^{5/2} = x$$
$$(4^{1/2})^5 = x$$
$$2^5 = x$$
$$32 = x.$$

The solution set is $\{32\}$. ■

Logarithms were originally important as an aid for numerical calculations, but the availability of inexpensive calculators has greatly reduced the need for this application of logarithms. Yet the principles behind the use of logarithms for calculation are important in other applications; these principles are based on the properties of logarithms discussed in the next theorem.

**Properties of
Logarithms**

If x and y are any positive real numbers, r is any real number, and a is any positive real number, $a \neq 1$, then

(a) $\log_a xy = \log_a x + \log_a y$ (b) $\log_a \dfrac{x}{y} = \log_a x - \log_a y$

(c) $\log_a x^r = r \cdot \log_a x$ (d) $\log_a a = 1$

(e) $\log_a 1 = 0$.

To prove part (a) of the properties of logarithms, let $m = \log_a x$ and $n = \log_a y$. Then, by the definition of logarithm,

$$a^m = x \quad \text{and} \quad a^n = y.$$

Multiplication gives $\qquad\qquad a^m \cdot a^n = xy.$

By a property of exponents, $\qquad a^{m+n} = xy.$

Now use the definition of logarithm to write this statement as

$$\log_a xy = m + n.$$

Since $m = \log_a x$ and $n = \log_a y$,

$$\log_a xy = \log_a x + \log_a y.$$

To prove part (b) of the properties, use m and n as defined above. Then

$$\frac{a^m}{a^n} = \frac{x}{y}.$$

Since $\qquad\qquad\qquad\qquad \dfrac{a^m}{a^n} = a^{m-n},$

then $\qquad\qquad\qquad\qquad a^{m-n} = \dfrac{x}{y}.$

By the definition of logarithm, this statement can be written

$$\log_a \frac{x}{y} = m - n,$$

or $\qquad\qquad\qquad \log_a \dfrac{x}{y} = \log_a x - \log_a y.$

For part (c),

$$(a^m)^r = x^r \quad \text{or} \quad a^{mr} = x^r.$$

Again using the definition of logarithm

$$\log_a x^r = mr,$$

or $\qquad\qquad\qquad \log_a x^r = r \cdot \log_a x.$

Finally, (d) and (e) follow directly from the definition of logarithm since $a^1 = a$ and $a^0 = 1$.

The properties of logarithms are useful for rewriting expressions with logarithms in different forms, as shown in the next examples.

Example 6   Assuming all variables represent positive real numbers, use the properties of logarithms to write each of the following in a different form.

(a) $\log_6 7 \cdot 9 = \log_6 7 + \log_6 9$

(b) $\log_9 \dfrac{15}{7} = \log_9 15 - \log_9 7$

(c) $\log_5 \sqrt{8} = \log_5 8^{1/2} = \dfrac{1}{2} \log_5 8$

(d) $\log_a \dfrac{mnq}{p^2} = \log_a m + \log_a n + \log_a q - 2 \log_a p$

(e) $\log_a \sqrt[3]{m^2} = \dfrac{2}{3} \log_a m$

(f) $\log_b \sqrt[n]{\dfrac{x^3 y^5}{z^m}} = \dfrac{1}{n} \log_b \dfrac{x^3 y^5}{z^m}$

$$= \dfrac{1}{n} (\log_b x^3 + \log_b y^5 - \log_b z^m)$$

$$= \dfrac{3}{n} \log_b x + \dfrac{5}{n} \log_b y - \dfrac{m}{n} \log_b z \quad \blacksquare$$

Example 7   Use the properties of logarithms to write each of the following as a single logarithm with a coefficient of 1. Assume all variables represent positive real numbers.

(a) $\log_3 (x + 2) + \log_3 x - \log_3 2 = \log_3 \dfrac{(x + 2)x}{2}$

(b) $2 \log_a m - 3 \log_a n = \log_a m^2 - \log_a n^3 = \log_a \dfrac{m^2}{n^3}$

(c) $\dfrac{1}{2} \log_b m + \dfrac{3}{2} \log_b 2n - \log_b m^2 n$

$$= \log_b m^{1/2} + \log_b (2n)^{3/2} - \log_b m^2 n$$

$$= \log_b \dfrac{m^{1/2}(2n)^{3/2}}{m^2 n}$$

$$= \log_b \dfrac{2^{3/2} n^{1/2}}{m^{3/2}} \quad \blacksquare$$

Example 8   Assume $\log_{10} 2 = .3010$. Find the base 10 logarithms of 4 and 5.

By the properties of logarithms,

$$\log_{10} 4 = \log_{10} 2^2 = 2 \log_{10} 2 = 2(.3010) = .6020$$

$$\log_{10} 5 = \log_{10} \dfrac{10}{2} = \log_{10} 10 - \log_{10} 2 = 1 - .3010 = .6990. \quad \blacksquare$$

Compositions of the exponential and logarithmic functions can be used to get two more useful properties. If $f(x) = a^x$ and $g(x) = \log_a x$, then

$$f[g(x)] = a^{\log_a x}$$

and
$$g[f(x)] = \log_a a^x.$$

The study of inverse functions shows that if functions f and g are inverses of each other, $f[g(x)] = g[f(x)] = x$. Since exponential and logarithmic functions of the same base are inverses of each other, this result gives the following theorem.

Theorem

If $a > 0$, $a \neq 1$, and $x > 0$, then

$$a^{\log_a x} = x \quad \text{and} \quad \log_a a^x = x.$$

Example 9   Simplify each of the following.

   (a) $\log_5 5^3 = 3$ **(b)** $7^{\log_7 10} = 10$ **(c)** $\log_r r^{k+1} = k + 1$ ∎

7.2 Exercises

For each of the following statements, write an equivalent statement in logarithmic form.

1. $3^4 = 81$ **2.** $2^5 = 32$ **3.** $10^4 = 10,000$ **4.** $8^2 = 64$

5. $(1/2)^{-4} = 16$ **6.** $(2/3)^{-3} = 27/8$ **7.** $10^{-4} = .0001$ **8.** $(1/100)^{-2} = 10,000$

For each of the following statements, write an equivalent statement in exponential form.

9. $\log_6 36 = 2$ **10.** $\log_5 5 = 1$ **11.** $\log_{\sqrt{3}} 81 = 8$ **12.** $\log_4 (1/64) = -3$

13. $\log_{10} .0001 = -4$ **14.** $\log_3 \sqrt[3]{9} = 2/3$ **15.** $\log_m k = n$ **16.** $\log_2 r = y$

Find the value of each of the following. Assume all variables represent positive real numbers.

17. $\log_5 25$ **18.** $\log_3 81$ **19.** $\log_8 8$ **20.** $\log_7 1$

21. $\log_{10} 0.001$ **22.** $\log_6 \dfrac{1}{216}$ **23.** $\log_{25} 5$ **24.** $\log_{16} 2$

25. $\log_4 \dfrac{\sqrt[3]{4}}{2}$ **26.** $\log_9 \dfrac{\sqrt[4]{27}}{3}$ **27.** $\log_{1/3} \dfrac{9^{-4}}{3}$ **28.** $\log_{1/4} \dfrac{16^2}{2^{-3}}$

29. $\log_6 36^4$ **30.** $\log_5 125^2$ **31.** $\log_e e^4$ **32.** $\log_e \dfrac{1}{e}$

33. $\log_e \sqrt{e}$ **34.** $\log_e e^x$ **35.** $e^{\log_e 2}$ **36.** $e^{\log_e 5}$

37. $e^{\log_e x}$ **38.** $e^{\log_e (x+2)}$ **39.** $2^{\log_2 9}$ **40.** $8^{\log_8 11}$

Solve each of the following equations.

41. $\log_x 25 = -2$ **42.** $\log_x \dfrac{1}{16} = -2$ **43.** $\log_9 27 = m$ **44.** $\log_8 4 = z$

45. $\log_y 8 = \dfrac{3}{4}$ **46.** $\log_r 7 = 1/2$ **47.** $\log_e x = 0$ **48.** $\log_e x = 1$

Write each of the following as a sum, difference, or product of logarithms. Simplify the result if possible. Assume all variables represent positive real numbers.

49. $\log_3 (2/5)$

50. $\log_4 (6/7)$

51. $\log_2 \dfrac{6x}{y}$

52. $\log_3 \dfrac{4p}{q}$

53. $\log_5 \dfrac{5\sqrt{7}}{3}$

54. $\log_2 \dfrac{2\sqrt{3}}{5}$

55. $\log_4 (2x + 5y)$

56. $\log_6 (7m + 3q)$

57. $\log_k \dfrac{pq^2}{m}$

58. $\log_z \dfrac{x^5 y^3}{3}$

59. $\log_m \sqrt{\dfrac{5r^3}{z^5}}$

60. $\log_p \sqrt[3]{\dfrac{m^5 n^4}{t^2}}$

Write each of the following expressions as a single logarithm with a coefficient of 1. Assume that all variables represent positive real numbers.

61. $\log_a x + \log_a y - \log_a m$

62. $(\log_b k - \log_b m) - \log_b a$

63. $2 \log_m a - 3 \log_m b^2$

64. $\dfrac{1}{2} \log_y p^3 q^4 - \dfrac{2}{3} \log_y p^4 q^3$

65. $-\dfrac{3}{4} \log_x a^6 b^8 + \dfrac{2}{3} \log_x a^9 b^3$

66. $\log_a (pq^2) + 2 \log_a (p/q)$

67. $\log_b (x + 2) + \log_b 7x - \log_b 8$

68. $\log_h (4m + 1) + \log_h 2m - \log_h 3$

69. $2 \log_a (z + 1) + \log_a (3z + 2)$

70. $\log_b (2y + 5) - \dfrac{1}{2} \log_b (y + 3)$

71. $-\dfrac{2}{3} \log_5 5m^2 + \dfrac{1}{2} \log_5 25m^2$

72. $-\dfrac{3}{4} \log_3 16p^4 - \dfrac{2}{3} \log_3 8p^3$

73. Graph $f(x) = \log_3 x$ and $f(x) = 3^x$ on the same axes.

74. Graph $f(x) = \log_4 x$ and $f(x) = 4^x$ on the same axes.

75. Graph each of the following equations. Compare the graphs to that of $y = \log_2 x$.
   (a) $y = (\log_2 x) + 3$ (b) $y = \log_2 (x + 3)$ (c) $y = |\log_2 (x + 3)|$

76. Graph each of the following equations. Compare the graphs to that of $y = \log_{1/2} x$.
   (a) $y = (\log_{1/2} x) - 2$ (b) $y = \log_{1/2} (x - 2)$ (c) $y = |\log_{1/2} (x - 2)|$

Graph each of the functions defined as follows.

77. $f(x) = \log_5 x$

78. $f(x) = \log_{10} x$

79. $f(x) = \log_{1/2} (1 - x)$

80. $f(x) = \log_{1/3} (3 - x)$

81. $f(x) = \log_2 x^2$

82. $f(x) = \log_3 (x - 1)$

83. $f(x) = x \cdot \log_{10} x$

84. $f(x) = x^2 \cdot \log_{10} x$

Given $\log_{10} 2 = .3010$ and $\log_{10} 3 = .4771$, find each of the following without using calculators or tables.

85. $\log_{10} 6$

86. $\log_{10} 12$

87. $\log_{10} 9$

88. $\log_{10} 20$

89. $\log_{10} 30$

90. $\log_{10} 36$

91. The population of an animal species that is introduced into a certain area may grow rapidly at first but then grow more slowly as time goes on. A logarithmic function can provide an excellent description of such growth. Suppose that the population of foxes, $F(t)$, in an area t months after the foxes were introduced there is

$$F(t) = 500 \log_{10} (2t + 3).$$

Use a calculator with a log key to find the population of foxes at the following times: **(a)** when they are first released into the area (that is, when $t = 0$)   **(b)** after 3 months   **(c)** after 15 months.   **(d)** Graph $y = F(t)$.

The loudness of sounds is measured in a unit called a *decibel*. To measure with this unit, we first assign an intensity of I_0 to a very faint sound, called the *threshold sound*. If a particular sound has intensity I, then the decibel rating of this louder sound is

$$d = 10 \cdot \log_{10} \frac{I}{I_0}.$$

92. Find the decibel ratings of sounds having the following intensities: **(a)** $100I_0$ **(b)** $1000I_0$ **(c)** $100{,}000I_0$ **(d)** $1{,}000{,}000I_0$.

93. Find the decibel ratings of the following sounds, having intensities as given. (You will need a calculator with a log key.) Round answers to the nearest whole number.
 (a) Whisper, $115I_0$
 (b) Busy street, $9{,}500{,}000I_0$
 (c) Heavy truck, 20 m away, $1{,}200{,}000{,}000I_0$
 (d) Rock music, $895{,}000{,}000{,}000I_0$
 (e) Jetliner at takeoff, $109{,}000{,}000{,}000{,}000I_0$

94. The intensity of an earthquake, measured on the *Richter Scale,* is given by

$$\log_{10} \frac{I}{I_0},$$

where I_0 is the intensity of an earthquake of a certain (small) size. Find the Richter Scale ratings of earthquakes having intensity   **(a)** $1000I_0$ **(b)** $1{,}000{,}000I_0$ **(c)** $100{,}000{,}000I_0$.

95. The San Francisco earthquake of 1906 had a Richter Scale rating of 8.6. Use a calculator with a y^x key to express the intensity of this earthquake as a multiple of I_0 (see Exercise 94).

96. How much more powerful is an earthquake with a Richter Scale rating of 8.6 than one with a rating of 8.2?

97. Using a calculator*, evaluate $(3^{.003})^{1001}$ and $3^{(.003 \times 1001)}$ by computing the expression within parentheses first. Did you get the same results? If not, can you explain the difference? What does this tell you about the laws of exponents as applied to calculator arithmetic?

98. Using a calculator, evaluate $\log_{10} (2^{.0001})$ and $.0001 \times (\log_{10} 2)$ by computing the expression within parentheses first. Did you get the same results? If not, can you explain the difference? What does this tell you about the properties of logarithms as applied to calculator arithmetic?

99. Let $f(x) = e^x$. Graph $y = f^{-1}(x)$.

100. Prove that $\log_e (x - \sqrt{x^2 - 1}) = -\log_e (x + \sqrt{x^2 - 1})$.

*Exercises 97 and 98 from *Calculus and Analytic Geometry* by Abe Mizrahi and Michael Sullivan. Copyright © 1982 by Wadsworth, Inc., Belmont, CA 94002. Reprinted by permission.

7.3 Natural Logarithms

Since our number system uses base 10, logarithms to base 10 are most convenient for numerical calculation, historically the main application of logarithms. Base 10 logarithms are called **common logarithms.** The common logarithm of the number x, or $\log_{10} x$, is often abbreviated as just $\log x$. Common logarithms are discussed in Section 7.6.

In most other practical applications of logarithms, however, the number $e \approx 2.718281828$ is used as base. Logarithms to base e are called **natural logarithms,** since they occur in many natural-world applications, such as those involving growth and decay. The abbreviation $\ln x$ is used for the natural logarithm of x, so that $\log_e x = \ln x$. The graph of $f(x) = e^x$ was given in Figure 5. Since the functions defined by $f(x) = e^x$ and $f(x) = \ln x$ are inverses, the graph of $f(x) = \ln x$, shown in Figure 9, can be found by reflecting the graph of $f(x) = e^x$ about the line $y = x$.

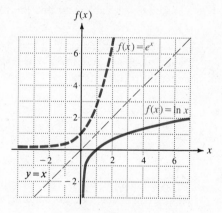

Figure 9

The following results for the natural logarithm function are direct applications of the properties of logarithms.

Properties of Natural Logarithms		
$e^{\ln x} = x$ if $x > 0$		$\ln 1 = 0$
$\ln e^x = x$		$\ln e = 1$

Example 1   Evaluate each expression.

(a) $e^{\ln 5} = 5$

(b) $e^{\ln \sqrt{3}} = \sqrt{3}$

(c) $\ln e^2 = 2$

(d) $\ln e^{-1.5} = -1.5$

(e) $8 \ln 1 - 16 \ln e + 2 \ln e^4$

Since $\ln 1 = 0$, $\ln e = 1$, and $\ln e^4 = 4$,

$$8 \ln 1 - 16 \ln e + 2 \ln e^4 = 8(0) - 16(1) + 2(4) = -8. \quad \blacksquare$$

Values of natural logarithms can be found with a calculator which has a ln key or with a table of natural logarithms. A table of natural logarithms is given in Table 1. Reading directly from this table,

$$\ln 55 = 4.0073,$$

$$\ln 1.9 = .6419,$$

and $$\ln .4 = -.9163.$$

Example 2 Use a calculator or Table 1 to find the following logarithms.

(a) $\ln 85$

With a calculator, enter 85, press the ln key, and read the result, 4.4427.

Table 1 does not give ln 85. However, the value of ln 85 can be found using the properties of logarithms.

$$\ln 85 = \ln (8.5 \times 10)$$
$$= \ln 8.5 + \ln 10$$
$$\approx 2.1401 + 2.3026$$
$$= 4.4427$$

A result found in this way is sometimes slightly different from the answer found using a calculator, due to rounding error.

(b) $\ln 36$

A calculator gives $\ln 36 = 3.5835$. To use the table, first use properties of logarithms, since 36 is not listed in Table 1.

$$\ln 36 = \ln 6^2$$
$$= 2 \ln 6$$
$$\approx 2(1.7918)$$
$$= 3.5836$$

Alternatively, ln 36 can be found as follows.

$$\ln 36 = \ln 9 \cdot 4 = \ln 9 + \ln 4 = 2.1972 + 1.3863 = 3.5835 \quad \blacksquare$$

Logarithms to other bases A calculator or a table will give the values of either natural logarithms (base e) or common logarithms (base 10). However, sometimes it is convenient to use logarithms to other bases. The following theorem can be used to convert logarithms from one base to another.

| **Change of Base Theorem** | If x is any positive number and if a and b are positive real numbers, $a \neq 1$, $b \neq 1$, then $$\log_a x = \frac{\log_b x}{\log_b a}.$$ |

To prove this result, use the definition of logarithm to write $y = \log_a x$ as $x = a^y$ or $x = a^{\log_a x}$ (for positive x and positive a, $a \neq 1$). Now take base b logarithms of both sides of this last equation.

$$\log_b x = \log_b a^{\log_a x}$$

or

$$\log_b x = (\log_a x)(\log_b a),$$

from which

$$\log_a x = \frac{\log_b x}{\log_b a}.$$

Example 3 Use natural logarithms to find each of the following. Round to the nearest hundredth.

(a) $\log_5 27$

Let $x = 27$, $a = 5$, and $b = e$. Substituting into the change of base theorem gives

$$\log_5 27 = \frac{\log_e 27}{\log_e 5}$$

$$= \frac{\ln 27}{\ln 5}.$$

Now use a calculator or Table 1.

$$\log_5 27 \approx \frac{3.2958}{1.6094}$$

$$\approx 2.05$$

To check, use a calculator with a y^x key, along with the definition of logarithm, to verify that $5^{2.05} \approx 27$.

(b) $\log_2 .1$

$$\log_2 .1 = \frac{\ln .1}{\ln 2} \approx \frac{-2.3026}{.6931} \approx -3.32 \quad \blacksquare$$

Example 4 One measure of the diversity of the species in an ecological community is given by

$$H = -[P_1 \log_2 P_1 + P_2 \log_2 P_2 + \cdots + P_n \log_2 P_n],$$

where $P_1, P_2, \ldots, P_n$ are the proportions of a sample belonging to each of n species found in the sample. For example, in a community with two species, where there are 90 of one species and 10 of the other, $P_1 = 90/100 = .9$ and $P_2 = 10/100 = .1$, with

$$H = -[.9 \log_2 .9 + .1 \log_2 .1].$$

The value of $\log_2 .1$ was found in Example 3(b) above. Now find $\log_2 .9$.

$$\log_2 .9 = \frac{\ln .9}{\ln 2} \approx \frac{-.1054}{.6931} \approx -.152$$

Therefore, $\quad H \approx -[(.9)(-.152) + (.1)(-3.32)] \approx .469.$ ∎

7.3 Exercises

Evaluate each expression.

1. $\ln e$ **2.** $\ln 1$ **3.** $\ln e^4$ **4.** $\ln e^2$

5. $\ln e^{-3}$ **6.** $\ln e^{-.017}$ **7.** $e^{\ln 1.2}$ **8.** $e^{\ln .003}$

Find each of the following logarithms to four decimal places. Use a calculator or Table 1.

9. $\ln 4$ **10.** $\ln 6$ **11.** $\ln 17$ **12.** $\ln 29$

13. $\ln 350$ **14.** $\ln 900$ **15.** $\ln 49$ **16.** $\ln 64$

17. $\ln 42$ **18.** $\ln 72$ **19.** $\ln 81,000$ **20.** $\ln 121,000$

Find each of the following logarithms to the nearest hundredth.

21. $\log_5 10$ **22.** $\log_9 12$ **23.** $\log_{15} 5$ **24.** $\log_6 8$

25. $\log_{1/2} 3$ **26.** $\log_{12} 62$ **27.** $\log_{100} 83$ **28.** $\log_{200} 175$

29. $\log_{2.9} 7.5$ **30.** $\log_{5.8} 12.7$ **31.** $\log_{.6} 13.2$ **32.** $\log_{.9} 5.77$

Graph each of the functions defined as follows.

33. $f(x) = \ln x$ **34.** $f(x) = |\ln x|$ **35.** $f(x) = x \ln x$ **36.** $f(x) = x^2 \ln x$

Work the following problems. Refer to Example 4 for Exercises 37 and 38.

37. Suppose a sample of a small community shows two species with 50 individuals each. Find the index of diversity H.

38. A virgin forest in northwestern Pennsylvania has 4 species of large trees with the following proportions of each: hemlock, .521; beech, .324; birch, .081; maple, .074. Find the index of diversity H.

39. The number of species in a sample is given by

$$S(n) = a \ln \left(1 + \frac{n}{a}\right).$$

Here n is the number of individuals in the sample and a is a constant that indicates the diversity of species in the community. If $a = .36$, find $S(n)$ for the following values of n. **(a)** 100 **(b)** 200 **(c)** 150 **(d)** 10

40. In Exercise 39, find n if $S(n) = 9$ and $a = .36$.

41. Suppose the number of rabbits in a colony is

$$y = y_0 e^{.4t},$$

where t represents time in months and y_0 is the rabbit population when $t = 0$.
(a) If $y_0 = 100$, find the number of rabbits present at time $t = 4$.
(b) Find y_0 if there are 30 rabbits after 4 months.

42. A Midwestern city finds its residents moving to the suburbs. Its population is declining according to the relationship

$$P = P_0 e^{-.04t},$$

where t is time measured in years and P_0 is the population at time $t = 0$. Assume $P_0 = 1,000,000$.

(a) Find the population at time $t = 1$.

(b) If the decline continues at the same rate, what will the population be after 10 years?

43. If an object is fired vertically upward and is subject only to the force of gravity, g, and to air resistance, then the maximum height, H, attained by the object is

$$H = \frac{1}{K} \left(V_0 - \frac{g}{K} \ln \frac{g + V_0 K}{g} \right),$$

where V_0 is the initial velocity of the object and K is a constant. Find H if $K = 2.5$, $V_0 = 1000$ feet per second, and $g = 32$ feet per second per second.

44. The pull, P, of a tracked vehicle on dry sand under certain conditions is approximated by

$$P = W \left[.2 + .16 \ln \frac{G(bl)^{3/2}}{W} \right]$$

where G is an index of sand strength, W is the load on the vehicle, b is the width of the track, and l is the length of the track.* Find P if $W = 10$, $G = 5$, $b = 30.5$ cm, and $l = 61.0$ cm.

45. The following formula† can be used to estimate the population of the United States, where t is time in years measured from 1914 (times before 1914 are negative):

$$N = \frac{197,273,000}{1 + e^{-.03134t}}$$

(a) Complete the following chart.

Year	Observed population	Predicted population
1790	3,929,000	3,929,000
1810	7,240,000	
1860	31,443,000	30,412,000
1900	75,995,000	
1970	204,000,000	
1980	224,000,000	

(b) Estimate the number of years after 1914 that it would take for the population to increase to 197,273,000.

---

*Gerald W. Turnage, *Prediction of Track Pull Performance in Desert Sand*, unpublished MS thesis, The Florida State University, 1971. †The formula is given in *Elements of Mathematical Biology*, by Alfred J. Lotka, 1957, p. 67. Reprinted by permission of Dover Publications, Inc.

To find the maximum permitted levels of certain pollutants in fresh water, the EPA has established the functions defined in Exercises 46–47, where $M(h)$ is the maximum permitted level of pollutant for a water hardness of h milligrams per liter. Find $M(h)$ in each case. (These results give the maximum permitted average concentration in micrograms per liter for a 24-hour period.)

46. Pollutant: copper; $M(h) = e^r$, where $r = .65 \cdot \ln h - 1.94$ and $h = 9.7$

47. Pollutant: lead; $M(h) = e^r$, where $r = 1.51 \cdot \ln h - 3.37$ and $h = 8.4$

In the central Sierra Nevada Mountains of California, the percent of moisture that falls as snow rather than rain is approximated reasonably well by $p = 86.3 \ln h - 680$, where p is the percent of snow at an altitude h (in feet). (Assume $h \geq 3000$.)

48. Find the percent of moisture that falls as snow at the following altitudes: **(a)** 3000 ft **(b)** 4000 ft **(c)** 7000 ft. **(d)** Graph p.

In Exercises 49 and 50, assume a and b represent positive numbers other than 1, and x is any real number.

49. Show that $\dfrac{1}{\log_a b} = \log_b a$.

50. Show that $a^x = e^{x \ln a}$.

7.4 Exponential and Logarithmic Equations

In Section 7.1 we solved exponential equations such as $(1/3)^x = 81$ by writing each side of the equation as a power of 3. However, that method cannot be used to solve an equation such as $7^x = 12$, since 12 cannot easily be written as a power of 7. However, the equation $7^x = 12$ can be solved by taking the logarithm of each side, a process that depends on the fact that a logarithmic function is one-to-one.

> If $x > 0$, $y > 0$, $b > 0$, $b \neq 1$, then
>
> $$x = y \quad \text{if and only if} \quad \log_b x = \log_b y.$$

Example 1 Solve the equation $7^x = 12$.

While the result above is valid for any appropriate base b, the best practical base to use is often base e. Taking base e (natural) logarithms of both sides gives

$$\ln 7^x = \ln 12$$
$$x \cdot \ln 7 = \ln 12$$
$$x = \frac{\ln 12}{\ln 7}.$$

To get a decimal approximation for x, use Table 1 or a calculator.

$$x = \frac{\ln 12}{\ln 7} \approx \frac{2.4849}{1.9459}$$

Using a calculator to divide 2.4849 by 1.9459 gives

$$x \approx 1.277.$$

A calculator with a y^x key can be used to check this answer. Evaluate $7^{1.277}$; the result should be approximately 12. This step verifies that, to the nearest thousandth, the solution set is $\{1.277\}$.   ■

Example 2   Solve $3^{2x-1} = 4^{x+2}$.

Taking natural logarithms on both sides gives

$$\ln 3^{2x-1} = \ln 4^{x+2}$$

Now use properties of logarithms.

$$(2x - 1) \ln 3 = (x + 2) \ln 4$$
$$2x \ln 3 - \ln 3 = x \ln 4 + 2 \ln 4$$
$$2x \ln 3 - x \ln 4 = 2 \ln 4 + \ln 3.$$

Factor out x on the left to get

$$x(2 \ln 3 - \ln 4) = 2 \ln 4 + \ln 3$$

or

$$x = \frac{2 \ln 4 + \ln 3}{2 \ln 3 - \ln 4}.$$

Using properties of logarithms,

$$x = \frac{\ln 16 + \ln 3}{\ln 9 - \ln 4}$$

or, finally,

$$x = \frac{\ln 48}{\ln \dfrac{9}{4}}.$$

This quotient could be approximated by a decimal if desired.

$$x = \frac{\ln 48}{\ln 2.25} \approx \frac{3.8712}{.8109} \approx 4.774$$

To the nearest thousandth, the solution set is $\{4.774\}$. To find $\ln 2.25$ with Table 1, write $\ln 2.25$ as $\ln 1.5^2 = 2 \ln 1.5$.   ■

Example 3   Solve $e^{x^2} = 200$.

Take natural logarithms on both sides; then use properties of logarithms.

$$e^{x^2} = 200$$
$$\ln e^{x^2} = \ln 200$$
$$x^2 \ln e = \ln 200$$

Since $\ln e = 1$,

$$x^2 = \ln 200$$
$$x = \pm \sqrt{\ln 200}$$
$$x \approx \pm 2.302.$$

The solution set is $\{\pm 2.302\}$, rounding to the nearest thousandth.   ■

Example 4   Solve $3 = 5(1 - e^x)$.

First solve for e^x.

$$3 = 5(1 - e^x)$$

$$\frac{3}{5} = 1 - e^x$$

$$e^x = 1 - \frac{3}{5} = \frac{2}{5}$$

Now take the natural logarithm on each side.

$$\ln e^x = \ln \frac{2}{5}$$

$$x \ln e = \ln \frac{2}{5}$$

Replace $\ln e$ with 1 to get

$$x = \ln \frac{2}{5} = \ln .4 \approx -.916.$$

To the nearest thousandth, the solution set is $\{-.916\}$.   ■

Logarithmic Equations   The properties of logarithms given in Section 7.2 are useful in solving logarithmic equations, as shown in the next examples.

Example 5   Solve $\log_a (x + 4) - \log_a (x + 2) = \log_a x$.

Using a property of logarithms, rewrite the equation as

$$\log_a \frac{x + 4}{x + 2} = \log_a x.$$

Then

$$\frac{x + 4}{x + 2} = x$$

$$x + 4 = x(x + 2)$$

$$x + 4 = x^2 + 2x$$

$$x^2 + x - 4 = 0.$$

By the quadratic formula,

$$x = \frac{-1 \pm \sqrt{1 + 16}}{2}$$

so that

$$x = \frac{-1 + \sqrt{17}}{2} \quad \text{or} \quad x = \frac{-1 - \sqrt{17}}{2}.$$

$\text{Log}_a x$ cannot be evaluated for $x = (-1 - \sqrt{17})/2$, since this number is negative and not in the domain of $\log_a x$. By substitution verify that $x = (-1 + \sqrt{17})/2$ is a solution, giving the solution set $\{(-1 + \sqrt{17})/2\}$.   ■

Example 6   Solve $\log (3x + 2) + \log (x - 1) = 1$.

Since $\log x$ is an abbreviation for $\log_{10} x$, and $1 = \log_{10} 10$, the properties of logarithms give

$$\log (3x + 2)(x - 1) = \log 10$$
$$(3x + 2)(x - 1) = 10$$
$$3x^2 - x - 2 = 10$$
$$3x^2 - x - 12 = 0.$$

Now use the quadratic formula to arrive at

$$x = \frac{1 \pm \sqrt{1 + 144}}{6}.$$

If $x = (1 - \sqrt{145})/6$, then $x - 1 < 0$ and $\log (x - 1)$ does not exist. For this reason, $(1 - \sqrt{145})/6$ must be discarded as a solution. A calculator can help to show that $(1 + \sqrt{145})/6$ is a solution, so the solution set is $\{(1 + \sqrt{145})/6\}$. ∎

Example 7   Solve $\log_3 (3m^2)^{1/4} - 1 = 2$

Solve for the logarithm first.

$$\log_3 (3m^2)^{1/4} - 1 = 2$$
$$\log_3 (3m^2)^{1/4} = 3$$

Now write the expression in exponential form as

$$(3m^2)^{1/4} = 3^3$$
$$3^{1/4} m^{1/2} = 3^3$$
$$m^{1/2} = 3^{11/4}$$

Square both sides to get

$$m = (3^{11/4})^2 = 3^{11/2}$$

or

$$m \approx 420.888.$$

To the nearest thousandth the solution set is $\{420.888\}$. ∎

Example 8   Suppose

$$P(t) = 10{,}000 e^{.4t}$$

gives the population of a city at time t (in years). In how many years will the population double?

Replace t with 0 to show that the population of the city is 10,000 when $t = 0$. We want to find the value of t for which $P(t)$ doubles to 20,000. Do this by substituting 20,000 for $P(t)$ and solving for $e^{.4t}$.

$$20{,}000 = 10{,}000 e^{.4t}$$
$$2 = e^{.4t}.$$

Now take natural logarithms on both sides.

$$\ln 2 = \ln e^{.4t}$$

Since $\ln e^{.4t} = (.4t)\ln e = (.4t)(1) = .4t$,

$$\ln 2 = .4t,$$

with $$t = \frac{\ln 2}{.4}.$$

Using a calculator or Table 1 to find $\ln 2$ and then using a calculator to find the quotient gives $t = 1.733$ to the nearest thousandth. The population of the city will double in about 1.733 years.   ■

Example 9   Solve the equation $y = \dfrac{1 - e^x}{1 - e^{-x}}$ for x.

Begin by multiplying both sides by $1 - e^{-x}$.

$$y(1 - e^{-x}) = 1 - e^x$$

or $$y - ye^{-x} = 1 - e^x.$$

Get 0 alone on one side of the equation.

$$e^x + y - 1 - ye^{-x} = 0.$$

Multiply both sides by e^x.

$$e^{2x} + (y - 1)e^x - y = 0.$$

Rewrite this equation as

$$(e^x)^2 + (y - 1)e^x - y = 0,$$

a quadratic equation in e^x. Solve this equation by using the quadratic formula with $a = 1$, $b = y - 1$, and $c = -y$.

$$e^x = \frac{-(y - 1) \pm \sqrt{(y - 1)^2 - 4(1)(-y)}}{2(1)}$$

$$= \frac{-(y - 1) \pm \sqrt{y^2 - 2y + 1 + 4y}}{2}$$

$$= \frac{-(y - 1) \pm \sqrt{y^2 + 2y + 1}}{2}$$

$$= \frac{-(y - 1) \pm \sqrt{(y + 1)^2}}{2}$$

$$e^x = \frac{-y + 1 \pm (y + 1)}{2}.$$

First, use the $+$ sign:

$$e^x = \frac{-y + 1 + y + 1}{2} = \frac{2}{2} = 1.$$

If $e^x = 1$, $x = 0$ and $e^{-x} = 1$ also. This leads to a zero denominator in the original equation, so the $+$ sign leads to no solution.

Try the $-$ sign.

$$e^x = \frac{-y + 1 - y - 1}{2} = \frac{-2y}{2} = -y$$

If $e^x = -y$, take natural logarithms on both sides to get

$$\ln e^x = \ln(-y)$$
$$x \cdot \ln e = \ln(-y)$$
$$x \cdot 1 = \ln(-y)$$
$$x = \ln(-y).$$

This result will be satisfied by all values of y less than 0. ■

Solving Exponential or Logarithmic Equations

In summary, to solve an exponential or logarithmic equation, first use the properties of algebra to change the given equation into one of the following forms, where a and b are real numbers.

1. $a^{f(x)} = b$

To solve, take logarithms to base a on both sides.

2. $\log_a f(x) = b$

Solve by changing to the exponential form $a^b = f(x)$.

3. $\log_a f(x) = \log_a g(x)$

From the given equation, obtain the equation $f(x) = g(x)$, then solve algebraically.

4. In a more complicated equation, such as the one in Example 9, it is necessary to first solve for $e^{f(x)}$ or $\log_a f(x)$ and then solve the resulting equation using one of the methods given above.

7.4 Exercises

Solve the following equations. Give answers as decimals rounded to the nearest thousandth.

1. $3^x = 6$ **2.** $4^x = 12$ **3.** $7^x = 8$ **4.** $13^p = 55$

5. $3^{a+2} = 5$ **6.** $5^{2-x} = 12$ **7.** $6^{1-2k} = 8$ **8.** $2^{k-3} = 11$

9. $4^{3m-1} = 12^{m+2}$ **10.** $3^{2m-5} = 13^{m-1}$ **11.** $e^{k-1} = 4$ **12.** $e^{2-y} = 12$

13. $2e^{5a+2} = 8$ **14.** $10e^{3z-7} = 5$ **15.** $2^x = -3$ **16.** $(1/4)^p = -4$

17. $\left(1 + \dfrac{r}{2}\right)^5 = 9$ **18.** $\left(1 + \dfrac{n}{4}\right)^3 = 12$

19. $100(1 + .02)^{3+n} = 150$ **20.** $500(1 + .05)^{p/4} = 200$ **21.** $2^{x^2-1} = 12$

22. $3^{2-x^2} = 4$ **23.** $2(e^x + 1) = 10$ **24.** $5(e^{2x} - 2) = 15$

25. $\log(t - 1) = 1$ **26.** $\log q^2 = 1$ **27.** $\log(x - 3) = 1 - \log x$

28. $\log(z - 6) = 2 - \log(z + 15)$ **29.** $\ln(y + 2) = \ln(y - 7) + \ln 4$

30. $\ln p - \ln(p + 1) = \ln 5$ **31.** $\ln(3x - 1) - \ln(2 + x) = \ln 2$

32. $\ln(8k - 7) - \ln(3 + 4k) = \ln(9/11)$ **33.** $\ln(5 + 4y) - \ln(3 + y) = \ln 3$

34. $\ln m + \ln(2m + 5) = \ln 7$ **35.** $\ln x + 1 = \ln(x - 4)$

36. $\ln(4x - 2) = \ln 4 - \ln(x - 2)$ **37.** $2\ln(x - 3) = \ln(x + 5) + \ln 4$

38. $\ln (k + 5) + \ln (k + 2) = \ln 14k$

39. $\log_5 (r + 2) + \log_5 (r - 2) = 1$

40. $\log_4 (z + 3) + \log_4 (z - 3) = 1$

41. $\log_3 (a - 3) = 1 + \log_3 (a + 1)$

42. $\log w + \log (3w - 13) = 1$

43. $\log_2 \sqrt{2y^2} - 1 = 1/2$

44. $\log_2 (\log_2 x) = 1$

45. $\log z = \sqrt{\log z}$

46. $\log x^2 = (\log x)^2$

47. $\log_x 5.87 = 2$

48. $\log_x 11.9 = 3$

49. $1.8^{p+4} = 9.31$

50. $3.7^{5z-1} = 5.88$

51. The amount of a radioactive specimen present at time t (measured in seconds) is $A(t) = 5000(10)^{-.02t}$, where $A(t)$ is measured in grams. Find the half-life of the specimen, that is, the time it will take for exactly half the specimen to remain.

A large cloud of radioactive debris from a nuclear explosion has floated over the Pacific Northwest, contaminating much of the hay supply. Consequently, farmers in the area are concerned that the cows who eat this hay will give contaminated milk. (The tolerance level for radioactive iodine in milk is 0.) The percent of the initial amount of radioactive iodine still present in the hay after t days is approximated by $P(t) = 100\, e^{-.1t}$, where t is time measured in days.

52. Some scientists feel that the hay is safe after the percent of radioactive iodine has declined to 10% of the original amount. Find the number of days before the hay could be used.

53. Other scientists believe that the hay is not safe until the level of radioactive iodine has declined to only 1% of the level. Find the number of days this would take.

Solve the following equations for x. (*Hint:* In Exercises 56–57 multiply by e^x.)

54. $2^{2x} - 3 \cdot 2^x + 2 = 0$ **55.** $5^{2x} + 2 \cdot 5^x - 3 = 0$ **56.** $e^x - 5 + 6e^{-x} = 0$ **57.** $e^x - 6 + 5e^{-x} = 0$

58. $y = \dfrac{1 + e^{-x}}{1 + e^x}$ **59.** $y = \dfrac{e^x}{1 - e^x}$ **60.** $y = \dfrac{e^x + e^{-x}}{2}$ **61.** $y = \dfrac{e^x - e^{-x}}{2}$

Solve each of the following equations for the indicated variables. Use logarithms to the appropriate bases.

62. $P = P_0 e^{kt/1000}$, for t

63. $I = \dfrac{E}{R} (1 - e^{-Rt/2})$, for t

64. $T = T_0 + (T_1 - T_0)\, 10^{-kt}$, for t

65. $A = \dfrac{Pi}{1 - (1 + i)^{-n}}$, for n

66. $(\log_{10} x) - y = \log_{10} (3x - 1)$, for x

67. $\log_{10} (x - y) = \log_{10} (3x - 1)$, for x

Solve each inequality for x.

68. $\log_2 x < 4$ **69.** $\log_3 x > -1$ **70.** $\log_x 16 < 4$ **71.** $\log_x .1 < -1$

72. $\log_2 x < \log_3 x$ **73.** $\log_5 x < \log_{10} x$ **74.** $\log_6 |5 - 2x| < 0$ **75.** $\log_2 |x^6 - 1| \le 1$

76. Recall (from the exercises for Section 7.2) the formula for the decibel rating of a sound:

$$d = 10 \log \frac{I}{I_0}.$$

Solve this formula for I.

77. A few years ago, there was a controversy about a proposed government limit on factory noise—one group wanted a maximum of 89 decibels, while another group wanted 86. This difference seemed very small to many people. Find the percent by which the 89 decibel intensity exceeds that for 86 decibels.

The formula for compound interest is

$$A = P\left(1 + \frac{i}{m}\right)^{nm}.$$

Use natural logarithms and solve for the following.

78. i **79.** n

80. The turnover of legislators is a problem of interest to political scientists. One model of legislative turnover in the U.S. House of Representatives is given by

$$M = 434e^{-.08t},$$

where M is the number of continuously serving members at time t.* This model is based on the 1965 membership of the House. Find the number of continuously serving members in each of the following years: **(a)** 1969   **(b)** 1973   **(c)** 1979.

81. Solve the formula in Exercise 80 for t.

82. The growth of bacteria in food products makes it necessary to time-date some products (such as milk) so that they will be sold and consumed before the bacteria count is too high. Suppose for a certain product that the number of bacteria present is given by

$$f(t) = 500e^{.1t},$$

under certain storage conditions, where t is time in days after packing of the product and the value of $f(t)$ is in millions. Find the number of bacteria present at each of the following times: **(a)** 2 days   **(b)** 1 week   **(c)** 2 weeks.
(d) Suppose the product cannot be safely eaten after the bacteria count reaches 3,000,000,000. How long will this take?
(e) If $t = 0$ corresponds to January 1, what date should be placed on the product?

7.5   **Exponential Growth and Decay (Optional)**

In many situations that occur in biology, economics, and the social sciences, a quantity changes at a rate proportional to the amount present. In such cases the amount present at time t is a function of t called the **exponential growth function.**

Exponential Growth Function

> Let y_0 be the amount or number present at time $t = 0$. Then, under certain conditions, the amount present at any time t is given by
>
> $$y = y_0 e^{kt},$$
>
> where k is a constant.

*From "Exponential Models of Legislative Turnover" by Thomas W. Casstevens, *UMAP*, Unit 296. Reprinted by permission of COMAP, Inc.

This section shows how to determine the equation from given data.

Radioactive decay is an important application; it has been shown that radioactive substances decay exponentially—that is, according to

$$y = y_0 e^{kt},$$

where k is a negative number.

Example 1 If 600 grams of a radioactive substance are present initially and 3 years later only 300 grams remain, how much of the substance will be present after 6 years?

From the statement of the problem, $y = 600$ when $x = 0$ (that is, initially), so

$$600 = y_0 e^{k(0)}$$

$$600 = y_0,$$

giving the exponential decay equation

$$y = 600 \, e^{kt}.$$

Since there are 300 grams after 3 years, use the fact that $y = 300$ when $x = 3$ to find k.

$$300 = 600 \, e^{3k}$$

$$\frac{1}{2} = e^{3k}$$

Take natural logarithms on both sides, then solve for k.

$$\ln \frac{1}{2} = \ln e^{3k}$$

$$\ln .5 = 3k \qquad \text{(Since } \ln e^x = x \text{)}$$

$$\frac{\ln .5}{3} = k$$

$$k \approx -.231,$$

giving
$$y = 600 e^{-.231t}$$

as the exponential decay equation. To find the amount present after 6 years let $t = 6$.

$$y = 600 \, e^{-.231(6)}$$

$$\approx 600 \, e^{-1.386}$$

$$\approx 150.$$

After 6 years, 150 grams of the substance remain. ∎

As mentioned in Exercise 51 of Section 7.4, the *half-life* of a radioactive substance is the time it takes for half of a given amount of the substance to decay.

Example 2 The amount in grams of a certain radioactive substance at time t is given by

$$y = y_0 \, e^{-.1t},$$

where t is in days. Find the half-life of the substance.

Find the time t when y will equal $y_0/2$. That is, solve the equation

$$\frac{y_0}{2} = y_0 \, e^{-.1t}.$$

Divide both sides by y_0 to get

$$\frac{1}{2} = e^{-.1t}$$

Now take natural logarithms on both sides and solve for t.

$$\ln \frac{1}{2} = \ln e^{-.1t}$$

$$\ln \frac{1}{2} = -.1t \quad (\text{Using } \ln e^x = x)$$

$$t = \frac{-\ln 1/2}{.1}$$

From Table 1 or a calculator, $\ln 1/2 = \ln .5 = -.6931$, so that

$$t \approx 6.9 \text{ days.} \quad \blacksquare$$

Example 3   Carbon 14 is a radioactive isotope of carbon which has a half-life of about 5600 years. The atmosphere contains much carbon, mostly in the form of carbon dioxide, with small traces of carbon 14. Most of this atmospheric carbon is in the form of the nonradioactive isotope carbon 12. The ratio of carbon 14 to carbon 12 is virtually constant in the atmosphere. However, as a plant absorbs carbon dioxide from the air in the process of photosynthesis, the carbon 12 stays in the plant while the carbon 14 decays by conversion to nitrogen. Thus, the ratio of carbon 14 to carbon 12 is smaller in the plant than it is in the atmosphere. Even when the plant is eaten by an animal, this ratio will continue to decrease. Based on these facts, a method of dating objects called carbon 14 dating has been developed.

Let R be the (nearly constant) ratio of carbon 14 to carbon 12 found in the atmosphere, and let r be the ratio found in a fossil. It can be shown that the relationship between R and r is given by

$$\frac{R}{r} = e^{(t \ln 2)/5600},$$

where t is the age of the fossil in years.

(a) Verify the formula for $t = 0$.
   If $t = 0$,

$$\frac{R}{r} = e^0 = 1.$$

The quotient R/r can equal 1 only when $R = r$, so that the ratio in the fossil is the same as the ratio in the atmosphere. This is true only when $t = 0$.

(b) Verify the formula for $t = 5600$.

Substitute 5600 for t. Then

$$\frac{R}{r} = e^{(5600 \ln 2)/5600}$$

$$\frac{R}{r} = e^{\ln 2} = 2 \qquad \text{(Recall that } e^{\ln x} = x\text{)}$$

$$r = \frac{1}{2}R.$$

From this last result, the ratio in the fossil is half the ratio in the atmosphere. Since the half-life of carbon 14 is 5600 years, only half of it would remain at the end of that time. The formula gives the correct result for $t = 5600$.

(c) If the ratio in a fossil is 40% of R, how old is the fossil?

Let $r = .40R$, so

$$\frac{R}{r} = \frac{R}{.4R} = \frac{1}{.4} = 2.5.$$

Now use the equation $R/r = e^{t \ln 2/5600}$ and solve for t.

$$\frac{R}{r} = e^{t \ln 2/5600}$$

$$2.5 = e^{t \ln 2/5600}$$

$$\ln 2.5 = \ln e^{t \ln 2/5600}$$

$$\ln 2.5 = \frac{t \ln 2}{5600} \qquad \text{(Since } \ln e^x = x\text{)}$$

$$\frac{5600}{\ln 2} (\ln 2.5) = t$$

$$t \approx 7400$$

The fossil is about 7400 years old. ∎

The compound interest formula

$$A = P\left(1 + \frac{i}{m}\right)^{nm}$$

was discussed in Section 7.1. The table presented there shows that increasing the frequency of compounding makes smaller and smaller differences in the amount of interest earned. In fact, it can be shown that even if interest is compounded at intervals of time as small as one chooses (such as each hour, each minute, or each second), the total amount of interest earned will be only slightly more than for daily compounding. This is true even for a process called **continuous compounding**, which can be described loosely as compounding every instant. It turns out (although we shall not prove it) that the formula for continuous compounding involves the number e.

Continuous
Compounding

> If P dollars is deposited at a rate of interest i compounded continuously for n years, the final amount on deposit is
>
> $$A = Pe^{ni}$$
>
> dollars.

Example 4   Suppose $5000 is deposited in an account paying 8% compounded continuously for five years. Find the total amount on deposit at the end of five years.

Let $P = 5000$, $n = 5$, and $i = .08$. Then

$$A = 5000e^{5(.08)} = 5000e^{.4}.$$

From Table 1 or a calculator, $e^{.4} \approx 1.49182$, and

$$A = 5000(1.49182) = 7459.10,$$

or $7459.10. Check that daily compounding would have produced a compound amount about 30¢ less. ∎

Example 5   How long will it take for the money in an account that is compounded continuously at 8% interest to double?

Use the formula for continuous compounding, $A = Pe^{ni}$, to find the time n that makes $A = 2P$. Substitute $2P$ for A and .08 for i; then solve for n.

$$A = Pe^{ni}$$
$$2P = Pe^{.08n}$$
$$2 = e^{.08n}$$

Taking natural logarithms on both sides gives

$$\ln 2 = \ln e^{.08n}.$$

Use the property $\ln e^x = x$ to get $\ln e^{.08n} = .08n$, and

$$\ln 2 = .08n$$
$$\frac{\ln 2}{.08} = n$$
$$8.664 = n.$$

It will take about 8 2/3 years for the amount to double. ∎

Example 6   Suppose inflation is about 10% per year. Assuming this rate continues, in how many years would prices double?

Use the equation for continuous compounding, with $A = 2P$ and $i = .10$.

$$A = Pe^{ni}$$
$$2P = Pe^{.1n}$$
$$2 = e^{.1n}$$

Solve for n by taking natural logarithms on both sides.

$$\ln 2 = \ln e^{.1n}$$

$$\ln 2 = .1n$$

$$n = \frac{\ln 2}{.1} \approx 6.9$$

Prices will double in about 7 years. ■

7.5 Exercises

1. A population of lice is growing exponentially. After 2 months the population has increased from 100 to 125. In how many months will the population reach 500 lice?

2. A population of bacteria in a culture is increasing exponentially. The original culture of 25,000 bacteria contains 40,000 bacteria after 10 hours. How long will it be until there are 60,000 bacteria in the culture?

3. A radioactive substance is decaying exponentially. The substance is reduced from 800 grams to 400 grams after 4 days. How much remains after 6 days?

4. When a bactericide is introduced into a culture of bacteria, the number of bacteria decreases exponentially. After 9 hours there are only 20,000 bacteria. In how many hours will the original population of 50,000 bacteria be reduced to half?

5. The amount of a certain chemical that will dissolve in a solution increases exponentially as the temperature is increased. At 0°C 1 gram dissolved and at a temperature of 10°C 11 grams dissolved. At what temperature will 15 grams dissolve?

6. The amount of a certain radioactive specimen present at time t (in days) decreases exponentially. If 5000 grams decreased to 4000 grams in 5 days, find the half-life of the specimen.

Exercises 7–12 refer to the carbon dating process of Example 3 in the text.

7. Suppose an Egyptian mummy is discovered in which the ratio of carbon 14 to carbon 12 is only about half the ratio found in the atmosphere. About how long ago did the Egyptian die?

8. If the ratio of carbon 14 to carbon 12 in an object is 1/4 the atmospheric ratio, how old is the object? How old if the ratio is 1/8?

9. Verify the formula of Example 3 for $t = 11,200$.

10. Solve the formula for t.

11. In the Lascaux caves of France, the ratio of carbon 14 to carbon 12 is only about 15% of the ratio found in the atmosphere. Estimate the age of the caves.

12. Suppose a specimen is found in which $r = (2/3)R$. Estimate the age of the specimen.

Find each of the amounts in Exercises 13–18, assuming continuous compounding.

13. $2,000 at 8% for 1 year

14. $2,000 at 8% for 5 years

15. $12,700 at 10% for 3 years

16. $175.25 at 11% for 8 years

17. $5,800 at 13% for 6 years

18. $10,000 at 12% for 3 years

19. Assuming an inflation rate of 5% compounded continuously, how long will it take for prices to double?

20. In Exercise 19, how long will it take if the inflation rate is 9%?

Solve $A = Pe^{ni}$ for the following.

21. i **22.** n

Newton's law of cooling says that the rate at which a body cools is proportional to the difference in temperature between the body and the environment into which it is introduced. The temperature $f(t)$ of the body at time t in hours after being introduced into an environment having constant temperature T_0 is

$$f(t) = T_0 + Ce^{-kt},$$

where C and k are constants. Use this result in Exercises 23–26.

23. A piece of metal is heated to 300°C and then placed in a cooling liquid at 50°C. After 4 minutes, the metal has cooled to 175°C. Find its temperature after 12 minutes.

24. Boiling water, at 100°C, is placed in a freezer at 0°C. The temperature of the water is 50°C after 24 minutes. Find the temperature of the water after 96 minutes.

25. A volcano discharges lava at 800°C. The surrounding air has a temperature of 20°C. The lava cools to 410°C in five hours. Find its temperature after 15 hours.

26. Paisley refuses to drink coffee cooler than 95°F. She makes coffee with a temperature of 170°F in a room with a temperature of 70°F. The coffee cools to 120°F in 10 minutes. What is the longest time she can let the coffee sit before she drinks the coffee?

Many environmental situations place effective limits on the growth of the number of an organism in an area. Many such limited growth situations are described by the *logistic function*, defined by

$$G(t) = \frac{m \cdot G_0}{G_0 + (m - G_0)e^{-kmt}},$$

where G_0 is the initial number present, m is the maximum possible size of the population, and k is a positive constant. Assume $G_0 = 1000$, $m = 2500$, $k = .0004$, and t is time in decades (10-year periods).

27. Find $G(1)$.   **28.** Find $G(2)$.

29. When will the population double?   **30.** When will the population reach 5000?

7.6 Common Logarithms (Optional)

As we said earlier, base 10 logarithms are called **common logarithms.** It is customary to abbreviate $\log_{10} x$ as simply $\log x$. This convention started when base 10 logarithms were used extensively for calculation. (But be careful: some advanced books use $\log x$ as an abbreviation for $\log_e x$.) Examples of common logarithms include

$$\log 1000 = \log 10^3 = 3$$
$$\log 100 = \log 10^2 = 2$$

$$\log 10 = \log 10^1 = 1$$
$$\log 1 = \log 10^0 = 0$$
$$\log .1 = \log 10^{-1} = -1$$
$$\log .001 = \log 10^{-3} = -3.$$

Though it can be shown that there is no rational number x such that $10^x = 6$ (and thus no rational number x such that $x = \log 6$), a table of common logarithms or a calculator can be used to find a decimal approximation for log 6. From Table 2 in the back of the book, a decimal approximation of log 6 is

$$\log 6 \approx .7782,$$

or, equivalently, $10^{.7782} \approx 6$. Since most logarithms are approximations anyway, it is common to replace $\approx$ with $=$ and write

$$\log 6 = .7782.$$

Table 2 gives the logarithms of numbers between 1 and 10. Since every positive number can be written in scientific notation as the product of numbers between 1 and 10 and a power of 10, the logarithm of any positive number can be found by using the table and the properties of logarithms.

Example 1   Find log 6.24.

Locate the first two digits, 6.2, in the left column of the table. Then find the third digit, 4, across the top of the table. You should find

$$\log 6.24 = .7952. \quad \blacksquare$$

Example 2   Find log 6240.

Write 6240 using scientific notation, as

$$6240 = 6.24 \times 10^3.$$

Then use the properties of logarithms.

$$\log 6240 = \log (6.24 \times 10^3)$$
$$= \log 6.24 + \log 10^3$$
$$= \log 6.24 + 3 \log 10$$
$$= \log 6.24 + 3$$

From Example 1, log 6.24 = .7952, so

$$\log 6240 = .7952 + 3 = 3.7952. \quad \blacksquare$$

The decimal part of the logarithm, .7952 in Example 2, is called the **mantissa,** and the integer part, 3 here, is the **characteristic.** When using a table of logarithms, always make sure the mantissa is positive. The characteristic can be any integer, positive, negative, or zero.

Example 3   Find log .00587.
Use scientific notation and the properties of logarithms to get

$$\log .00587 = \log (5.87 \times 10^{-3})$$
$$= \log 5.87 + \log 10^{-3}$$
$$= .7686 + (-3)$$
$$= .7686 - 3.$$

The logarithm is usually left in this form. A calculator would give the answer as -2.2314, the algebraic sum of .7686 and -3. The decimal portion in the calculator answer is a negative number. This is not the best form for the logarithm when using tables, since it is not clear which number is the mantissa.

It is possible to write the characteristic in other forms. For example, log .00587 could be written as

$$\log .00587 = 7.7686 - 10.$$

The best choice depends on the anticipated use of the logarithm. ∎

Example 4   Find each of the following.
(a) $\log (2.73)^4$
Use a property of logarithms to get

$$\log (2.73)^4 = 4 \log 2.73$$
$$= 4(.4362)$$
$$= 1.7448.$$

(b) $\log \sqrt[3]{.0762}$
Here

$$\log \sqrt[3]{.0762} = \log (.0762)^{1/3}$$
$$= \frac{1}{3} \log .0762 = \frac{1}{3}(.8820 - 2).$$

To preserve the characteristic as an integer, change the characteristic to a multiple of 3 before multiplying by 1/3. One way to do this is to add and subtract 1 (which adds to 0) as follows.

$$.8820 - 2 = 1 - 1 + .8820 - 2$$
$$= 1.8820 - 3$$

Now complete the work above.

$$\log \sqrt[3]{.0762} = \frac{1}{3}(1.8820 - 3)$$
$$= .6273 - 1 \quad ∎$$

Sometimes the logarithm of a number is known and the number itself must be found. The number is called the **antilogarithm,** sometimes abbreviated **antilog.** For example, .756 is the antilogarithm of .8785 − 1, since

$$\log .756 = .8785 - 1.$$

To find the antilogarithm, look for .8785 in the body of the logarithm table. You should find 7.5 at the left and 6 at the top. Since the characteristic of .8785 − 1 is −1, the antilogarithm is

$$7.56 \times 10^{-1} = .756.$$

In exponential notation, since

$$\log .756 = .8785 - 1 = -.1215,$$

then

$$.756 = 10^{-.1215}.$$

Example 5   Find each of the following antilogarithms.

(a) $\log x = 2.5340$

Find .5340 in the body of the table; 3.4 is at the left and 2 is at the top.

$$\begin{aligned}
\log x &= 2 + .5340 \\
&= \log 10^2 + \log 3.42 \qquad \text{(From Table 2)} \\
&= \log (3.42 \times 10^2) = \log 342
\end{aligned}$$

Since

$$\begin{aligned}
\log x &= \log 342, \\
x &= 342,
\end{aligned}$$

and 342 is the antilogarithm of 2.5340.

(b) $\log x = .7536 - 3$

Table 2 shows that the antilogarithm of .7536 − 3 is

$$5.67 \times 10^{-3} = .00567.$$

(c) $\log x = -4.0670 = -4 + (-.0670)$

$$= -4 + (-1) + [1 + (-.0670)] = -5 + .9330$$

$$\log x = \log .0000857$$

$$x = .0000857 \quad \blacksquare$$

The next example uses Table 2 and the properties of logarithms for a numerical calculation.

Example 6   Find $(\sqrt[3]{42})(76.9)(.00283)$.

Use the properties of logarithms and Table 2.

$$\log (\sqrt[3]{42})(76.9)(.00283) = \frac{1}{3} \log 42 + \log 76.9 + \log .00283$$

$$= \frac{1}{3}(1.6232) + 1.8859 + (.4518 - 3)$$

$$= .5411 + 1.8859 + (.4518 - 3)$$

$$= 2.8788 - 3 = .8788 - 1.$$

From the logarithm table, .756 is the antilogarithm of .8788 − 1, and so

$$(\sqrt[3]{42})(76.9)(.00283) \approx .756. \quad \blacksquare$$

Example 7 The cost in dollars, $C(x)$, of manufacturing x picture frames, where x is measured in thousands, is

$$C(x) = 5000 + 2000 \log (x + 1).$$

Find the cost of manufacturing 19,000 frames.

To find the cost of producing 19,000 frames, let $x = 19$. This gives

$$
\begin{aligned}
C(19) &= 5000 + 2000 \log (19 + 1) \\
&= 5000 + 2000 \log 20 \\
&= 5000 + 2000(1.3010) \\
&= 7602.
\end{aligned}
$$

Thus, 19,000 frames cost a total of about $7600 to produce. $\blacksquare$

Example 8 In chemistry, the pH of a solution is defined as

$$pH = -\log [H_3O^+],$$

where $[H_3O^+]$ is the hydronium ion concentration in moles per liter. The number pH is a measure of the acidity or alkalinity of solutions. Pure water has a pH of 7.0 with values greater than that indicating alkalinity and values less than 7.0 indicating acidity. Find the following.

(a) The pH of a solution with $[H_3O^+] = 2.5 \times 10^{-4}$

$$
\begin{aligned}
pH &= -\log [H_3O^+] \\
pH &= -\log (2.5 \times 10^{-4}) \\
&= -(\log 2.5 + \log 10^{-4}) \\
&= -(.3979 - 4) \\
&= -.3979 + 4 \\
&\approx 3.6
\end{aligned}
$$

It is customary to round pH values to the nearest tenth.

(b) The hydronium ion concentration of a solution with pH = 7.1

$$
\begin{aligned}
pH &= -\log [H_3O^+] \\
7.1 &= -\log [H_3O^+] \\
-7.1 &= \log [H_3O^+]
\end{aligned}
$$

To find the antilogarithm of the number −7.1, first write −7.1 as −7.1 + 8 − 8 = .9 − 8. Look up the mantissa .9 in Table 2. Write the antilogarithm in scientific notation, rounding to the nearest tenth.

$$H_3O^+ = 7.9 \times 10^{-8} \quad \blacksquare$$

7.6 Exercises

Find the characteristic of the logarithms of each of the following.

1. 875   **2.** 9462   **3.** 2,400,000   **4.** 875,000

5. .00023   **6.** .098   **7.** .000042   **8.** .000000257

Find the common logarithms of the following numbers. A calculator is needed for Exercises 15–18.

9. .000893   **10.** .00376   **11.** 68,200   **12.** 103,000

13. 7.63   **14.** 9.37   **15.** 42,967   **16.** 1.51172

17. .008094   **18.** .00004397

Find the common antilogarithms of each of the following logarithms to three significant digits.

19. 1.5366   **20.** 2.9253   **21.** .8733 − 2   **22.** .2504 − 1

23. 3.4947   **24.** 4.6863   **25.** .8039 − 3   **26.** .8078 − 2

Use the change-of-base formula from Section 7.3 with common logarithms and Table 2 to find each of the following logarithms. Round to the nearest hundredth.

27. $\ln 125$ **28.** $\ln 63.1$ **29.** $\ln .98$ **30.** $\ln 1.53$

31. $\ln 275$ **32.** $\ln 39.8$ **33.** $\log_2 10$ **34.** $\log_5 8$

35. $\log_3 180$ **36.** $\log_2 9.65$ **37.** $\log_4 12$ **38.** $\log_7 4$

Use common logarithms to solve the following equations. Round to the nearest hundredth.

39. $10^x = 2.5$ **40.** $10^{x-1} = 143$ **41.** $100^{2x+1} = 17$ **42.** $1000^{5x} = 2010$

43. $50^{x+2} = 8$ **44.** $200^{x+3} = 12$ **45.** $.01^{3x} = .005$ **46.** $.001^{2x} = .0004$

Use common logarithms to find approximations to three significant digits for each of the following.

47. $(8.16)^{1/3}$ **48.** $\sqrt{276}$ **49.** $\dfrac{(7.06)^3}{(31.7)(\sqrt{1.09})}$ **50.** $\dfrac{(2.51)^2}{(\sqrt{1.52})(3.94)}$

51. $(115)^{1/2} + (35.2)^{2/3}$ **52.** $(778)^{1/3} + (159)^{3/4}$

Use logarithms to solve each of the following applications.

53. The number of years, n, since two independently evolving languages split off from a common ancestral language is approximated by

$$n \approx -7600 \log r,$$

where r is the proportion of words from the ancestral language common to both languages. Find n if   **(a)** $r = .9$;   **(b)** $r = .3$;   **(c)** How many years have elapsed since the split if half of the words of the ancestral language are common to both languages?

54. Midwest Creations finds that its total sales, $T(x)$, from the distribution of x catalogs, measured in thousands, is approximated by $T(x) = 5000 \log (x + 1)$. Find the total sales resulting from the distribution of   **(a)** 0 catalogs   **(b)** 5000 catalogs   **(c)** 24,000 catalogs   **(d)** 49,000 catalogs.

Find the pH of each of the following substances, using the given hydronium ion concentration.

55. Grapefruit, 6.3×10^{-4}

56. Crackers, 3.9×10^{-9}

57. Limes, 1.6×10^{-2}

58. Sodium hydroxide (lye), 3.2×10^{-14}

Find $[H_3O^+]$ for each of the following substances, using the given pH.

59. Soda pop, 2.7   **60.** Wine, 3.4   **61.** Beer, 4.8   **62.** Drinking water, 6.5

The area of a triangle having sides of length a, b, and c is given by

$$A = \sqrt{s(s - a)(s - b)(s - c)},$$

where $s = (a + b + c)/2$. Use logarithms to find the areas of the following triangles.

63. $a = 114$, $b = 196$, $c = 153$

64. $a = .0941$, $b = .0873$, $c = .0896$

65. A common problem in archaeology is to determine estimates of populations at a particular site. Several methods have been proposed to do this. One method relates the total surface area of a site to the number of occupants. If P represents the population of a site which covers an area of a square units, then

$$\log P = k \log a,$$

where k is an appropriate constant which varies for hilly, coastal, or desert environments, or for sites with single family dwellings or multiple family dwellings.* Find the population of sites with the following areas (use .8 for k): **(a)** 230 m^2 **(b)** 95 m^2 **(c)** $20,000 \text{ m}^2$.

66. In Exercise 65, find the population of a site with an area $100,000 \text{ m}^2$ for the following values of k: **(a)** 1.2 **(b)** .5 **(c)** .7.

Appendix: Interpolation

The table of common logarithms included in this text contains decimal approximations of common logarithms to four places of accuracy. If greater accuracy is necessary, a calculator or more accurate tables can be used. However, if desired, more accuracy can be obtained from the table included in this text by the process of **linear interpolation.** As an example, use linear interpolation to approximate log 75.37. First,

$$\log 75.3 < \log 75.37 < \log 75.4$$

Figure 10 on the next page shows the portion of the curve $y = \log x$ between $x = 75.3$ and $x = 75.4$. We shall use the line segment PR to approximate the logarithm curve (this approximation is usually adequate for values of x relatively close to one another). From the figure, log 75.37 is given by the length of segment MQ, which

*From "The Quantitative Investigation of Indian Mounds" by S. F. Cook, Berkeley: *University of California Publications in American Archaeology and Ethnology*, 40: 231–33, 1950. Reprinted by permission of the University of California Press.

cannot be found directly from Table 2. We can, however, find MN, which we shall use as our approximation to log 75.37. By properties of similar triangles,

$$\frac{PS}{PT} = \frac{SN}{TR}.$$

In this case, $PS = 75.37 - 75.3 = .07$, $PT = 75.4 - 75.3 = .1$, and $RT = \log 75.4 - \log 75.3 = 1.8774 - 1.8768 = .0006$. Hence,

$$\frac{.07}{.1} = \frac{SN}{.0006}$$

or

$$SN = .7(.0006) \approx .0004.$$

(This is 7/10 of the difference of the two logarithms.) Since $\log 75.37 = MN = MS + SN$, and since $MS = \log 75.3 = 1.8768$,

$$\log 75.37 = 1.8768 + .0004 = 1.8772.$$

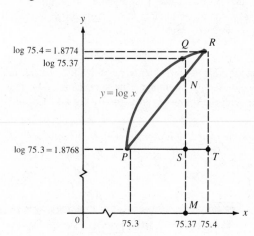

Figure 10

Example 1   Find log 8726.

All the work of the example above can be condensed as follows.

$$10\left\{ 6\left\{ \begin{array}{l} \log 8720 = 3.9405 \\ \log 8726 = \end{array} \right\}x \atop \log 8730 = 3.9410 \right\}.0005$$

From this display,

$$\frac{6}{10} = \frac{x}{.0005}$$

$$x = \frac{6(.0005)}{10}$$

$$x = .0003,$$

and log 8726 = 3.9405 + .0003 = 3.9408. ∎

Example 2 Find log .0005958.

Work as shown above.

$$10\left\{\begin{array}{l}8\left\{\begin{array}{l}\log .0005950 = -4 + .7745\\ \log .0005958 = \\ \log .0005960 = -4 + .7752\end{array}\right\}x\end{array}\right\}.0007$$

Then
$$\frac{8}{10} = \frac{x}{.0007},$$

or
$$x = .0006,$$

and
$$\log .0005958 = (-4 + .7745) + .0006$$
$$= -4 + .7751 \quad \blacksquare$$

Interpolation also can be used when finding an antilogarithm, as shown in the next example.

Example 3 Find x such that log $x = .3275$.

From the logarithm table,

$$\log 2.12 = .3263 < .3275 < .3284 = \log 2.13.$$

Set up the work as follows.

$$.01\left\{\begin{array}{l}y\left\{\begin{array}{l}\log 2.12 = .3263\\ \log x \quad = .3275\\ \log 2.13 = .3284\end{array}\right\}.0012\end{array}\right\}.0021$$

$$\frac{y}{.01} = \frac{.0012}{.0021}$$

$$y = \frac{(.01)(.0012)}{.0021}$$

$$y = .006$$

Thus, $x = 2.12 + .006 = 2.126$, and

$$\log 2.126 = .3275. \quad \blacksquare$$

Appendix Exercises

Interpolate to find common logarithms of the following numbers.

1. 2345   **2.** 1.732   **3.** 48.2ʗ   **4.** 351.9

5. .06273   **6.** .003471   **7.** 27.ʗ5   **8.** 342.6

Use interpolation to find the antilogarithms of the following common logarithms.

9. 1.7942   **10.** 3.9225   **11.** 7.6565 − 10   **12.** 8.7296 − 10

13. 5.6930   **14.** 12.6268   **15.** −3.7778   **16.** −4.1323

Use common logarithms (and interpolation) to find approximations to four-digit accuracy for each of the following computations.

17. $\dfrac{(26.13)(5.427)}{101.6}$ **18.** $\dfrac{(32.68)(142.8)}{973.4}$ **19.** $\sqrt{\dfrac{6.532}{2.718}}$ **20.** $\left(\dfrac{27.46}{58.29}\right)^3$

21. $(.2374)^{.05}$ **22.** $(1.792)^{.23}$ **23.** $(49.83)^{1/2} + (2.917)^2$ **24.** $(38.42)^{1/3} + (86.13)^{1/4}$

25. The maximum load L that a cylindrical column can hold is given by

$$L = \frac{d^4}{h^2},$$

where d is the diameter of the cylindrical cross section and h is the height of the column. Find L if $d = 2.143$ feet and $h = 12.25$ feet.

26. Two electrons repel each other with a force $F = kd^{-2}$, where d is the distance between the electrons. Find F if $d = .0005241$ cm and $k = 1$.

Chapter 7 Summary

Key Words

exponential function
logarithm
logarithmic function

common logarithm
natural logarithm
mantissa

characteristic
antilogarithm
interpolation

Review Exercises

Graph each of the functions defined as follows.

1. $f(x) = 2^x$ **2.** $f(x) = 2^{-x}$ **3.** $f(x) = (1/2)^{x+1}$ **4.** $f(x) = 4^x + 4^{-x}$

5. $f(x) = (x - 1)e^{-x}$ **6.** $f(x) = \log_2 (x - 1)$ **7.** $f(x) = \log_3 2x$ **8.** $f(x) = \ln x^{-2}$

Solve each of the following equations.

9. $8^p = 32$ **10.** $9^{2y-1} = 27^y$ **11.** $\dfrac{8}{27} = b^{-3}$ **12.** $\dfrac{1}{2} = \left(\dfrac{b}{4}\right)^{1/4}$

The amount of a certain radioactive material, in grams, present after t days, is given by

$$A(t) = 800e^{-.04t}.$$

Find $A(t)$ if

13. $t = 0$. **14.** $t = 5$.

How much would $1200 amount to at 10% compounded continuously for the following number of years?

15. 4 years **16.** 10 years

17. Historically, the consumption of electricity has increased at a continuous rate of 6% per year. If it continued to increase at this rate, find the number of years before exactly twice as much electricity would be needed.

18. Suppose a conservation campaign together with higher rates caused demand for electricity to increase at only 2% per year. (See Exercise 17.) Find the number of years before twice as much electricity would be needed as is needed today.

Write each of the following expressions in logarithmic form.

19. $2^5 = 32$ **20.** $100^{1/2} = 10$ **21.** $(1/16)^{1/4} = 1/2$ **22.** $(3/4)^{-1} = 4/3$

23. $10^{.4771} = 3$ **24.** $e^{2.4849} = 12$ **25.** $e^{.1} = 1.1052$ **26.** $2^{2.322} = 5$

Write each of the following logarithms in exponential form.

27. $\log_{10} .001 = -3$ **28.** $\log_2 \sqrt{32} = 5/2$ **29.** $\log 3.45 = .537819$ **30.** $\ln 45 = 3.806662$

Use properties of logarithms to write each of the following as a sum, difference, or product of logarithms. Assume all variables represent positive real numbers.

31. $\log_3 \dfrac{mn}{p}$ **32.** $\log_2 \dfrac{\sqrt{5}}{3}$ **33.** $\log_5 x^2 y^4 \sqrt[5]{m^3 p}$ **34.** $\log_7 (7k + 5r^2)$

Find the common logarithm of each of the following numbers.

35. 8.47   **36.** .00421   **37.** 1050   **38.** 69,800

Find the antilogarithm of each of the following numbers to three significant digits.

39. 3.4983   **40.** 9.7243   **41.** $.6493 - 2$ **42.** $.4232 - 3$

Use common logarithms to approximate each of the following.

43. $\dfrac{6^{2.1}}{\sqrt{52}}$ **44.** $2.43^{3.2}$ **45.** $\sqrt[5]{\dfrac{27.1}{4.33}}$ **46.** $(2^{5.71})(5.43^3)$

The height, in meters, of the members of a certain tribe is approximated by

$$h(t) = .5 + \log t,$$

where t is the tribe member's age in years, and $1 \le t \le 20$. Find the height of a tribe member of each of the following ages.

47. 2 years   **48.** 5 years   **49.** 10 years   **50.** 20 years

51. A person learning certain skills involving repetition tends to learn quickly at first. Then learning tapers off and approaches some upper limit. Suppose the number of symbols per minute a keypunch operator can produce is given by

$$p(t) = 250 - 120(2.8)^{-.5t}$$

where t is the number of months the operator has been in training. Find each of the following: **(a)** $p(2)$ **(b)** $p(4)$ **(c)** $p(10)$. **(d)** Graph $p(t)$.

52. The concentration of pollutants, in grams per liter, in the east fork of the Big Weasel River is approximated by

$$P(x) = .04 \, e^{-4x},$$

where x is the number of miles downstream from a paper mill that the measurement is taken. Find **(a)** $P(.5)$ **(b)** $P(1)$ **(c)** the concentration of pollutants 2 miles downstream.

Solve each of the following equations. Round to the nearest thousandth.

53. $5^r = 11$ **54.** $10^{2r-3} = 17$ **55.** $e^{p+1} = 10$

56. $(1/2)^{3k+1} = 3$ **57.** $6^{2-m} = 2^{3m+1}$ **58.** $4(1 + e^x) = 8$

59. $\log_{64} y = 1/3$ **60.** $\log_2 (y + 3) = 5$ **61.** $\ln 6x - \ln (x + 1) = \ln 4$

62. $\log_{16} \sqrt{x + 1} = 1/4$ **63.** $\log (3p - 1) = 1 - \log p$ **64.** $\log_2 (2b - 1)^2 = 4$

Solve for the indicated variable.

65. $y = \dfrac{5^x - 5^{-x}}{2}$,   for x **66.** $y = \dfrac{1}{2(5^x - 5^{-x})}$,   for x

67. $2 \log_a (x - 2) = 1 + \log_a (x + 1)$,   for x **68.** $r = r_0 e^{nt}$,   for t

69. $N = a + b \ln \dfrac{c}{d}$,   for c **70.** $P = \dfrac{k}{1 + e^{-rt}}$,   for t

Find each of the following logarithms.

71. $\ln e^{-5.3}$ **72.** $\ln e^{.04}$ **73.** $\ln 89$ **74.** $\ln .000050$

75. $\ln 8$ **76.** $\log_{3.4} 15.8$ **77.** $\log_{1/2} 9.45$ **78.** $\log_3 769$

Use interpolation to find each of the following logarithms.

79. $\log 18.99$ **80.** $\log 4.763$ **81.** $\log .009814$

The formula

$$A = P\left(1 + \frac{i}{m}\right)^{mn}$$

in Section 7.1 gives the amount of money in an account after n years if P dollars are deposited at a rate of interest i compounded m times per year. If A is known, then P is called the *present value* of A. (Think of present value as the value today of a sum of money to be received at some time in the future.) Find the present value of the following sums.

82. $4500 at 6% compounded annually for 9 years   **83.** $11,500 at 4% compounded annually for 12 years

84. $2000 at 4% compounded semiannually for 11 years   **85.** $2000 at 6% compounded quarterly for 8 years

86. How long would it take for $1000 at 5% compounded quarterly to double?

87. In Exercise 86, how long would it take at 10%?

88. If the inflation rate were 10%, use the formula for continuous compounding to find the number of years for a $1 item to cost $2.

89. In Exercise 88, find the number of years if the rate of inflation were 13%.

If R dollars is deposited at the end of each year in an account paying a rate of interest of i per year compounded annually, then after n years the account will contain a total of

$$R\left[\frac{(1 + i)^n - 1}{i}\right]$$

dollars. Find the final amount on deposit for each of the following. Use logarithms or a calculator. (Such a sequence of payments is called an *annuity*.)

90. $800, 12%, 10 years   **91.** $1500, 14%, 7 years   **92.** $375, 10%, 12 years

93. Manual deposits $10,000 at the end of each year for 12 years in an account paying 12% compounded annually. He then puts this total amount on deposit in another account paying 10% compounded semiannually for another 9 years. Find the total amount on deposit after the entire 21-year period.

94. Scott Hardy deposits $12,000 at the end of each year for 8 years in an account paying 14% compounded annually. He then leaves the money alone with no further deposits for an additional 6 years. Find the total amount on deposit after the entire 14-year period.

95. A radioactive substance is decaying exponentially. A sample of the substance is reduced from 500 grams to 400 grams after 4 days.
   (a) How much is left after 10 days?
   (b) How long will it take for the substance to decay to 100 grams?

96. The population of a boomtown is increasing exponentially. There were 10,000 people in town when the boom began. Two years later the population had reached 12,000. Assume this growth rate continues.
   (a) What will be the population after 5 years?
   (b) How long will it take for the population to double?

97. Natural logarithms can be calculated on a hand calculator even if the calculator does not have an ln key. If $1/2 \leq x \leq 3/2$ and if $A = (x - 1)/(x + 1)$, then a good approximation to $\ln x$ is given by

$$\ln x \approx ((3A^2/5 + 1)A^2/3 + 1)2A.$$

   (a) Use this formula to calculate ln .8 and ln 1.2.
   (b) Using facts about logarithms, use the formula to calculate (approximately)

$$\ln 2 = \ln ((3/2)(4/3)).$$

   (c) Using (b), calculate ln 3 and ln 8.

98. The exponential e^x can be estimated for x in $[-1/2, 1/2]$ by the formula

$$e^x \approx \left(\left(\left(\left(\frac{x}{5} + 1\right)\frac{x}{4} + 1\right)\frac{x}{3} + 1\right)\frac{x}{2} + 1\right)x + 1.$$

   (a) Calculate an approximate value for $e^{.13}$.
   (b) Calculate an approximate value for $e^{-.37}$.
   (c) Calculate an approximate value for $e^{4.13}$.
   (d) Calculate an approximate value for $e^{-2.63}$.
      (*Hint:* Use part (b).)

99. The quantity $n! = n(n - 1)(n - 2) \cdots 3 \cdot 2 \cdot 1$ grows very rapidly as n increases. According to *Stirling's Formula,* when n is large,

$$n! \approx \sqrt{2\pi n}\left(\frac{n}{e}\right)^n.$$

Use Stirling's Formula to estimate 100! and 200! (*Hint:* Use common logarithms.)

100. Watch carefully. Suppose $0 < A < B$. Because the logarithm is an increasing function, we have
   (a) $\log A < \log B$.   Then
   (b) $10A \cdot \log A < 10B \cdot \log B$,
   (c) $\log A^{10A} < \log B^{10B}$,
   (d) $A^{10A} < B^{10B}$.
   On the other hand, we run into trouble with particular choices of A and B. For instance, choose $A = 1/10$ and $B = 1/2$. Clearly $0 < A < B$, but $A^{10A} = (1/10)^1 = 1/10$ is greater than $B^{10B} = (1/2)^5 = 1/32$. Where was the first false step made?

8

Analytic Geometry.

Analytic geometry, as its name implies, is a study of geometric figures using algebraic (analytic) methods. This chapter develops some useful equations of geometric figures from their definitions.

8.1 Lines

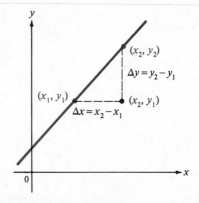

Figure 1

An important characteristic of a straight line is its *slope*, a numerical measure of the steepness of the line. This measure is found by choosing two distinct points, (x_1, y_1) and (x_2, y_2), on the line, as shown in Figure 1.

Slope

The **slope** m of the line through (x_1, y_1) and (x_2, y_2) is

$$m = \frac{\Delta y}{\Delta x} = \frac{y_2 - y_1}{x_2 - x_1}, \quad \Delta x \neq 0.$$

The slope of a line can be found only if the line is nonvertical. This guarantees that $x_2 \neq x_1$ so that the denominator $x_2 - x_1 \neq 0$. It is not possible to define the slope of a vertical line.

The slope of a vertical line is undefined.

Example 1 Find the slope of the line through each of the following pairs of points.

(a) $(-4, 8)$, $(2, -3)$

Let $x_1 = -4$, $y_1 = 8$, $x_2 = 2$ and $y_2 = -3$. Then $\Delta y = -3 - 8 = -11$ and $\Delta x = 2 - (-4) = 6$. The slope $m = \Delta y/(\Delta x) = -11/6$.

(b) $(2, 7)$, $(2, -4)$

A sketch shows that the line through $(2, 7)$ and $(2, -4)$ is vertical. As mentioned above, the slope of a vertical line is not defined. (An attempt to use the definition of slope would produce a zero denominator.)

(c) $(5, -3)$ and $(-2, -3)$

By the definition of slope,

$$m = \frac{-3 - (-3)}{-2 - 5} = \frac{0}{-7} = 0. \quad \blacksquare$$

As in Example 1(c), the slope of a horizontal line is 0.

The **angle of inclination** of a line L is the smallest positive angle α between the positive x-axis and the line. Figures 2 and 3 show the angle of inclination for two lines. The figures suggest that $0° \leq \alpha < 180°$. The slope of a line has been defined as the change in y divided by the change in x. Using this and the definition of the tangent of an angle,

$$m = \frac{y_2 - y_1}{x_2 - x_1} = \tan \alpha,$$

where m is the slope of the line and α is the angle of inclination of the line.

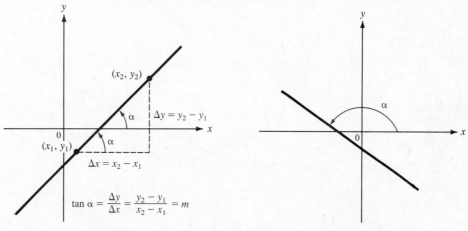

Figure 2   Figure 3

For horizontal lines, $\alpha = 0°$ and $m = 0$. For vertical lines, $\alpha = 90°$ but the slope is undefined (since $\tan 90°$ is undefined).

Example 2   Find the angle of inclination of the line through $(5, 8)$ and $(8, 11)$.

The tangent of the angle of inclination equals the slope of the line. Thus, if α is the angle of inclination,

$$\tan \alpha = \frac{11 - 8}{8 - 5} = \frac{3}{3} = 1.$$

Since $\tan \alpha = 1$ and since $0° \le \alpha < 180°$,

$$\alpha = 45°. \quad \blacksquare$$

Example 3   Find the angle of inclination of the line through $(-4, 3)$ and $(2, -5)$.

Here

$$\tan \alpha = \frac{-5 - 3}{2 - (-4)} = \frac{-8}{6} = \frac{-4}{3}.$$

Use a calculator or Table 3 to get $\alpha \approx 127°$. $\quad \blacksquare$

The vertical line through the point $(k, 0)$ goes through all points of the form (k, y); this gives the following equation of a vertical line.

Equation of a Vertical Line	An equation of the vertical line through $(k, 0)$ is $$x = k.$$

For example, the vertical line through the point $(-6, 0)$ has equation $x = -6$, or $x + 6 = 0$.

Suppose now that a line has slope m and goes through the fixed point (x_1, y_1), as in Figure 4. Let (x, y) be any other point on this line. By the definition of slope, the

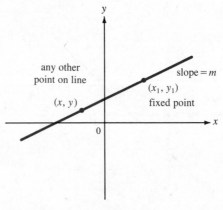

any other
point on line

(x, y)

slope $= m$

(x_1, y_1)
fixed point

0

Figure 4

slope of this line is

$$\frac{y - y_1}{x - x_1}.$$

Since the slope of the line is m,

$$\frac{y - y_1}{x - x_1} = m.$$

Multiplying both sides by $x - x_1$ gives

$$y - y_1 = m(x - x_1). \qquad (*)$$

Substituting x_1 for x and y_1 for y shows that this equation also holds for the point (x_1, y_1), showing that any point (x, y) whose coordinates satisfy equation $(*)$ above lies on the given line, and also that any point on the line has coordinates satisfying the equation. The equation of this discussion, called the *point-slope form* of the equation of a line, is summarized as follows.

Point-Slope Form

> The line with slope m passing through the point (x_1, y_1) has an equation
>
> $$y - y_1 = m(x - x_1),$$
>
> the **point-slope form** of the equation of a line.

Example 4   Write an equation of the line through $(-4, 1)$ with slope -3.

Here $x_1 = -4$, $y_1 = 1$, and $m = -3$. Use the point-slope form of the equation of a line to get

$$y - 1 = -3[x - (-4)]$$
$$y - 1 = -3(x + 4)$$
$$y - 1 = -3x - 12$$
$$3x + y = -11. \quad \blacksquare$$

Example 5   Find an equation of the line through $(-3, 2)$ and $(2, -4)$.

First find the slope, using the definition.

$$m = \frac{-4 - 2}{2 - (-3)} = -\frac{6}{5}.$$

Either $(-3, 2)$ or $(2, -4)$ can be used for (x_1, y_1). Choosing $x_1 = -3$ and $y_1 = 2$, the point-slope form gives

$$y - 2 = -\frac{6}{5}[x - (-3)]$$
$$5(y - 2) = -6(x + 3)$$
$$5y - 10 = -6x - 18$$
$$6x + 5y = -8.$$

Verify that this same equation is obtained if $(2, -4)$ is used instead of $(-3, 2)$ in the point-slope form. $\blacksquare$

As a special case of the point-slope form of the equation of a line, suppose that a line passes through the point $(0, b)$, so the line has y-intercept b. If the line has slope m, then using the point-slope form with $x_1 = 0$ and $y_1 = b$ gives

$$y - y_1 = m(x - x_1)$$
$$y - b = m(x - \mathbf{0})$$
$$y = mx + b,$$

as an equation of the line. This result is called the *slope-intercept form* of the equation of a line since it shows the slope and the y-intercept.

Slope-Intercept Form

The line with slope m and y-intercept b has an equation

$$y = mx + b,$$

the **slope-intercept form** of the equation of a line.

Example 6   Find the slope and y-intercept of $3x - y = 2$. Graph the line.

First write $3x - y = 2$ in the slope-intercept form, $y = mx + b$. Do this by solving for y, getting $y = 3x - 2$. From this form of the equation, the slope is $m = 3$ and the y-intercept is $b = -2$. Draw the graph by first locating the y-intercept, as in Figure 5. Then use the slope of 3, or 3/1, to get a second point on the graph. The line through these two points is the graph of $3x - y = 2$.   ■

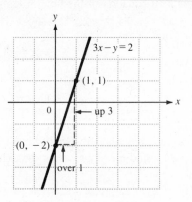

Figure 5

As mentioned earlier, a horizontal line has a slope of 0. Letting $m = 0$ in the slope-intercept form gives $y = 0x + b$, or $y = b$, an equation of a horizontal line.

Horizontal Line

An equation of the horizontal line through $(0, b)$ is

$$y = b.$$

The slope-intercept form, together with the vertical-line equation $x = k$, show that every line has an equation of the form $ax + by + c = 0$, where a and b are not both 0. Conversely, assuming $b \neq 0$, solving $ax + by + c = 0$ for y gives $y = (-a/b)x - c/b$, the equation of a line with slope $-a/b$ and y-intercept $-c/b$. If $b = 0$, solving for x gives $x = -c/a$, a vertical line. In any case, the equation $ax + by + c = 0$ has a line for its graph.

If a and b are not both 0, then the equation $ax + by + c = 0$ has a line for its graph. Also, any line has an equation of the form $ax + by + c = 0$.

Parallel and Perpendicular Lines One application of slope involves deciding whether or not two lines are parallel. Since two parallel lines are equally "steep," they should have the same slope. Also, two lines with the same "steepness" are parallel.

Parallel Lines

Two nonvertical lines are parallel if and only if they have the same slope.

A proof of this statement is requested in Exercises 72 and 73.

Example 7 Find an equation of the line through the point (3, 5) and parallel to the line $2x + 5y = 4$.

The point-slope form of the equation of a line requires a point that the line goes through and the slope of the line. The line here goes through (3, 5). Find the slope by writing the given equation in slope-intercept form (that is, solve for y).

$$2x + 5y = 4$$

$$y = -\frac{2}{5}x + \frac{4}{5}$$

The slope is $-2/5$. Since the lines are parallel, the slope of the line whose equation is needed must also be $-2/5$. Substituting $m = -2/5$, $x_1 = 3$, and $y_1 = 5$ into the point-slope form gives

$$y - y_1 = m(x - x_1)$$

$$y - 5 = -\frac{2}{5}(x - 3)$$

$$5(y - 5) = -2(x - 3)$$

$$5y - 25 = -2x + 6$$

$$2x + 5y = 31. \quad \blacksquare$$

Two lines with the same slope are parallel. As stated below, two lines having slopes with a product of -1 are perpendicular.

Perpendicular Lines

Two lines, neither of which is vertical, are perpendicular if and only if their slopes have a product of -1.

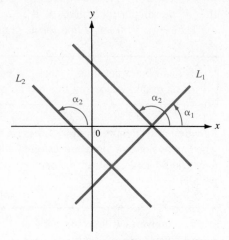

Figure 6

As shown in Figure 6, the angles of inclination of two perpendicular lines differ by 90°. If α_1 and α_2 are the angles of inclination of two perpendicular lines, then

$$\alpha_2 = \alpha_1 + 90°$$

$$\tan \alpha_2 = \tan (\alpha_1 + 90°).$$

The identity for $\tan (A + B)$ cannot be used to change the right side of the equation to an expression with $\tan \alpha_1$, since $\tan 90°$ is undefined. Instead, change the tangent function to the quotient of sine and cosine, then use the identities for $\sin (A + B)$ and $\cos (A + B)$.

$$\tan (\alpha_1 + 90°) = \frac{\sin (\alpha_1 + 90°)}{\cos (\alpha_1 + 90°)}$$

$$= \frac{\sin \alpha_1 \cos 90° + \cos \alpha_1 \sin 90°}{\cos \alpha_1 \cos 90° - \sin \alpha_1 \sin 90°}$$

Substitute 0 for $\cos 90°$ and 1 for $\sin 90°$ to get

$$\tan (\alpha_1 + 90°) = \frac{\cos \alpha_1}{-\sin \alpha_1} = -\cot \alpha_1.$$

Substituting $-\cot \alpha_1$ for $\tan (\alpha_1 + 90°)$ in the equation $\tan \alpha_2 = \tan (\alpha_1 + 90°)$ gives

$$\tan \alpha_2 = -\cot \alpha_1 = -\frac{1}{\tan \alpha_1}.$$

This result shows that the corresponding slopes m_1 and m_2 satisfy the relationship

$$m_2 = -\frac{1}{m_1}, \quad \text{or} \quad m_2 m_1 = -1.$$

Thus, two perpendicular lines have slopes with a product of -1, so the slopes are negative reciprocals of each other. Repeating these steps in reverse order and again using the fact that each angle of inclination α satisfies the relationship $0° \leq \alpha < 180°$, shows that if two slopes have a product of -1, then the corresponding lines are perpendicular.

Example 8   Find the slope of the line L perpendicular to the line with equation $5x - y = 4$. To find the slope of line L, solve $5x - y = 4$ for y to get

$$y = 5x - 4.$$

By the slope-intercept form, the slope is 5. The lines are perpendicular so that if m is the slope of line L, then

$$5m = -1$$

$$m = -\frac{1}{5}. \quad \blacksquare$$

Angle Between Two Lines   A formula for finding the angle between two nonparallel lines, given their slopes, can now be derived. Let α_1 be the angle of inclination of line L_1 with slope m_1, and α_2 be the angle of inclination of line L_2 with slope m_2, and let $\alpha_2 > \alpha_1$. If θ is the angle of smallest positive measure between lines L_1 and L_2, then, as Figure 7 suggests,

$$\theta = \alpha_2 - \alpha_1,$$

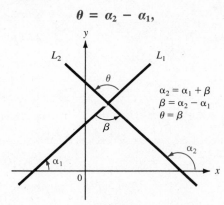

Figure 7

by properties from plane geometry. Furthermore,

$$\tan \theta = \tan (\alpha_2 - \alpha_1) = \frac{\tan \alpha_2 - \tan \alpha_1}{1 + \tan \alpha_2 \tan \alpha_1}.$$

Replacing $\tan \alpha_2$ with m_2 and $\tan \alpha_1$ with m_1 gives the following result.

Angle Between Two Lines

If θ is the angle of smallest positive measure between lines with slopes m_1 and m_2, with $m_2 > m_1$, then

$$\tan \theta = \frac{m_2 - m_1}{1 + m_2 m_1}.$$

If either α_1 or α_2 is $90°$, use $\theta = \alpha_2 - \alpha_1$ to find θ directly.

Example 9   Find the acute angle between the lines $2x + y = 5$ and $x - 3y = 8$.

First find the slope of each line. Writing each equation in the slope-intercept form shows that the slopes are -2 and $1/3$. Since m_2 must be greater than m_1, select $m_1 = -2$ and $m_2 = 1/3$. From the formula above,

$$\tan \theta = \frac{\dfrac{1}{3} - (-2)}{1 + \left(\dfrac{1}{3}\right)(-2)} = \frac{\dfrac{7}{3}}{\dfrac{1}{3}} = 7.$$

Use a calculator or Table 3 to find $\theta = 81.9°$ to the nearest tenth of a degree.

If $\tan \theta$ had been negative, then angle θ would be a second-quadrant (obtuse) angle. In that case, the acute angle between the lines is $180° - \theta$.   ■

8.1 Exercises

Find the slope of each of the following lines that has a slope.

1. Through $(4, 5)$ and $(-1, 2)$

2. Through $(5, -4)$ and $(1, 3)$

3. Through $(8, 4)$ and $(8, -7)$

4. Through $(1, 5)$ and $(-2, 5)$

5. $y = 2x$

6. $y = 3x - 2$

7. $5x - 9y = 11$

8. $4x + 7y = 1$

9. $x = -6$

10. The x-axis

11. The line parallel to $2y - 4x = 7$

12. The line perpendicular to $6x = y - 3$

13. Through $(-1.978, 4.806)$ and $(3.759, 8.125)$

14. Through $(11.72, 9.811)$ and $(-12.67, -5.009)$

Find the angle of inclination of the line through each of the following pairs of points.

15. $(1, 2)$, $(4, 5)$

16. $(-2, 8)$, $(1, 11)$

17. $(1, 2)$, $(2, 2 + \sqrt{3})$

18. $(-3, 5)$, $(-3 + \sqrt{3}, 6)$

19. $(2, -1)$, $(3, -4)$

20. $(-3, 2)$, $(-5, 4)$

21. $(2, -9)$, $(-5, 6)$

22. $(3, 7)$, $(-7, 2)$

Write an equation of each line.

23. Through $(1, 3)$, $m = -2$

24. Through $(2, 4)$, $m = -1$

25. Through $(6, 1)$, $m = 0$

26. Through $(-8, 1)$, no slope

27. Through $(4, 2)$ and $(1, 3)$

28. Through $(8, -1)$ and $(4, 3)$

29. Through $(0, 3)$ and $(4, 0)$

30. Through $(-3, 0)$ and $(0, -5)$

31. x-intercept 3, y-intercept -2

32. x-intercept -2, y-intercept 4

33. Vertical, through $(-6, 5)$

34. Horizontal, through $(8, 7)$

35. Through $(-1.76, 4.25)$, with slope -5.081

36. Through $(5.469, 11.08)$, with slope 4.723

Graph the following lines.

37. Through $(-1, 3)$, $m = 3/2$

38. Through $(-2, 8)$, $m = -1$

39. Through $(3, -4)$, $m = -1/3$

40. Through $(-2, -3)$, $m = -3/4$

41. $3x + 5y = 15$

42. $2x - 3y = 12$

43. $4x - y = 8$

44. $x + 3y = 9$

45. $x + 2y = 0$

46. $3x - y = 0$

47. $x = -1$

48. $y + 2 = 0$

49. $y = -3$

50. $x = 5$

Write an equation for each of the following lines.

51. Through $(-1, 4)$, parallel to $x + 3y = 5$

52. Through $(2, -5)$, parallel to $y - 4 = 2x$

53. Through $(3, -4)$, perpendicular to $x + y = 4$

54. Through $(-2, 6)$, perpendicular to $2x - 3y = 5$

55. x-intercept -2, parallel to $y = 2x$

56. y-intercept 3, parallel to $x + y = 4$

57. The perpendicular bisector of the segment connecting $(4, -6)$ and $(3, 5)$

58. The line with x-intercept $-2/3$ and perpendicular to $2x - y = 4$

59. Do the points $(4, 3)$, $(2, 0)$, and $(-18, -12)$ lie on the same line? (*Hint:* Find the equation of the line through two of the points.)

60. Do the points $(4, -5)$, $(3, -5/2)$, and $(-6, 18)$ lie on a line?

61. Find k so that the line through $(4, -1)$ and $(k, 2)$ is (a) parallel to $3y + 2x = 6$, and (b) perpendicular to $2y - 5x = 1$.

62. Use slopes to show that the quadrilateral with vertices at $(1, 3)$, $(-5/2, 2)$, $(-7/2, 4)$ and $(2, 1)$ is a parallelogram.

63. Use slopes to show that the square with vertices at $(-2, 3)$, $(4, 3)$, $(4, -3)$, and $(-2, -3)$ has diagonals that are perpendicular.

Give the acute angle between each of the following pairs of lines to the nearest tenth of a degree.

64. $5x + 3y = 12$; $2x - 4y = 8$

65. $3x - y = 7$, $x + 2y = 9$

66. $2x + y = 5$, $x - y = 3$

67. $4x + 3y = 12$, $2x + y = 6$

68. $x = 7$, $3x - 2y = 6$

69. $x = -2$, $x + y = 4$

70. Give, to the nearest degree, the angles of the triangle with vertices at $(2, -1)$, $(1, 4)$, and $(-3, 1)$.

71. Given the triangle with vertices at $(-5, 2)$, $(4, 1)$, and $(2, -3)$, find all angles of the triangle to the nearest degree.

Prove each of the following statements.

72. Two nonvertical parallel lines have the same slope.

73. Two lines with the same slope are parallel.

74. The line $y = x$ is the perpendicular bisector of the segment connecting (a, b) and (b, a), where $a \neq b$.

75. The line $ax + by = 0$, where $a \neq 0$, goes through the origin. If $b \neq 0$, the slope of the line is $-a/b$.

76. When discussing parallel and perpendicular lines, we were always careful to exclude vertical lines. Why do you think this is?

8.2 Ellipses and Hyperbolas

As the earth travels around the sun over a year's time, it traces out a curve called an *ellipse*. There are many applications of ellipses. In one interesting application, patients with kidney stones are treated by being placed in a water bath in a tub with an elliptical cross section. Several hundred spark discharges are produced at one focus of the ellipse (see F or F' in Figure 8), with the kidney stone at the other focus. The discharges go through the water, causing the stone to break up into small pieces which can readily be excreted from the body.

An ellipse is defined as follows.

Ellipse

> An **ellipse** is the set of all points in a plane the sum of whose distances from two distinct fixed points is constant. The two fixed points are called the **foci** of the ellipse.

For example, the ellipse in Figure 8 has foci at points F and F'. By the definition, the ellipse is made up of all points P such that the sum $d(P, F) + d(P, F')$ is constant. The ellipse in Figure 8 has its **center** at the origin. Points V and V' are the **vertices** of the ellipse, the line segment connecting V and V' is the **major axis,** and the line segment connecting B and B' is the **minor axis.**

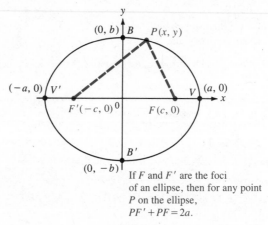

If F and F' are the foci of an ellipse, then for any point P on the ellipse, $PF' + PF = 2a$.

Figure 8

To obtain an equation for an ellipse centered at the origin, let the two foci have coordinates $(-c, 0)$ and $(c, 0)$, respectively. (See Figure 8.) Let the sum of the distances from any point $P(x, y)$ on the ellipse to the two foci be $2a$. By the distance formula, segment PF has length

$$d(P,F) = \sqrt{(x - c)^2 + y^2},$$

while segment PF' has length

$$d(P, F') = \sqrt{[x - (-c)]^2 + y^2} = \sqrt{(x + c)^2 + y^2}.$$

The sum of the lengths $d(P, F)$ and $d(P, F')$ must be $2a$, or

$$\sqrt{(x - c)^2 + y^2} + \sqrt{(x + c)^2 + y^2} = 2a.$$

Using algebra (see Exercise 33) and letting $a^2 - c^2 = b^2$ gives

$$\frac{x^2}{a^2} + \frac{y^2}{b^2} = 1,$$

the **standard form** of the equation of an ellipse centered at the origin with foci on the x-axis.

Letting $y = 0$ in the standard form gives

$$\frac{x^2}{a^2} + \frac{0^2}{b^2} = 1$$

$$\frac{x^2}{a^2} = 1$$

$$x^2 = a^2$$

$$x = \pm a$$

as the x-intercepts of the ellipse. The points $V'(-a, 0)$ and $V(a, 0)$ are the vertices of the ellipse; the segment VV' is the major axis. In a similar manner, letting $x = 0$ shows that the y-intercepts are $\pm b$; the segment connecting $(0, b)$ and $(0, -b)$ is the minor axis. It was assumed throughout the work above that the foci were on the x-axis. If the foci were on the y-axis, an almost identical proof could be used to get the standard form

$$\frac{y^2}{a^2} + \frac{x^2}{b^2} = 1.$$

Do not be confused by the two standard forms—in one case a^2 is associated with x^2; in the other case a^2 is associated with y^2. However, in practice it is necessary only to find the intercepts of the graph—if the positive x-intercept is larger than the positive y-intercept, the major axis is horizontal, and otherwise it is vertical. When using the relationship $a^2 - c^2 = b^2$, or $a^2 - b^2 = c^2$, choose a^2 and b^2 so that $a^2 - b^2 > 0$. A summary of this work with ellipses follows.

Equations for Ellipses

The ellipse with center at the origin and major axis along the x-axis has equation

$$\frac{x^2}{a^2} + \frac{y^2}{b^2} = 1 \quad (a > b),$$

while the ellipse centered at the origin with major axis along the y-axis has equation

$$\frac{y^2}{a^2} + \frac{x^2}{b^2} = 1 \quad (a > b).$$

Notice that, as shown in the summary, an ellipse always has $a > b$. Also, an ellipse is symmetric with respect to its major axis and its minor axis.

Example 1 Graph $4x^2 + 9y^2 = 36$.

To obtain the standard form for the equation of an ellipse, divide each side by 36 to get

$$\frac{x^2}{9} + \frac{y^2}{4} = 1.$$

The x-intercepts of this ellipse are ± 3, and the y-intercepts ± 2. Additional ordered pairs satisfying the equation of the ellipse may be found if desired. The graph of the ellipse is shown in Figure 9.

Since $9 > 4$, find the foci by letting $c^2 = 9 - 4 = 5$ so that $c = \pm\sqrt{5}$. The major axis is along the x-axis so the foci are at $(-\sqrt{5}, 0)$ and $(\sqrt{5}, 0)$. ∎

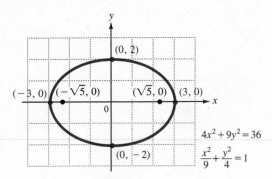

Figure 9

Example 2 Find the equation of the ellipse having center at the origin, foci at $(0, 3)$ and at $(0, -3)$, and major axis of length 8 units.

Since the major axis is 8 units long,

$$2a = 8$$

or

$$a = 4.$$

To find b^2, use the relationship $a^2 - b^2 = c^2$. Here $a = 4$ and $c = 3$. Substituting for a and c gives

$$a^2 - b^2 = c^2$$
$$4^2 - b^2 = 3^2$$
$$16 - b^2 = 9$$
$$b^2 = 7.$$

Since the foci are on the y-axis, the larger intercept, a, is used to find the denominator for y^2, giving the equation in standard form as

$$\frac{y^2}{16} + \frac{x^2}{7} = 1.$$

A graph of this ellipse is shown in Figure 10.   ■

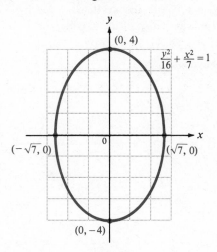

Figure 10

Example 3   Give the foci, vertices, and length of the major axis for the ellipse with equation

$$25x^2 + 9y^2 = 225.$$

Divide each term by 225 to get the equation in the standard form.

$$\frac{x^2}{9} + \frac{y^2}{25} = 1$$

We have $a^2 = 25$, and $b^2 = 9$. Since

$$c^2 = a^2 - b^2,$$

then $$c^2 = 25 - 9 = 16.$$

Thus $a = 5$, $b = 3$, and $c = 4$. The foci are located on the y-axis with coordinates $(0, 4)$ and $(0, -4)$. The vertices are also on the y-axis at $(0, 5)$ and $(0, -5)$. The length of the major axis is $2a = 2(5) = 10$.   ■

Hyperbolas   An ellipse was defined as the set of all points in a plane with the sum of the distances from two fixed points as a constant. A **hyperbola** is the set of all points in a plane for which the *difference* of the distances from two fixed points (called **foci**) is constant.

Hyperbola

Let $F'(-c, 0)$ and $F(c, 0)$ be two points on the x-axis. A **hyperbola** is the set of all points $P(x, y)$ in a plane such that the difference of the distances $d(P, F')$ and $d(P, F)$ is a constant.

The midpoint of the segment $F'F$ is the **center** of the hyperbola. The intercepts V and V' are the **vertices** of the hyperbola. See Figure 11. Segment $V'V$ is the **transverse axis** of the hyperbola.

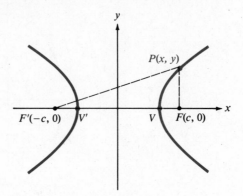

Figure 11

Suppose a hyperbola has center at the origin and foci at $F'(-c, 0)$ and $F(c, 0)$. Choosing $2a$ as the constant in the definition above gives

$$d(P, F') - d(P, F) = 2a.$$

The distance formula and algebraic manipulation (see Exercise 32) produce the result

$$\frac{x^2}{a^2} - \frac{y^2}{b^2} = 1,$$

where $b^2 = c^2 - a^2$. Letting $y = 0$ shows that the x-intercepts are $\pm a$. If $x = 0$ the equation becomes

$$\frac{0^2}{a^2} - \frac{y^2}{b^2} = 1$$

$$-\frac{y^2}{b^2} = 1$$

$$y^2 = -b^2,$$

which has no real number solutions, showing that this hyperbola has no y-intercepts.

Example 4 Graph $\dfrac{x^2}{16} - \dfrac{y^2}{9} = 1$.

This hyperbola has x-intercepts 4 and -4 and no y-intercepts. To sketch the graph, find some other points that lie on the graph. For example, letting $x = 6$ gives

$$\frac{6^2}{16} - \frac{y^2}{9} = 1$$

$$-\frac{y^2}{9} = 1 - \frac{6^2}{16}$$

$$\frac{y^2}{9} = \frac{20}{16}$$

Multiplying both sides by 9,

$$y^2 = \frac{180}{16} = \frac{45}{4}$$

$$y = \frac{\pm 3\sqrt{5}}{2} \approx \pm 3.4.$$

The graph includes the points $(6, 3.4)$ and $(6, -3.4)$. Also, letting $x = -6$ would still give $y \approx \pm 3.4$ with the points $(-6, 3.4)$ and $(-6, -3.4)$ also on the graph. These points, along with other points on the graph, were used to help sketch the final graph shown in Figure 12.   ■

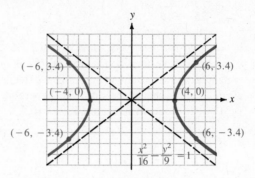

Figure 12

This basic information on hyperbolas is summarized as follows.

Equations for Hyperbolas

A hyperbola centered at the origin and having x-intercepts at a and $-a$ has an equation of the form

$$\frac{x^2}{a^2} - \frac{y^2}{b^2} = 1,$$

while a hyperbola centered at the origin and having y-intercepts at a and $-a$ has an equation of the form

$$\frac{y^2}{a^2} - \frac{x^2}{b^2} = 1.$$

By definition, $c^2 = a^2 + b^2$.

Starting with the equation for a hyperbola $(x^2/a^2) - (y^2/b^2) = 1$ and solving for y gives

$$\frac{x^2}{a^2} - 1 = \frac{y^2}{b^2}$$

$$\frac{x^2 - a^2}{a^2} = \frac{y^2}{b^2}$$

or

$$y = \pm \frac{b}{a}\sqrt{x^2 - a^2}. \tag{*}$$

If x^2 is very large in comparison to a^2, the difference $x^2 - a^2$ would be very close to x^2. If this happens, then the points satisfying equation (∗) above would be very close to one of the lines

$$y = \pm \frac{b}{a} x.$$

Thus, as $|x|$ gets larger and larger, the points of the hyperbola $(x^2/a^2) - (y^2/b^2) = 1$ come closer and closer to the lines $y = (\pm b/a)x$. These lines, called the **asymptotes** of the hyperbola, are very helpful when graphing the hyperbola.

Example 5   Graph $\dfrac{x^2}{25} - \dfrac{y^2}{49} = 1$.

For this hyperbola, $a = 5$ and $b = 7$. With these values, $y = (\pm b/a)x$ becomes $y = (\pm 7/5)x$. If $x = 5$, then $y = (\pm 7/5)(5) = \pm 7$, while $x = -5$ also gives $y = \pm 7$. These four points, $(5, 7)$, $(5, -7)$, $(-5, 7)$, and $(-5, -7)$, lead to the rectangle shown in Figure 13. The extended diagonals of this rectangle are the asymptotes of the hyperbola. The hyperbola crosses the x-axis at 5 and -5. The final graph is shown in Figure 13.   ■

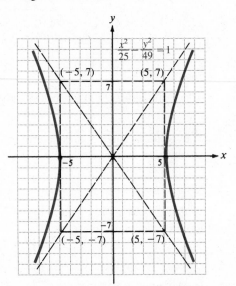

Figure 13

The rectangle used to graph the hyperbola in Example 5 is called the **fundamental rectangle.** While $a > b$ for an ellipse, the examples above show that for hyperbolas, it is possible that $a > b$ or $a < b$; other examples would show that a might equal b, also. If the foci of a hyperbola are on the y-axis, the equation of the hyperbola is of the form

$$\frac{y^2}{a^2} - \frac{x^2}{b^2} = 1, \quad \text{with asymptotes} \quad y = \pm \frac{a}{b} x.$$

If the foci of the hyperbola are on the x-axis, the asymptotes have equations $y = \pm(b/a)x$, while foci on the y-axis lead to asymptotes $y = \pm(a/b)x$. There is an obvious chance for confusion here; to avoid mistakes write the equation of the hyperbola in either the form

$$\frac{x^2}{a^2} - \frac{y^2}{b^2} = 1 \quad \text{or} \quad \frac{y^2}{a^2} - \frac{x^2}{b^2} = 1,$$

and replace 1 with 0. Solving the resulting equation for y produces the proper equations for the asymptotes. (The reason why this process works is explained in more advanced courses.)

Example 6   Graph $25y^2 - 4x^2 = 100$.
Divide each side by 100 to get

$$\frac{y^2}{4} - \frac{x^2}{25} = 1.$$

This hyperbola is centered at the origin, has foci on the y-axis, and has y-intercepts 2 and -2. To find the asymptotes, replace 1 with 0, getting

$$\frac{y^2}{4} - \frac{x^2}{25} = 0$$

$$\frac{y^2}{4} = \frac{x^2}{25}$$

$$y^2 = \frac{4x^2}{25}$$

$$y = \pm\frac{2}{5}x.$$

Use the points $(5, 2)$, $(5, -2)$, $(-5, 2)$, and $(-5, -2)$ to get the fundamental rectangle shown in Figure 14. Use the diagonals of this rectangle to determine the asymptotes for the graph, as shown in Figure 14.   ■

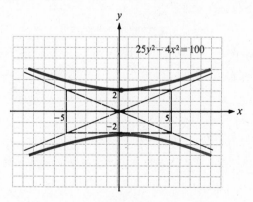

Figure 14

Example 7 Find an equation of the hyperbola with transverse axis of length 10 units, center at the origin, and a focus at (6, 0).

Since the transverse axis has length $2a = 10$, we have $a = 5$. A focus is at (6, 0), so $c = 6$. For this hyperbola,

$$b^2 = c^2 - a^2,$$

so that
$$b^2 = 36 - 25$$
$$= 11.$$

The hyperbola has intercepts on the x-axis, since the foci are on the x-axis. The equation is

$$\frac{x^2}{25} - \frac{y^2}{11} = 1. \quad \blacksquare$$

8.2 Exercises

Sketch the graph of each of the following equations. Give the endpoints of the major axis for ellipses and the equations of the asymptotes for hyperbolas.

1. $\dfrac{x^2}{9} + \dfrac{y^2}{4} = 1$ **2.** $\dfrac{x^2}{16} + \dfrac{y^2}{36} = 1$ **3.** $\dfrac{x^2}{9} + y^2 = 1$ **4.** $\dfrac{y^2}{16} - \dfrac{x^2}{9} = 1$

5. $\dfrac{x^2}{6} + \dfrac{y^2}{9} = 1$ **6.** $\dfrac{x^2}{8} - \dfrac{y^2}{12} = 1$ **7.** $x^2 + 4y^2 = 16$ **8.** $25x^2 + 9y^2 = 225$

9. $x^2 = 9 + y^2$ **10.** $y^2 = 16 + x^2$ **11.** $2x^2 + y^2 = 8$ **12.** $9x^2 - 25y^2 = 225$

13. $25x^2 - 4y^2 = -100$ **14.** $4x^2 - y^2 = -16$

Find equations for each of the following ellipses that have centers at the origin.

15. x-intercepts ± 4; foci at $(-2, 0)$ and $(2, 0)$

16. y-intercepts ± 3; foci at $(0, \sqrt{3})$, $(0, -\sqrt{3})$

17. Endpoints V and V' of major axis at $(6, 0)$, $(-6, 0)$; $c = 4$

18. Vertices $(0, 5)$, $(0, -5)$; $b = 2$

19. $a = 5$, $c = 3$; major axis vertical

20. Minor axis of length 6; major axis horizontal, of length 9

Find equations for each of the following hyperbolas with centers at the origin.

21. x-intercepts ± 3; foci at $(-4, 0)$, $(4, 0)$ **22.** y-intercepts ± 5; foci at $(0, 3\sqrt{3})$, $(0, -3\sqrt{3})$

23. A focus at $(-4, 0)$; a vertex at $(-3, 0)$ **24.** Transverse axis of length 8; a focus at $(0, 5)$

25. Asymptotes $y = \pm(3/5)x$; y-intercepts $(0, 3)$, $(0, -3)$

26. Passing through $(5, 3)$ and $(-10, 2\sqrt{21})$; no y-intercepts

The orbit of Mars is an ellipse, with the sun at one focus. An approximate equation for the orbit is

$$\frac{x^2}{5013} + \frac{y^2}{4970} = 1,$$

where x and y are measured in millions of miles.

27. Find the length of the major axis.

28. Find the length of the minor axis.

29. Draftspeople often use the method shown in the figure below to draw an ellipse. Explain why the method works.

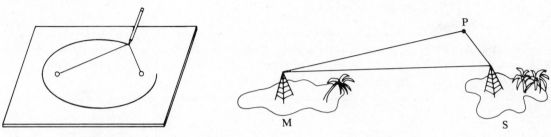

Exercise 29   **Exercise 30**

30. Ships and planes often use a location-finding system called LORAN. With this system, a radio transmitter at M in the figure below sends out a series of pulses. When each pulse is received at transmitter S, it then sends out a pulse. A ship at P receives pulses from both M and S. A receiver on the ship measures the difference in the arrival times of the pulses. The navigator then consults a special map, showing certain curves according to the differences in arrival times. In this way, the ship can be located as lying on a portion of which curve? (The method requires three transmitters operating as two pairs.)

31. Microphones are placed at points $(-c, 0)$ and $(c, 0)$. An explosion occurs at point $P(x, y)$ having positive x-coordinate. (See the figure.) The sound is detected at the closer microphone t seconds before being detected at the farther microphone. Assume that sound travels at a speed of 330 m per sec, and show that P must be on the hyperbola

$$\frac{x^2}{330^2 t^2} - \frac{y^2}{4c^2 - 330^2 t^2} = \frac{1}{4}.$$

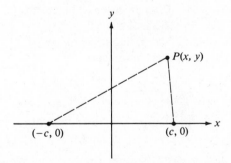

32. Suppose that a hyperbola has center at the origin, foci at $F'(-c, 0)$ and $F(c, 0)$, and $PF' - PF = 2a$. Let $b^2 = c^2 - a^2$, and show that the equation of the hyperbola is

$$\frac{x^2}{a^2} - \frac{y^2}{b^2} = 1.$$

33. Derive the standard form of the equation of an ellipse centered at the origin.

34. A rod of fixed length in the xy-coordinate plane is moved so that one end is always on the x-axis and the other end is always on the y-axis. Let P be any fixed point on the rod. Show that the path of P is an ellipse.

8.3 Conic Sections

As with the circle, ellipse, and hyperbola, the equation of a parabola can be determined from its geometric definition as a set of points.

Parabola

> A **parabola** is the set of points in a plane equidistant from a fixed point and a fixed line. The fixed point is called the **focus** and the fixed line the **directrix** of the parabola.

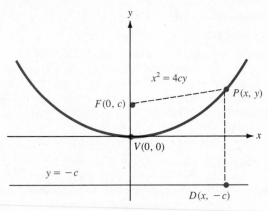

Figure 15

To find an equation of a parabola, let the directrix be the line $y = -c$ and the focus point F with coordinates $(0, c)$ as shown in Figure 15. Find the equation of the set of points that are the same distance from the line $y = -c$ and the point $(0, c)$ by choosing one such point P and giving it coordinates (x, y). Then, $d(P, F)$ and $d(P, D)$ are the same. By the distance formula,

$$d(P,F) = d(P,D)$$
$$\sqrt{x^2 + (y - c)^2} = \sqrt{(y + c)^2}$$
$$x^2 + y^2 - 2yc + c^2 = y^2 + 2yc + c^2$$
$$x^2 = 4cy.$$

Vertical Parabola

> The parabola with focus at $(0, c)$ and directrix $y = -c$ has an equation of the form
>
> $$x^2 = 4cy.$$

The parabola opens upward if $c > 0$ and downward if $c < 0$.

The point V in Figure 15 is the **vertex** of the parabola. The line through V perpendicular to the directrix, in this case the y-axis, is called the **axis** of the parabola.

The parabola is symmetric about its axis, that is, if folded on that line, the two halves of the figure would match.

If the directrix in Figure 15 had been drawn as a vertical line, going through the same algebraic process would lead to the equation of a parabola opening in a horizontal, rather than vertical, direction.

Horizontal Parabola

> The parabola with focus at $(c, 0)$ and directrix $x = -c$ has an equation of the form
>
> $$y^2 = 4cx.$$

The proof of this case is given as Exercise 47.

Parabolas have many practical applications. For example, if a light source is placed at the focus of a parabolic reflector, as in Figure 16, light rays reflect parallel to the axis, making a spotlight or flashlight.

The process also works in reverse. Light rays from a distant source come in parallel to the axis and are reflected to a point at the focus. (If such a reflector is aimed at the sun, a temperature of several thousand degrees may be obtained.)

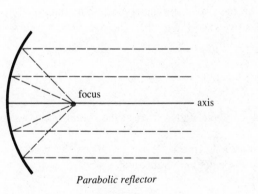

Parabolic reflector

Figure 16

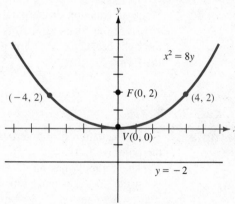

Figure 17

Example 1 Find the focus, directrix, vertex, and axis of the following parabolas.

(a) $x^2 = 8y$

The equation is in the form $x^2 = 4cy$, so set $4c = 8$, from which $c = 2$. Since the x-term is squared, the parabola is vertical, with focus at $(0, c) = (0, 2)$ and directrix $y = -2$. The vertex is $(0, 0)$, and the axis of the parabola is the y-axis. See Figure 17.

(b) $y^2 = -28x$

This parabola is of the form $y^2 = 4cx$, with $4c = -28$, so that $c = -7$. The parabola is horizontal, with focus $(-7, 0)$, directrix $x = 7$, vertex $(0, 0)$, and *x*-axis as axis of the parabola. Since *c* is negative, the graph opens to the left, as shown in Figure 18.   ■

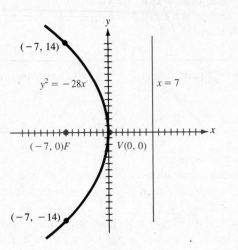

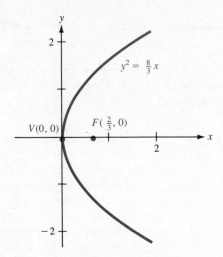

Figure 18

Figure 19

Example 2   Write an equation of the parabola with focus (2/3, 0) and vertex at the origin.

Since the focus (2/3, 0) is on the *x*-axis, the parabola is horizontal and opens to the right because $c = 2/3$ is positive. See Figure 19. The equation, which will have the form $y^2 = 4cx$, is

$$y^2 = 4\left(\frac{2}{3}\right)x \quad \text{or} \quad y^2 = \frac{8}{3}x. \quad ■$$

Example 3   Find an equation of the vertical parabola with vertex at the origin that goes through the point (−2, 12).

The parabola has an equation of the form $x^2 = 4cy$. Since the point (−2, 12) is on the graph, it must satisfy the equation. Substitute $x = -2$ and $y = 12$ into $x^2 = 4cy$ to get

$$(-2)^2 = 4c(12)$$
$$4 = 48c$$
$$c = \frac{1}{12},$$

which gives an equation of the parabola,

$$x^2 = 4\left(\frac{1}{12}\right)y \quad \text{or} \quad y = 3x^2. \quad ■$$

In summary, a parabola with vertex at the origin has the following characteristics.

	Equation	Characteristics
Parabola: Vertex at the Origin	$x^2 = 4cy$	Opens upward if $c > 0$
		Opens downward if $c < 0$
		Focus $(0, c)$
		Directrix $y = -c$
		Axis y-axis
	$y^2 = 4cx$	Opens to right if $c > 0$
		Opens to left if $c < 0$
		Focus $(c, 0)$
		Directrix $x = -c$
		Axis x-axis

Conic Sections   Parabolas, hyperbolas, ellipses, and circles are called *conic sections,* since each can be obtained by cutting a doublenapped cone with a plane, as shown in Figure 20.

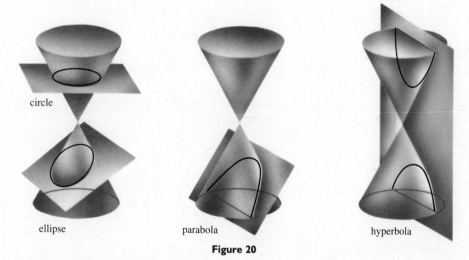

circle

ellipse

parabola

hyperbola

Figure 20

A general definition of any conic figure can be given as a set of points in a plane.

Conic	A **conic** is the set of all points $P(x, y)$ in a plane such that the ratio of the distance from P to a fixed line and the distance from P to a fixed point is constant.

As before, the fixed line is the **directrix** and the fixed point is the **focus.** In Figure 21 on the following page, the focus is $F(c, 0)$ and the directrix is the line

$x = -c$. The constant ratio is called the **eccentricity** of the conic, written e. (This is not the same e as the base of natural logarithms.) By definition, $d(P, F)$ and $d(P, D)$ in Figure 21 are equal if the conic is a parabola. Thus, a parabola always has eccentricity $e = 1$.

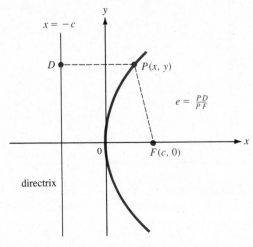

Figure 21

For the ellipse, it can be shown that the constant ratio is

$$e = \frac{c}{a},$$

where c is the distance from the center of the figure to a focus and a is the distance from the center to a vertex.

For an ellipse, it is always the case that $a^2 > b^2$. Since $b^2 = a^2 - c^2$, then $c^2 = a^2 - b^2$ and $c = \sqrt{a^2 - b^2}$, so that

$$0 < c < a,$$

$$0 < \frac{c}{a} < 1,$$

and, for an ellipse,

$$0 < e < 1.$$

If a is constant, letting c approach 0 will force the ratio c/a to approach 0, which also forces b to approach a (so that $\sqrt{a^2 - b^2} = c$ can approach 0). Since b leads to the y-intercepts, this means that the x- and y-intercepts are almost the same, producing an ellipse very close in shape to a circle when e is very close to 0. In a similar manner, if e approaches 1, then b will approach 0, which gives a very flat ellipse. The path of the earth around the sun is an ellipse that is very nearly circular. In fact, for this ellipse, $e \approx .017$. On the other hand, the path of Halley's comet is a very flat ellipse, with $e \approx .98$.

Example 4 Find the eccentricity of each ellipse.

(a) $\dfrac{x^2}{9} + \dfrac{y^2}{16} = 1$

Since $16 > 9$, let $a^2 = 16$, which gives $a = 4$. Also,

$$c = \sqrt{a^2 - b^2}$$
$$c = \sqrt{16 - 9} = \sqrt{7}.$$

Finally,
$$e = \frac{c}{a} = \frac{\sqrt{7}}{4}.$$

(b) $5x^2 + 10y^2 = 50$

Divide by 50 to get

$$\frac{x^2}{10} + \frac{y^2}{5} = 1.$$

Here $a^2 = 10$, with $a = \sqrt{10}$. Now find c:

$$c = \sqrt{10 - 5} = \sqrt{5},$$

and
$$e = \frac{c}{a} = \frac{\sqrt{5}}{\sqrt{10}} = \frac{1}{\sqrt{2}} = \frac{\sqrt{2}}{2}. \quad \blacksquare$$

The hyperbola

$$\frac{x^2}{a^2} - \frac{y^2}{b^2} = 1 \qquad \text{or} \qquad \frac{y^2}{a^2} - \frac{x^2}{b^2} = 1,$$

where $c = \sqrt{a^2 + b^2}$, also has eccentricity

$$e = \frac{c}{a}.$$

By the definition of c for a hyperbola, $c = \sqrt{a^2 + b^2} > a$, so that $c/a > 1$. Thus, for a hyperbola,

$$e > 1.$$

Example 5 Find the eccentricity for the hyperbola

$$\frac{x^2}{9} - \frac{y^2}{4} = 1.$$

Here $a^2 = 9$; thus $a = 3$ and $c = \sqrt{9 + 4} = \sqrt{13}$. The eccentricity is

$$e = \frac{c}{a} = \frac{\sqrt{13}}{3}. \quad \blacksquare$$

The chart on the next page summarizes the above discussion of eccentricity.

	Conic	Eccentricity
Eccentricity of Conics	Parabola	$e = 1$
	Ellipse	$e = \dfrac{c}{a}$ and $0 < e < 1$
	Hyperbola	$e = \dfrac{c}{a}$ and $e > 1$

Example 6 Find an equation for a conic with focus at (3, 0) and eccentricity 2.

Since $e = 2$, which is greater than 1, the conic is a hyperbola with $c = 3$. From $e = c/a$, find a by substituting $e = 2$ and $c = 3$.

$$e = \frac{c}{a}$$

$$2 = \frac{3}{a}$$

$$a = \frac{3}{2}$$

Now find b:

$$b^2 = c^2 - a^2 = 9 - \frac{9}{4} = \frac{27}{4}.$$

The given focus is on the x-axis, so the x^2 term is positive and the equation is

$$\frac{x^2}{9/4} - \frac{y^2}{27/4} = 1,$$

or

$$\frac{4x^2}{9} - \frac{4y^2}{27} = 1. \quad \blacksquare$$

8.3 Exercises

Graph each of the following.

1. $3y = x^2$

2. $\frac{1}{5}y = x^2$

3. $y = -x^2$

4. $\frac{1}{2}y = -x^2$

5. $4x = y^2$

6. $9x = y^2$

7. $16y^2 = -x$

8. $4y^2 = -x$

Give the focus, directrix, and axis for each of the following parabolas.

9. $x^2 = 16y$

10. $x^2 = 4y$

11. $y = -2x^2$

12. $y = 9x^2$

13. $x = 16y^2$

14. $x = -32y^2$

15. $x = -\frac{1}{16}y^2$

16. $x = -\frac{1}{4}y^2$

Write an equation for each of the following parabolas with vertex at the origin.

17. Focus $(0, -2)$ **18.** Focus $(5, 0)$ **19.** Focus $\left(-\dfrac{1}{2}, 0\right)$ **20.** Focus $\left(0, \dfrac{1}{4}\right)$

21. Through $(2, -2\sqrt{2})$, opening to the right   **22.** Through $(\sqrt{3}, 3)$, opening upward

23. Through $(\sqrt{10}, -5)$, opening downward   **24.** Through $(-3, 3)$, opening to the left

25. Through $(2, -4)$, symmetric to the y-axis   **26.** Through $(3, 2)$, symmetric to the x-axis

Find the eccentricity of each of the following ellipses or hyperbolas.

27. $12x^2 + 9y^2 = 36$ **28.** $8x^2 - y^2 = 16$ **29.** $x^2 - y^2 = 4$ **30.** $x^2 + 2y^2 = 8$

31. $4x^2 + 7y^2 = 28$ **32.** $9x^2 - y^2 = 1$ **33.** $x^2 - 9y^2 = 18$ **34.** $x^2 + 10y^2 = 10$

35. $2x^2 + y^2 = 32$ **36.** $5x^2 - 4y^2 = 20$

Write the equation for each of the following conics. Assume that each parabola has vertex at the origin, and each ellipse or hyperbola is centered at the origin.

37. The conic with focus at $(0, 8)$ and $e = 1$ **38.** The conic with focus at $(-2, 0)$ and $e = 1$

39. The conic with focus at $(3, 0)$ and $e = 1/2$ **40.** The conic with focus at $(0, -2)$ and $e = 2/3$

41. The conic with vertex at $(-6, 0)$ and $e = 2$ **42.** The conic with vertex at $(0, 4)$ and $e = 5/3$

43. The conic with focus at $(0, -1)$ and $e = 1$ **44.** The conic with focus at $(2, 0)$ and $e = 6/5$

45. The conic with a vertical major axis of length 6 and $e = 4/5$

46. The conic with a vertical transverse axis of length 8 and $e = 7/3$

47. Prove that the parabola with focus $(c, 0)$ and directrix $x = -c$ has equation $y^2 = 4cx$.

48. Use the definition of a parabola to find an equation of the parabola with vertex at (h, k) and axis $y = k$. Let the distance from the vertex to the focus and the distance from the vertex to the directrix be c, where $c > 0$.

49. What is the eccentricity of a circle? (Think of a circle as an ellipse with $a = b$.)

50. The orbit of Mars around the sun is an ellipse with equation

$$\frac{x^2}{5013} + \frac{y^2}{4970} = 1,$$

where x and y are measured in millions of miles. Find the eccentricity of this ellipse.

8.4 Translation of Axes

It is sometimes easier to draw a graph on a coordinate system that has been transformed from the usual one. In this section we discuss one such transformation, the translation of axes, and in the next section we discuss another, the rotation of axes.

Figure 22 on the next page shows two coordinate systems. One has an x-axis and a y-axis and is called the xy-system. The other has an x'-axis and a y'-axis and is called the $x'y'$-system. The x'-axis is parallel to the x-axis, and the y'-axis is parallel to the y-axis. The $x'y'$-system is a **translation** of the xy-system. The origin of the $x'y'$-system has coordinates $(0, 0)$ with respect to the $x'y'$-system, but coordinates (h, k) with respect to the xy-system.

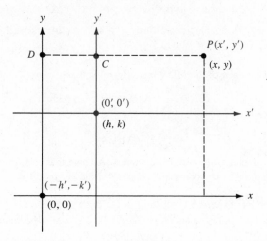

Figure 22

To find formulas for converting coordinates from one system to another, look at point P of Figure 22. Point P has coordinates (x, y) with respect to the xy-system and coordinates (x', y') with respect to the $x'y'$-system. From Figure 22, $d(P, D) = x$, $d(C, D) = h$, and $d(P, C) = x'$. Since $d(P, D) = d(P, C) + d(C, D)$,

$$x = x' + h.$$

In the same way,

$$y = y' + k.$$

Solving for x' and y' gives

$$x' = x - h \quad \text{and} \quad y' = y - k.$$

Although the figure shows the origin of the $x'y'$-system in the first quadrant of the xy-system, the results summarized below are true no matter where the origin lies.

Translation Equations

> If an xy-coordinate system is translated so that the origin of the $x'y'$-system has coordinates (h, k) with respect to the xy-system, and if (x, y) and (x', y') are the corresponding coordinates of the same point, then the **translation equations** are
>
> $$x' = x - h \quad \text{and} \quad y' = y - k.$$

These translation equations are used to change an equation in x and y to one involving x' and y', as shown in the next example. (The equations $x = x' + h$ and $y = y' + k$ can be used to convert in the opposite direction, but they are seldom used.)

Example 1   Graph $\dfrac{(x - 5)^2}{9} + \dfrac{(y + 3)^2}{25} = 1$.

Let $x' = x - 5$ and $y' = y + 3$. Then the given equation becomes

$$\frac{x'^2}{9} + \frac{y'^2}{25} = 1$$

with respect to the new $x'y'$-system. This equation represents an ellipse with x'-intercepts 3 and -3 and y'-intercepts 5 and -5. Since $x' = x - 5$ and $y' = y + 3 = y - (-3)$, the origin of the $x'y'$-system has coordinates $(h, k) = (5, -3)$ in the xy-system. The graph is shown in Figure 23.   ■

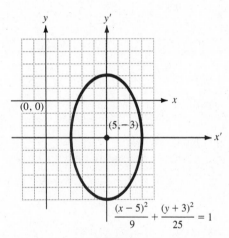

Figure 23

Sometimes an equation must be written in a form similar to the one in Example 1 before substituting from the translation equations.

Example 2   Graph $2x + y^2 - 8y = -26$.

Collect the terms with y on one side of the equation, and complete the square on y.

$$y^2 - 8y = -2x - 26$$
$$y^2 - 8y + \mathbf{16} = -2x - 26 + \mathbf{16}$$
$$(y - 4)^2 = -2x - 10$$
$$(y - 4)^2 = -2(x + 5)$$

Now let $y' = y - 4$ and $x' = x + 5$. The equation becomes

$$y'^2 = -2x',$$

the equation of a parabola with vertex at the origin of the $x'y'$-system and opening to the left. The origin of the $x'y'$-system has coordinates $(h, k) = (-5, 4)$ in the xy-system. The graph of the parabola is shown in Figure 24. ■

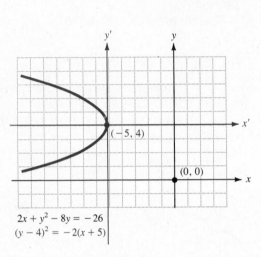

$2x + y^2 - 8y = -26$
$(y - 4)^2 = -2(x + 5)$

Figure 24

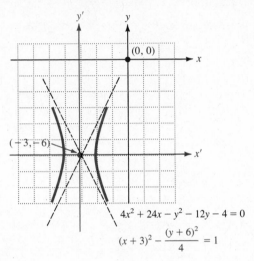

$4x^2 + 24x - y^2 - 12y - 4 = 0$

$(x + 3)^2 - \dfrac{(y + 6)^2}{4} = 1$

Figure 25

Example 3   Graph $4x^2 + 24x - y^2 - 12y - 4 = 0$.

Complete the square on both x and y.

$$4(x^2 + 6x +) - (y^2 + 12y +) = 4$$
$$4(x^2 + 6x + 9) - (y^2 + 12y + 36) = 4 + 36 - 36$$
$$4(x + 3)^2 - (y + 6)^2 = 4$$
$$\frac{(x + 3)^2}{1} - \frac{(y + 6)^2}{4} = 1$$

To graph, translate the axes by letting $x' = x + 3$ and $y' = y + 6$. This gives

$$\frac{x'^2}{1} - \frac{y'^2}{4} = 1,$$

a hyperbola with x'-intercepts of 1 and -1. Here the origin of the $x'y'$-system has xy-coordinates $(h, k) = (-3, -6)$. The asymptotes are given by

$$y' = \pm\frac{b}{a}x' = \pm 2x'.$$

These results lead to the graph in Figure 25. ■

The examples of this section suggest the following result.

General Second-Degree Equation

If the **general second-degree equation**

$$Ax^2 + Cy^2 + Dx + Ey + F = 0$$

has a graph, it will be one of the following:

(a) a circle (or a point) if $A = C \neq 0$;
(b) an ellipse (or a point) if $AC > 0$, and $A \neq C$;
(c) a hyperbola (or two intersecting lines) if $AC < 0$;
(d) a parabola (or one line or two parallel lines) if either $A = 0$ or $C = 0$, but not both.

The coefficient B is reserved for an xy-term, which will be discussed in the next section.

8.4 Exercises

For each of the following equations, use the general second-degree equation to decide which conic section the equation represents without translating the axes.

1. $x^2 + 3y^2 - 4x + 2y = 8$

2. $5x^2 - 2y^2 + 3x + 6y - 10 = 0$

3. $4x^2 + 6x - 9y = 10$

4. $3y^2 + 2x - y + 7 = 0$

5. $2x^2 + 2y^2 - 5x - 5y = 16$

6. $x^2 + 4y^2 + 6x + 8y = 24$

7. $-x^2 + 2y^2 - 3x + 2y - 12 = 0$

8. $-x^2 - 10x + 3y = 9$

For each of the following equations, translate the axes and sketch the graph.

9. $(x - 5)^2 = 4y$

10. $(y + 2)^2 = -6x$

11. $(y - 5)^2 = 3(x + 2)$

12. $(x + 1) = -4(y - 3)^2$

13. $\dfrac{(x + 3)^2}{9} + \dfrac{(y - 5)^2}{4} = 1$

14. $\dfrac{(x - 4)^2}{16} + \dfrac{(y + 6)^2}{25} = 1$

15. $\dfrac{(x - 6)^2}{9} - \dfrac{(y + 2)^2}{25} = 1$

16. $\dfrac{(y + 5)^2}{16} - \dfrac{(x - 7)^2}{25} = 1$

17. $x^2 - 2x - 6y = 11$

18. $2x - y^2 - 6y - 11 = 0$

19. $x - 2y^2 - 20y - 47 = 0$

20. $3x^2 - 30x - 2y + 79 = 0$

21. $x^2 - 2x + y^2 + 6y + 1 = 0$

22. $x^2 + 10x + y^2 + 4y + 13 = 0$

23. $x^2 - 6x + y^2 - 10y + 42 = 0$

24. $x^2 + 4x - y^2 + 10y - 20 = 0$

25. $9x^2 - 90x + 25y^2 - 200y + 400 = 0$

26. $x^2 + 6x + 5y^2 - 20y - 51 = 0$

27. $x^2 - 6x - 4y^2 - 32y - 71 = 0$

28. $16x^2 + 96x - 9y^2 - 36y - 36 = 0$

29. $5y^2 - 20y - 3x^2 - 24x - 28 = 0$

30. $25x^2 - 150x + 4y^2 + 32y + 189 = 0$

8.5 Rotation of Axes

If an xy-coordinate system having origin O has the axes rotated about O through an angle θ, the new coordinate system is called a **rotation** of the xy-system. Trigonometric identities can be used to obtain equations for converting the coordinates of a point from the xy-system to the rotated $x'y'$-system. Let P be any point other than the origin, with coordinates (x, y) in the xy-system and (x', y') in the $x'y'$-system. See Figure 26. Let $OP = r$, and let α represent the angle made by OP and the x'-axis. From Figure 26,

$$\cos(\theta + \alpha) = \frac{OA}{r} = \frac{x}{r}$$

$$\sin(\theta + \alpha) = \frac{AP}{r} = \frac{y}{r}$$

$$\cos \alpha = \frac{OB}{r} = \frac{x'}{r}$$

$$\sin \alpha = \frac{PB}{r} = \frac{y'}{r}.$$

These four statements can be rewritten as

$$x = r \cos(\theta + \alpha), \qquad y = r \sin(\theta + \alpha),$$
$$x' = r \cos \alpha, \qquad y' = r \sin \alpha.$$

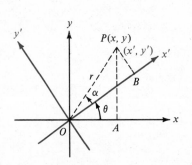

Figure 26

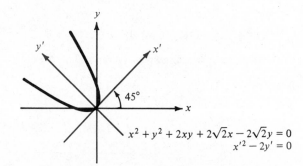

$$x^2 + y^2 + 2xy + 2\sqrt{2}x - 2\sqrt{2}y = 0$$
$$x'^2 - 2y' = 0$$

Figure 27

From the trigonometric identity for the cosine of the sum of two angles,

$$x = r \cos(\theta + \alpha)$$
$$= r(\cos \theta \cos \alpha - \sin \theta \sin \alpha)$$
$$= (r \cos \alpha) \cos \theta - (r \sin \alpha) \sin \theta$$
$$x = x' \cos \theta - y' \sin \theta.$$

In the same way, using the identity for the sine of the sum of two angles gives $y = x' \sin \theta + y' \cos \theta$, proving the following result.

Rotation Equations	If the rectangular coordinate axes are rotated about the origin through an angle θ, and if the coordinates of a point P are (x, y) and (x', y') with respect to the xy-system and the $x'y'$-system, then the **rotation equations** are

$$x = x' \cos \theta - y' \sin \theta$$

and

$$y = x' \sin \theta + y' \cos \theta.$$

Example 1   The equation of a curve is $x^2 + y^2 + 2xy + 2\sqrt{2}x - 2\sqrt{2}y = 0$. Find the resulting equation if the axes are rotated 45°. Graph the equation.

If $\theta = 45°$, then $\sin \theta = \sqrt{2}/2$ and $\cos \theta = \sqrt{2}/2$, and the rotation equations become

$$x = \frac{\sqrt{2}}{2} x' - \frac{\sqrt{2}}{2} y'$$

and

$$y = \frac{\sqrt{2}}{2} x' + \frac{\sqrt{2}}{2} y'.$$

Substituting these values into the given equation yields

$$x^2 + y^2 + 2xy + 2\sqrt{2}x - 2\sqrt{2}y = 0$$

$$\left[\frac{\sqrt{2}}{2} x' - \frac{\sqrt{2}}{2} y' \right]^2 + \left[\frac{\sqrt{2}}{2} x' + \frac{\sqrt{2}}{2} y' \right]^2$$

$$+ 2 \left[\frac{\sqrt{2}}{2} x' - \frac{\sqrt{2}}{2} y' \right] \cdot \left[\frac{\sqrt{2}}{2} x' + \frac{\sqrt{2}}{2} y' \right]$$

$$+ 2\sqrt{2} \left[\frac{\sqrt{2}}{2} x' - \frac{\sqrt{2}}{2} y' \right] - 2\sqrt{2} \left[\frac{\sqrt{2}}{2} x' + \frac{\sqrt{2}}{2} y' \right] = 0.$$

Expanding these terms,

$$\frac{1}{2} x'^2 - x'y' + \frac{1}{2} y'^2 + \frac{1}{2} x'^2 + x'y' + \frac{1}{2} y'^2 + x'^2 - y'^2$$

$$+ 2x' - 2y' - 2x' - 2y' = 0.$$

Collecting terms gives

$$2x'^2 - 4y' = 0$$

$$x'^2 - 2y' = 0,$$

or, finally,

$$x'^2 = 2y',$$

the equation of a parabola. The graph is shown in Figure 27.   ∎

Equations written in the general form $Ax^2 + Cy^2 + Dx + Ey + F = 0$ were graphed in the previous section. In Example 1, the rotation of axes eliminated the xy-term resulting in a general second-degree equation. This suggests that to graph an equation that has an xy-term, it is necessary to find an appropriate angle of rotation to eliminate the xy-term. The appropriate angle of rotation can be determined by using the following result. The proof is quite lengthy and is not presented here.

Angle of Rotation

The xy-term is removed from the equation

$$Ax^2 + Bxy + Cy^2 + Dx + Ey + F = 0$$

by a rotation of the axes through an angle θ, $0° < \theta < 90°$, where

$$\cot 2\theta = \frac{A - C}{B}.$$

This result can be used to find the appropriate angle of rotation, θ. Then, to get the rotation equations, first find $\sin \theta$ and $\cos \theta$. The following example illustrates a way to obtain $\sin \theta$ and $\cos \theta$ from $\cot 2\theta$ without first identifying the angle θ.

Example 2   Rotate the axes and graph $52x^2 - 72xy + 73y^2 = 200$.

In this equation, $A = 52$, $B = -72$, and $C = 73$, so that

$$\cot 2\theta = \frac{52 - 73}{-72} = \frac{-21}{-72} = \frac{7}{24}.$$

To find $\sin \theta$ and $\cos \theta$, use the trigonometric identities

$$\sin \theta = \sqrt{\frac{1 - \cos 2\theta}{2}} \quad \text{and} \quad \cos \theta = \sqrt{\frac{1 + \cos 2\theta}{2}}.$$

Sketch a right triangle and label it as in Figure 28, to see that $\cos 2\theta = 7/25$. (Recall: For the two quadrants of concern, cosine and cotangent have the same sign.)

$$\sin \theta = \sqrt{\frac{1 - 7/25}{2}} = \sqrt{\frac{9}{25}} = \frac{3}{5}$$

$$\cos \theta = \sqrt{\frac{1 + 7/25}{2}} = \sqrt{\frac{16}{25}} = \frac{4}{5}$$

Use these values for $\sin \theta$ and $\cos \theta$ to get the rotation equations

$$x = \frac{4}{5}x' - \frac{3}{5}y' \quad \text{and} \quad y = \frac{3}{5}x' + \frac{4}{5}y'.$$

Substituting these values of x and y into the original equation yields

$$52\left[\frac{4}{5}x' - \frac{3}{5}y'\right]^2 - 72\left[\frac{4}{5}x' - \frac{3}{5}y'\right]\left[\frac{3}{5}x' + \frac{4}{5}y'\right] + 73\left[\frac{3}{5}x' + \frac{4}{5}y'\right]^2$$

$$= 200,$$

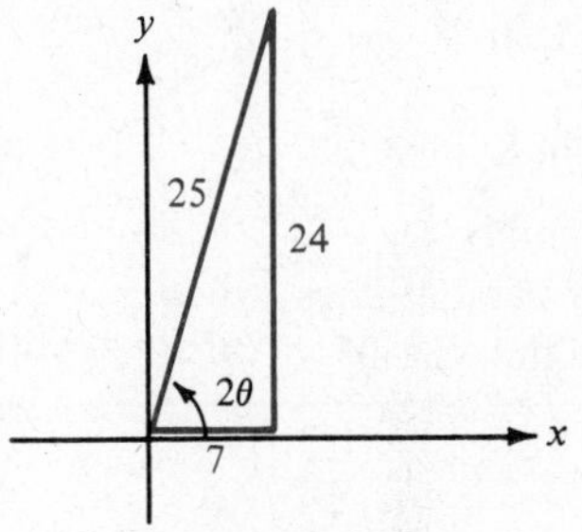

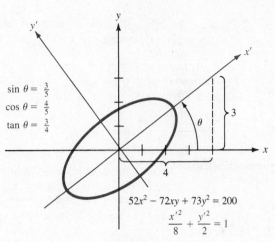

Figure 28

Figure 29

which becomes

$$52\left[\frac{16}{25}x'^2 - \frac{24}{25}x'y' + \frac{9}{25}y'^2\right] - 72\left[\frac{12}{25}x'^2 + \frac{7}{25}x'y' - \frac{12}{25}y'^2\right]$$

$$+ 73\left[\frac{9}{25}x'^2 + \frac{24}{25}x'y' + \frac{16}{25}y'^2\right] = 200.$$

Combining terms gives

$$25'^2 + 100y'^2 = 200.$$

Divide both sides by 200 to get

$$\frac{x'^2}{8} + \frac{y'^2}{2} = 1,$$

an equation of an ellipse having intercepts $(0, \sqrt{2})$, $(0, -\sqrt{2})$, $(2\sqrt{2}, 0)$, and $(-2\sqrt{2}, 0)$ with respect to the $x'y'$-system. The graph of this ellipse is shown in Figure 29. Find θ from the fact that

$$\frac{\sin \theta}{\cos \theta} = \frac{3/5}{4/5} = \frac{3}{4} = \tan \theta,$$

from which $\theta = 37°$. As shown in Figure 29, to locate the x'-axis without finding the value of θ, use the fact that

$$\tan \theta = \text{slope of } x'\text{-axis}$$

$$= \frac{\Delta y}{\Delta x}$$

$$= \frac{3}{4}. \quad \blacksquare$$

The following result tells how to decide from the equation which type of graph to expect.

<table>
<tr><td>

Equations of Conics with xy-Term

</td><td>

If the second-degree equation
$$Ax^2 + Bxy + Cy^2 + Dx + Ey + F = 0$$
has a graph, it will be one of the following:

(a) a circle or an ellipse (or a point)   if $B^2 - 4AC < 0$;
(b) a parabola (or a line or two parallel lines) if $B^2 - 4AC = 0$;
(c) a hyperbola (or two intersecting lines)   if $B^2 - 4AC > 0$.

</td></tr>
</table>

8.5 Exercises

Use the last result in this section to predict the graph of each of the following second-degree equations.

1. $4x^2 + 3y^2 + 2xy - 5x = 8$

2. $x^2 + 2xy - 3y^2 + 2y = 12$

3. $2x^2 + 3xy - 4y^2 = 0$

4. $x^2 - 2xy + y^2 + 4x - 8y = 0$

5. $4x^2 + 4xy + y^2 + 15 = 0$

6. $-x^2 + 2xy - y^2 + 16 = 0$

Find the angle of rotation θ that will remove the xy-term in each of the following equations.

7. $2x^2 + \sqrt{3}xy + y^2 + x = 5$

8. $4\sqrt{3}x^2 + xy + 3\sqrt{3}y^2 = 10$

9. $3x^2 + \sqrt{3}xy + 4y^2 + 2x - 3y = 12$

10. $4x^2 + 2xy + 2y^2 + x - 7 = 0$

11. $x^2 - 4xy + 5y^2 = 18$

12. $3\sqrt{3}x^2 - 2xy + \sqrt{3}y^2 = 25$

Use the given angle of rotation to remove the xy-term and graph each of the following equations.

13. $x^2 - xy + y^2 = 6; \quad \theta = 45°$

14. $2x^2 - xy + 2y^2 = 25; \quad \theta = 45°$

15. $8x^2 - 4xy + 5y^2 = 36; \quad \sin \theta = 2/\sqrt{5}$

16. $5y^2 + 12xy = 10; \quad \sin \theta = 3/\sqrt{13}$

Remove the xy-term from each of the following equations by performing a suitable rotation. Graph each equation.

17. $3x^2 - 2xy + 3y^2 = 8$

18. $x^2 + xy + y^2 = 3$

19. $x^2 - 4xy + y^2 = -5$

20. $x^2 + 2xy + y^2 + 4\sqrt{2}x - 4\sqrt{2}y = 0$

21. $7x^2 + 6\sqrt{3}xy + 13y^2 = 64$

22. $7x^2 + 2\sqrt{3}xy + 5y^2 = 24$

23. $3x^2 - 2\sqrt{3}xy + y^2 - 2x - 2\sqrt{3}y = 0$

24. $2x^2 + 2\sqrt{3}xy + 4y^2 = 5$

In each of the following equations, remove the xy-term by rotation. Then translate the axes and sketch the graph.

25. $x^2 + 3xy + y^2 - 5\sqrt{2}y = 15$

26. $x^2 - \sqrt{3}xy + 2\sqrt{3}x - 3y - 3 = 0$

27. $4x^2 + 4xy + y^2 - 24x + 38y - 19 = 0$

28. $12x^2 + 24xy + 19y^2 - 12x - 40y + 31 = 0$

29. $16x^2 + 24xy + 9y^2 - 130x + 90y = 0$

30. $9x^2 - 6xy + y^2 - 12\sqrt{10}x - 36\sqrt{10}y = 0$

Chapter 8 Summary

Key Words

slope	vertices	conic sections
angle of inclination	major axis	eccentricity
point-slope form	hyperbola	translation equations
slope-intercept form	asymptotes	general second-degree
angle between two lines	parabola	equation
ellipse	directrix	rotation equations
foci	axis	

Review Exercises

Find the slope and angle of inclination, to the nearest tenth of a degree, for each of the following lines.

1. Through $(5, 6)$ and $(5, -2)$

2. Through $(0, -7)$ and $(3, -7)$

3. $9x - 4y = 2$

4. $11x + 2y = 3$

5. $x - 5y = 0$

6. $y = x$

Write the equation for each of the following lines in the form $ax + by = c$.

7. Through $(-2, 4)$ and $(1, 3)$

8. Through $(-2/3, -1)$ and $(0, 4)$

9. Through $(3, -5)$, with slope -2

10. Through $(1/5, 1/3)$, with slope $-1/2$

11. x-intercept -3, y-intercept 5

12. No x-intercept, y-intercept $3/4$

13. Through $(2, -1)$, parallel to $3x - y = 1$

14. Through $(0, 5)$, perpendicular to $8x + 5y = 3$

Give the acute angle between each pair of lines to the nearest tenth of a degree.

15. $y = \sqrt{3}x - 1$, $x + y = 4$

16. $2y - 2x = 5$, $3y = 3x + 8$

17. $4y - x = 7$, $3x + y = 12$

18. $x - 2y = 5$, $5x + 3y = 7$

Graph each of the following equations. Give the focus, directrix, and axis of each parabola, the coordinates of the vertices and foci for each ellipse, and the equations of the asymptotes for each hyperbola.

19. $y^2 = 3x$

20. $y^2 = -\dfrac{1}{4}x$

21. $x^2 - 5y = 0$

22. $2x^2 = y$

23. $\dfrac{x^2}{4} + \dfrac{y^2}{9} = 36$

24. $9x^2 + 16y^2 = 144$

25. $25x^2 - 9y^2 = 225$

26. $\dfrac{y^2}{36} - \dfrac{x^2}{9} = 1$

Write an equation for each parabola with vertex at the origin.

27. Focus $(0, -4)$

28. Focus $(3, 0)$

29. Through $(1, 7)$, opening upward

30. Through $(-2, 3)$, opening to the left

Find an equation for each conic section with center at the origin.

31. Ellipse; x-intercept at $(4, 0)$, focus at $(-3, 0)$

32. Ellipse; vertex at $(0, 5)$, focus at $(0, 2)$

33. Hyperbola; y-intercept at $(0, -2)$, passing through $(2, 3)$

34. Hyperbola; focus at $(0, -7)$, transverse axis of length 10

For the following equations, name the conic, and give its eccentricity.

35. $4x^2 - 9y^2 = 36$ **36.** $y^2 + 4x^2 = 4$ **37.** $y^2 + x = 2$

38. $3x^2 - 4y^2 = 24$ **39.** $x^2 + y^2 = 16$ **40.** $2x^2 - y = 0$

Write an equation for each conic.

41. Center at the origin, focus $(0, 7)$, $e = 2/3$ **42.** Center at the origin, $e = 5/4$, focus $(-2, 0)$

43. Vertex $(0, 0)$, focus $(1, 0)$, $e = 1$ **44.** Vertex $(3, 0)$, center at the origin, $e = 1/4$

45. Vertical transverse axis 8 units long, $e = 3/2$, center at $(0, 0)$

46. Vertex $(0, 0)$, $e = 1$, focus $(0, -2)$

Translate the axes and sketch the graph of each conic.

47. $\dfrac{(x - 2)^2}{4} + \dfrac{(y - 3)^2}{25} = 1$ **48.** $\dfrac{(x + 3)^2}{9} - (y - 1)^2 = 1$

49. $(y + 2)^2 = 4(x - 3)$ **50.** $(x - 1)^2 = 2(y - 4)$

51. $9(x - 1)^2 + 4(y - 2)^2 = 36$ **52.** $16(x - 2)^2 - 4(y + 3)^2 = 64$

53. $x^2 + 2y^2 - 4x + 8y - 4 = 0$ **54.** $4x^2 - y^2 + 8x - 10y = 0$

55. $3x^2 - 6x + y + 1 = 0$ **56.** $2x - y^2 + 4y = 0$

Determine the type of graph for each equation.

57. $3xy - y^2 - 5 = 0$ **58.** $4x^2 - 2xy + y^2 + 2y - 6 = 0$

59. $x^2 - xy + 2x - 3y = 0$ **60.** $x^2 + 2xy - y^2 + 8 = 0$

Find the angle of rotation that will eliminate the xy-term.

61. $2\sqrt{3}x^2 + xy + \sqrt{3}y^2 + y = 2$ **62.** $5x^2 + 4xy + y^2 + 2x = 5$

Graph each conic by rotating the axis.

63. $24xy - 7y^2 + 36 = 0$ **64.** $5x^2 + 8xy + 5y^2 = 9$ **65.** $-3xy + 9\sqrt{2}x = 15$

Table IA Natural Logarithms and Powers of e

x	e^x	e^{-x}	ln x	x	e^x	e^{-x}	ln x
0.00	1.00000	1.00000		1.60	4.95302	0.20189	0.4700
0.01	1.01005	0.99004	−4.6052	1.70	5.47394	0.18268	0.5306
0.02	1.02020	0.98019	−3.9120	1.80	6.04964	0.16529	0.5878
0.03	1.03045	0.97044	−3.5066	1.90	6.68589	0.14956	0.6419
0.04	1.04081	0.96078	−3.2189	2.00	7.38905	0.13533	0.6931
0.05	1.05127	0.95122	−2.9957				
0.06	1.06183	0.94176	−2.8134	2.10	8.16616	0.12245	0.7419
0.07	1.07250	0.93239	−2.6593	2.20	9.02500	0.11080	0.7885
0.08	1.08328	0.92311	−2.5257	2.30	9.97417	0.10025	0.8329
0.09	1.09417	0.91393	−2.4079	2.40	11.02316	0.09071	0.8755
0.10	1.10517	0.90483	−2.3026	2.50	12.18248	0.08208	0.9163
				2.60	13.46372	0.07427	0.9555
0.11	1.11628	0.89583	−2.2073	2.70	14.87971	0.06720	0.9933
0.12	1.12750	0.88692	−2.1203	2.80	16.44463	0.06081	1.0296
0.13	1.13883	0.87810	−2.0402	2.90	18.17412	0.05502	1.0647
0.14	1.15027	0.86936	−1.9661	3.00	20.08551	0.04978	1.0986
0.15	1.16183	0.86071	−1.8971				
0.16	1.17351	0.85214	−1.8326	3.50	33.11545	0.03020	1.2528
0.17	1.18530	0.84366	−1.7720	4.00	54.59815	0.01832	1.3863
0.18	1.19722	0.83527	−1.7148	4.50	90.01713	0.01111	1.5041
0.19	1.20925	0.82696	−1.6607				
				5.00	148.41316	0.00674	1.6094
0.20	1.22140	0.81873	−1.6094	5.50	224.69193	0.00409	1.7047
0.30	1.34985	0.74081	−1.2040				
0.40	1.49182	0.67032	−0.9163	6.00	403.42879	0.00248	1.7918
0.50	1.64872	0.60653	−0.6931	6.50	665.14163	0.00150	1.8718
0.60	1.82211	0.54881	−0.5108				
0.70	2.01375	0.49658	−0.3567	7.00	1096.63316	0.00091	1.9459
0.80	2.22554	0.44932	−0.2231	7.50	1808.04241	0.00055	2.0149
0.90	2.45960	0.40656	−0.1054	8.00	2980.95799	0.00034	2.0794
				8.50	4914.76884	0.00020	2.1401
1.00	2.71828	0.36787	0.0000				
1.10	3.00416	0.33287	0.0953	9.00	8130.08392	0.00012	2.1972
1.20	3.32011	0.30119	0.1823	9.50	13359.72683	0.00007	2.2513
1.30	3.66929	0.27253	0.2624				
1.40	4.05519	0.24659	0.3365	10.00	22026.46579	0.00005	2.3026
1.50	4.48168	0.22313	0.4055				

Table IB Additional Natural Logarithms

x	ln x	x	ln x	x	ln x	x	ln x	x	ln x
		4.5	1.5041	6.0	1.7918	7.5	2.0149	9.0	2.1972
3.1	1.1314	4.6	1.5261	6.1	1.8083	7.6	2.0281	9.1	2.2083
3.2	1.1632	4.7	1.5476	6.2	1.8245	7.7	2.0412	9.2	2.2192
3.3	1.1939	4.8	1.5686	6.3	1.8405	7.8	2.0541	9.3	2.2300
3.4	1.2238	4.9	1.5892	6.4	1.8563	7.9	2.0669	9.4	2.2407
3.5	1.2528	5.0	1.6094	6.5	1.8718	8.0	2.0794	9.5	2.2513
3.6	1.2809	5.1	1.6292	6.6	1.8871	8.1	2.0919	9.6	2.2618
3.7	1.3083	5.2	1.6487	6.7	1.9021	8.2	2.1041	9.7	2.2721
3.8	1.3350	5.3	1.6677	6.8	1.9169	8.3	2.1163	9.8	2.2824
3.9	1.3610	5.4	1.6864	6.9	1.9315	8.4	2.1282	9.9	2.2925
4.0	1.3863	5.5	1.7047	7.0	1.9459	8.5	2.1401	10.0	2.3026
4.1	1.4110	5.6	1.7228	7.1	1.9601	8.6	2.1518	11.0	2.3979
4.2	1.4351	5.7	1.7405	7.2	1.9741	8.7	2.1633	12.0	2.4849
4.3	1.4586	5.8	1.7579	7.3	1.9879	8.8	2.1748	13.0	2.5649
4.4	1.4816	5.9	1.7750	7.4	2.0015	8.9	2.1861	14.0	2.6391

Table 2 Common Logarithms

n	0	1	2	3	4	5	6	7	8	9
1.0	.0000	.0043	.0086	.0128	.0170	.0212	.0253	.0294	.0334	.0374
1.1	.0414	.0453	.0492	.0531	.0569	.0607	.0645	.0682	.0719	.0755
1.2	.0792	.0828	.0864	.0899	.0934	.0969	.1004	.1038	.1072	.1106
1.3	.1139	.1173	.1206	.1239	.1271	.1303	.1335	.1367	.1399	.1430
1.4	.1461	.1492	.1523	.1553	.1584	.1614	.1644	.1673	.1703	.1732
1.5	.1761	.1790	.1818	.1847	.1875	.1903	.1931	.1959	.1987	.2014
1.6	.2041	.2068	.2095	.2122	.2148	.2175	.2201	.2227	.2253	.2279
1.7	.2304	.2330	.2355	.2380	.2405	.2430	.2455	.2480	.2504	.2529
1.8	.2553	.2577	.2601	.2625	.2648	.2672	.2695	.2718	.2742	.2765
1.9	.2788	.2810	.2833	.2856	.2878	.2900	.2923	.2945	.2967	.2989
2.0	.3010	.3032	.3054	.3075	.3096	.3118	.3139	.3160	.3181	.3201
2.1	.3222	.3243	.3263	.3284	.3304	.3324	.3345	.3365	.3385	.3404
2.2	.3424	.3444	.3464	.3483	.3502	.3522	.3541	.3560	.3579	.3598
2.3	.3617	.3636	.3655	.3674	.3692	.3711	.3729	.3747	.3766	.3784
2.4	.3802	.3820	.3838	.3856	.3874	.3892	.3909	.3927	.3945	.3962
2.5	.3979	.3997	.4014	.4031	.4048	.4065	.4082	.4099	.4116	.4133
2.6	.4150	.4166	.4183	.4200	.4216	.4232	.4249	.4265	.4281	.4298
2.7	.4314	.4330	.4346	.4362	.4378	.4393	.4409	.4425	.4440	.4456
2.8	.4472	.4487	.4502	.4518	.4533	.4548	.4564	.4579	.4594	.4609
2.9	.4624	.4639	.4654	.4669	.4683	.4698	.4713	.4728	.4742	.4757
3.0	.4771	.4786	.4800	.4814	.4829	.4843	.4857	.4871	.4886	.4900
3.1	.4914	.4928	.4942	.4955	.4969	.4983	.4997	.5011	.5024	.5038
3.2	.5051	.5065	.5079	.5092	.5105	.5119	.5132	.5145	.5159	.5172
3.3	.5185	.5198	.5211	.5224	.5237	.5250	.5263	.5276	.5289	.5302
3.4	.5315	.5328	.5340	.5353	.5366	.5378	.5391	.5403	.5416	.5428
3.5	.5441	.5453	.5465	.5478	.5490	.5502	.5514	.5527	.5539	.5551
3.6	.5563	.5575	.5587	.5599	.5611	.5623	.5635	.5647	.5658	.5670
3.7	.5682	.5694	.5705	.5717	.5729	.5740	.5752	.5763	.5775	.5786
3.8	.5798	.5809	.5821	.5832	.5843	.5855	.5866	.5877	.5888	.5899
3.9	.5911	.5922	.5933	.5944	.5955	.5966	.5977	.5988	.5999	.6010
4.0	.6021	.6031	.6042	.6053	.6064	.6075	.6085	.6096	.6107	.6117
4.1	.6128	.6138	.6149	.6160	.6170	.6180	.6191	.6201	.6212	.6222
4.2	.6232	.6243	.6253	.6263	.6274	.6284	.6294	.6304	.6314	.6325
4.3	.6335	.6345	.6355	.6365	.6375	.6385	.6395	.6405	.6415	.6425
4.4	.6435	.6444	.6454	.6464	.6474	.6484	.6493	.6503	.6513	.6522
4.5	.6532	.6542	.6551	.6561	.6571	.6580	.6590	.6599	.6609	.6618
4.6	.6628	.6637	.6646	.6656	.6665	.6675	.6684	.6693	.6702	.6712
4.7	.6721	.6730	.6739	.6749	.6758	.6767	.6776	.6785	.6794	.6803
4.8	.6812	.6821	.6830	.6839	.6848	.6857	.6866	.6875	.6884	.6893
4.9	.6902	.6911	.6920	.6928	.6937	.6946	.6955	.6964	.6972	.6981
5.0	.6990	.6998	.7007	.7016	.7024	.7033	.7042	.7050	.7059	.7067
5.1	.7076	.7084	.7093	.7101	.7110	.7118	.7126	.7135	.7143	.7152
5.2	.7160	.7168	.7177	.7185	.7193	.7202	.7210	.7218	.7226	.7235
5.3	.7243	.7251	.7259	.7267	.7275	.7284	.7292	.7300	.7308	.7316
5.4	.7324	.7332	.7340	.7348	.7356	.7364	.7372	.7380	.7388	.7396
n	0	1	2	3	4	5	6	7	8	9

Table 2   351

Table 2   Common Logarithms (continued)

n	0	1	2	3	4	5	6	7	8	9
5.5	.7404	.7412	.7419	.7427	.7435	.7443	.7451	.7459	.7466	.7474
5.6	.7482	.7490	.7497	.7505	.7513	.7520	.7528	.7536	.7543	.7551
5.7	.7559	.7566	.7574	.7582	.7589	.7597	.7604	.7612	.7619	.7627
5.8	.7634	.7642	.7649	.7657	.7664	.7672	.7679	.7686	.7694	.7701
5.9	.7709	.7716	.7723	.7731	.7738	.7745	.7752	.7760	.7767	.7774
6.0	.7782	.7789	.7796	.7803	.7810	.7818	.7825	.7832	.7839	.7846
6.1	.7853	.7860	.7868	.7875	.7882	.7889	.7896	.7903	.7910	.7917
6.2	.7924	.7931	.7938	.7945	.7952	.7959	.7966	.7973	.7980	.7987
6.3	.7993	.8000	.8007	.8014	.8021	.8028	.8035	.8041	.8048	.8055
6.4	.8062	.8069	.8075	.8082	.8089	.8096	.8102	.8109	.8116	.8122
6.5	.8129	.8136	.8142	.8149	.8156	.8162	.8169	.8176	.8182	.8189
6.6	.8195	.8202	.8209	.8215	.8222	.8228	.8235	.8241	.8248	.8254
6.7	.8261	.8267	.8274	.8280	.8287	.8293	.8299	.8306	.8312	.8319
6.8	.8325	.8331	.8338	.8344	.8351	.8357	.8363	.8370	.8376	.8382
6.9	.8388	.8395	.8401	.8407	.8414	.8420	.8426	.8432	.8439	.8445
7.0	.8451	.8457	.8463	.8470	.8476	.8482	.8488	.8494	.8500	.8506
7.1	.8513	.8519	.8525	.8531	.8537	.8543	.8549	.8555	.8561	.8567
7.2	.8573	.8579	.8585	.8591	.8597	.8603	.8609	.8615	.8621	.8627
7.3	.8633	.8639	.8645	.8651	.8657	.8663	.8669	.8675	.8681	.8686
7.4	.8692	.8698	.8704	.8710	.8716	.8722	.8727	.8733	.8739	.8745
7.5	.8751	.8756	.8762	.8768	.8774	.8779	.8785	.8791	.8797	.8802
7.6	.8808	.8814	.8820	.8825	.8831	.8837	.8842	.8848	.8854	.8859
7.7	.8865	.8871	.8876	.8882	.8887	.8893	.8899	.8904	.8910	.8915
7.8	.8921	.8927	.8932	.8938	.8943	.8949	.8954	.8960	.8965	.8971
7.9	.8976	.8982	.8987	.8993	.8998	.9004	.9009	.9015	.9020	.9025
8.0	.9031	.9036	.9042	.9047	.9053	.9058	.9063	.9069	.9074	.9079
8.1	.9085	.9090	.9096	.9101	.9106	.9112	.9117	.9122	.9128	.9133
8.2	.9138	.9143	.9149	.9154	.9159	.9165	.9170	.9175	.9180	.9186
8.3	.9191	.9196	.9201	.9206	.9212	.9217	.9222	.9227	.9232	.9238
8.4	.9243	.9248	.9253	.9258	.9263	.9269	.9274	.9279	.9284	.9289
8.5	.9294	.9299	.9304	.9309	.9315	.9320	.9325	.9330	.9335	.9340
8.6	.9345	.9350	.9355	.9360	.9365	.9370	.9375	.9380	.9385	.9390
8.7	.9395	.9400	.9405	.9410	.9415	.9420	.9425	.9430	.9435	.9440
8.8	.9445	.9450	.9455	.9460	.9465	.9469	.9474	.9479	.9484	.9489
8.9	.9494	.9499	.9504	.9509	.9513	.9518	.9523	.9528	.9533	.9538
9.0	.9542	.9547	.9552	.9557	.9562	.9566	.9571	.9576	.9581	.9586
9.1	.9590	.9595	.9600	.9605	.9609	.9614	.9619	.9624	.9628	.9633
9.2	.9638	.9643	.9647	.9652	.9657	.9661	.9666	.9671	.9675	.9680
9.3	.9685	.9689	.9694	.9699	.9703	.9708	.9713	.9717	.9722	.9727
9.4	.9731	.9736	.9741	.9745	.9750	.9754	.9759	.9763	.9768	.9773
9.5	.9777	.9782	.9786	.9791	.9795	.9800	.9805	.9809	.9814	.9818
9.6	.9823	.9827	.9832	.9836	.9841	.9845	.9850	.9854	.9859	.9863
9.7	.9868	.9872	.9877	.9881	.9886	.9890	.9894	.9899	.9903	.9908
9.8	.9912	.9917	.9921	.9926	.9930	.9934	.9939	.9943	.9948	.9952
9.9	.9956	.9961	.9965	.9969	.9974	.9978	.9983	.9987	.9991	.9996
n	0	1	2	3	4	5	6	7	8	9

Table 3 Trigonometric Functions in Degrees and Radians

θ (degrees)	θ (radians)	sin θ	cos θ	tan θ	cot θ	sec θ	csc θ	(radians)	(degrees)
0°00'	0000	0000	1.0000	0000	—	1.000	—	1.5708	90°00'
10	0029	0029	1.0000	0029	343.8	1.000	343.8	1.5679	50
20	0058	0058	1.0000	0058	171.9	1.000	171.9	1.5650	40
30	0087	0087	1.0000	0087	114.6	1.000	114.6	1.5621	30
40	0116	0116	9999	0116	85.94	1.000	85.95	1.5592	20
50	0145	0145	9999	0145	68.75	1.000	68.76	1.5563	10
1°00'	0175	0175	9998	0175	57.29	1.000	57.30	1.5533	89°00'
10	0204	0204	9998	0204	49.10	1.000	49.11	1.5504	50
20	0233	0233	9997	0233	42.96	1.000	42.98	1.5475	40
30	0262	0262	9997	0262	38.19	1.000	38.20	1.5446	30
40	0291	0291	9996	0291	34.37	1.000	34.38	1.5417	20
50	0320	0320	9995	0320	31.24	1.001	31.26	1.5388	10
2°00'	0349	0349	9994	0349	28.64	1.001	28.65	1.5359	88°00'
10	0378	0378	9993	0378	26.43	1.001	26.45	1.5330	50
20	0407	0407	9990	0407	24.54	1.001	24.56	1.5301	40
30	0436	0436	9990	0437	22.90	1.001	22.93	1.5272	30
40	0465	0465	9989	0466	21.47	1.001	21.49	1.5243	20
50	0495	0494	9988	0495	20.21	1.001	20.23	1.5213	10
3°00'	0524	0523	9986	0524	19.08	1.001	19.11	1.5184	87°00'
10	0553	0552	9985	0553	18.07	1.002	18.10	1.5155	50
20	0582	0581	9983	0582	17.17	1.002	17.20	1.5126	40
30	0611	0610	9981	0612	16.35	1.002	16.38	1.5097	30
40	0640	0640	9980	0641	15.60	1.002	15.64	1.5068	20
50	0669	0669	9978	0670	14.92	1.002	14.96	1.5039	10
4°00'	0698	0698	9976	0699	14.30	1.002	14.34	1.5010	86°00'
10	0727	0727	9974	0729	13.73	1.003	13.76	1.4981	50
20	0756	0756	9971	0758	13.20	1.003	13.23	1.4952	40
30	0785	0785	9969	0787	12.71	1.003	12.75	1.4923	30
40	0814	0814	9967	0816	12.25	1.003	12.29	1.4893	20
50	0844	0843	9964	0846	11.83	1.004	11.87	1.4864	10
5°00'	0873	0872	9962	0875	11.43	1.004	11.47	1.4835	85°00'
10	0902	0901	9959	0904	11.06	1.004	11.10	1.4806	50
20	0931	0929	9957	0934	10.71	1.004	10.76	1.4777	40
30	0960	0958	9954	0963	10.39	1.005	10.43	1.4748	30
40	0989	0987	9951	0992	10.08	1.005	10.13	1.4719	20
50	1018	1016	9948	1022	9.788	1.005	9.839	1.4690	10
6°00'	1047	1045	9945	1051	9.514	1.006	9.567	1.4661	84°00'
10	1076	1074	9942	1080	9.255	1.006	9.309	1.4632	50
20	1105	1103	9939	1110	9.010	1.006	9.065	1.4603	40
30	1134	1132	9936	1139	8.777	1.006	8.834	1.4573	30
40	1164	1161	9932	1169	8.556	1.007	8.614	1.4544	20
50	1193	1190	9929	1198	8.345	1.007	8.405	1.4515	10
	cos θ	sin θ	cot θ	tan θ	csc θ	sec θ	(radians)	(degrees)	

θ (degrees)	θ (radians)	sin θ	cos θ	tan θ	cot θ	sec θ	csc θ	(radians)	(degrees)
7°00'	1222	1219	9925	1228	8.144	1.008	8.206	1.4486	83°00'
10	1251	1248	9922	1257	7.953	1.008	8.016	1.4457	50
20	1280	1276	9918	1287	7.770	1.009	7.834	1.4428	40
30	1309	1305	9914	1317	7.596	1.009	7.661	1.4399	30
40	1338	1334	9911	1346	7.429	1.009	7.496	1.4370	20
50	1376	1363	9907	1376	7.269	1.009	7.337	1.4341	10
8°00'	1396	1392	9903	1405	7.115	1.010	7.185	1.4312	82°00'
10	1425	1421	9899	1435	6.968	1.010	7.040	1.4283	50
20	1454	1449	9894	1465	6.827	1.011	6.900	1.4254	40
30	1484	1478	9890	1495	6.691	1.011	6.765	1.4224	30
40	1513	1507	9886	1524	6.561	1.012	6.636	1.4195	20
50	1542	1536	9881	1554	6.435	1.012	6.512	1.4166	10
9°00'	1571	1564	9877	1584	6.314	1.012	6.392	1.4137	81°00'
10	1600	1593	9872	1614	6.197	1.013	6.277	1.4108	50
20	1629	1622	9868	1644	6.084	1.013	6.166	1.4079	40
30	1658	1650	9863	1673	5.976	1.014	6.059	1.4050	30
40	1687	1679	9858	1703	5.871	1.014	5.955	1.4021	20
50	1716	1708	9853	1733	5.769	1.015	5.855	1.3992	10
10°00'	1745	1736	9848	1763	5.671	1.015	5.759	1.3963	80°00'
10	1774	1765	9843	1793	5.576	1.016	5.665	1.3934	50
20	1804	1794	9838	1823	5.485	1.016	5.575	1.3904	40
30	1833	1822	9833	1853	5.396	1.017	5.487	1.3875	30
40	1862	1851	9827	1883	5.309	1.018	5.403	1.3846	20
50	1891	1880	9822	1914	5.226	1.018	5.320	1.3817	10
11°00'	1920	1908	9816	1944	5.145	1.019	5.241	1.3788	79°00'
10	1949	1937	9811	1974	5.066	1.019	5.164	1.3759	50
20	1978	1965	9805	2004	4.989	1.020	5.089	1.3730	40
30	2007	1994	9799	2035	4.915	1.020	5.016	1.3701	30
40	2036	2022	9793	2065	4.843	1.021	4.945	1.3672	20
50	2065	2051	9787	2095	4.773	1.022	4.876	1.3643	10
12°00'	2094	2079	9781	2126	4.705	1.022	4.810	1.3614	78°00'
10	2123	2108	9775	2156	4.638	1.023	4.745	1.3584	50
20	2153	2136	9769	2186	4.574	1.024	4.682	1.3555	40
30	2182	2164	9763	2217	4.511	1.024	4.620	1.3526	30
40	2211	2193	9757	2247	4.449	1.025	4.560	1.3497	20
50	2240	2221	9750	2278	4.390	1.026	4.502	1.3468	10
13°00'	2269	2250	9744	2309	4.331	1.026	4.445	1.3439	77°00'
10	2298	2278	9737	2339	4.275	1.027	4.390	1.3410	50
20	2327	2306	9730	2370	4.219	1.028	4.336	1.3381	40
30	2356	2334	9724	2401	4.165	1.028	4.284	1.3352	30
40	2385	2363	9717	2432	4.113	1.029	4.232	1.3323	20
50	2414	2391	9710	2462	4.061	1.030	4.182	1.3294	10
	cos θ	sin θ	cot θ	tan θ	csc θ	sec θ	(radians)	(degrees)	

Table 3   353

Table 3 Trigonometric Functions in Degrees and Radians (continued)

θ (degrees)	θ (radians)	sin θ	cos θ	tan θ	cot θ	sec θ	csc θ	θ (radians)	θ (degrees)
21°00'	3665	3584	9336	3839	2 605	1 071	2 790	1 2043	69°00'
10	3694	3611	9325	3872	2 583	1 072	2 769	1 2014	50
20	3723	3638	9315	3906	2 560	1 074	2 749	1 1985	40
30	3752	3665	9304	3939	2 539	1 075	2 729	1 1956	30
40	3782	3692	9293	3973	2 517	1 076	2 709	1 1926	20
50	3811	3719	9283	4006	2 496	1 077	2 689	1 1897	10
22°00'	3840	3746	9272	4040	2 475	1 079	2 669	1 1868	68°00'
10	3869	3773	9261	4074	2 455	1 080	2 650	1 1839	50
20	3898	3800	9250	4108	2 434	1 081	2 632	1 1810	40
30	3927	3827	9239	4142	2 414	1 082	2 613	1 1781	30
40	3956	3854	9228	4176	2 394	1 084	2 595	1 1752	20
50	3985	3881	9216	4210	2 375	1 085	2 577	1 1723	10
23°00'	4014	3907	9205	4245	2 356	1 086	2 559	1 1694	67°00'
10	4043	3934	9194	4279	2 337	1 088	2 542	1 1665	50
20	4072	3961	9182	4314	2 318	1 089	2 525	1 1636	40
30	4102	3987	9171	4348	2 300	1 090	2 508	1 1606	30
40	4131	4014	9159	4383	2 282	1 092	2 491	1 1577	20
50	4160	4041	9147	4417	2 264	1 093	2 475	1 1548	10
24°00'	4189	4067	9135	4452	2 246	1 095	2 459	1 1519	66°00'
10	4218	4094	9124	4487	2 229	1 096	2 443	1 1490	50
20	4247	4120	9112	4522	2 211	1 097	2 427	1 1461	40
30	4276	4147	9100	4557	2 194	1 099	2 411	1 1432	30
40	4305	4173	9088	4592	2 177	1 100	2 396	1 1403	20
50	4334	4200	9075	4628	2 161	1 102	2 381	1 1374	10
25°00'	4363	4226	9063	4663	2 145	1 103	2 366	1 1345	65°00'
10	4392	4253	9051	4699	2 128	1 105	2 352	1 1316	50
20	4422	4279	9038	4734	2 112	1 106	2 337	1 1286	40
30	4451	4305	9026	4770	2 097	1 108	2 323	1 1257	30
40	4480	4331	9013	4806	2 081	1 109	2 309	1 1228	20
50	4509	4358	9001	4841	2 066	1 111	2 295	1 1199	10
26°00'	4538	4384	8988	4877	2 050	1 113	2 281	1 1170	64°00'
10	4567	4410	8975	4913	2 035	1 114	2 268	1 1141	50
20	4596	4436	8962	4950	2 020	1 116	2 254	1 1112	40
30	4625	4462	8949	4986	2 006	1 117	2 241	1 1083	30
40	4654	4488	8936	5022	1 991	1 119	2 228	1 1054	20
50	4683	4514	8923	5059	1 977	1 121	2 215	1 1025	10
27°00'	4712	4540	8910	5095	1 963	1 122	2 203	1 0996	63°00'
10	4741	4566	8897	5132	1 949	1 124	2 190	1 0966	50
20	4771	4592	8884	5169	1 935	1 126	2 178	1 0937	40
30	4800	4617	8870	5206	1 921	1 127	2 166	1 0908	30
40	4829	4643	8857	5243	1 907	1 129	2 154	1 0879	20
50	4858	4669	8843	5280	1 894	1 131	2 142	1 0850	10
		cos θ	sin θ	cot θ	tan θ	csc θ	sec θ	θ (radians)	θ (degrees)

θ (degrees)	θ (radians)	sin θ	cos θ	tan θ	cot θ	sec θ	csc θ	θ (radians)	θ (degrees)
14°00'	2443	2419	9703	2493	4 011	1 031	4 134	1 3265	76°00'
10	2473	2447	9696	2524	3 962	1 031	4 086	1 3235	50
20	2502	2476	9689	2555	3 914	1 032	4 039	1 3206	40
30	2531	2504	9681	2586	3 867	1 033	3 994	1 3177	30
40	2560	2532	9674	2617	3 821	1 034	3 950	1 3148	20
50	2589	2560	9667	2648	3 776	1 034	3 906	1 3119	10
15°00'	2618	2588	9659	2679	3 732	1 035	3 864	1 3090	75°00'
10	2647	2616	9652	2711	3 689	1 036	3 822	1 3061	50
20	2676	2644	9644	2742	3 647	1 037	3 782	1 3032	40
30	2705	2672	9636	2773	3 606	1 038	3 742	1 3003	30
40	2734	2700	9628	2805	3 566	1 039	3 703	1 2974	20
50	2763	2728	9621	2836	3 526	1 039	3 665	1 2945	10
16°00'	2793	2756	9613	2867	3 487	1 040	3 628	1 2915	74°00'
10	2822	2784	9605	2899	3 450	1 041	3 592	1 2886	50
20	2851	2812	9596	2931	3 412	1 042	3 556	1 2857	40
30	2880	2840	9588	2962	3 376	1 043	3 521	1 2828	30
40	2909	2868	9580	2994	3 340	1 044	3 487	1 2799	20
50	2938	2896	9572	3026	3 305	1 045	3 453	1 2770	10
17°00'	2967	2924	9563	3057	3 271	1 046	3 420	1 2741	73°00'
10	2996	2952	9555	3089	3 237	1 047	3 388	1 2712	50
20	3025	2979	9546	3121	3 204	1 048	3 356	1 2683	40
30	3054	3007	9537	3153	3 172	1 049	3 326	1 2654	30
40	3083	3035	9528	3185	3 140	1 049	3 295	1 2625	20
50	3113	3062	9520	3217	3 108	1 050	3 265	1 2595	10
18°00'	3142	3090	9511	3249	3 078	1 051	3 236	1 2566	72°00'
10	3171	3118	9502	3281	3 047	1 052	3 207	1 2537	50
20	3200	3145	9492	3314	3 018	1 053	3 179	1 2508	40
30	3229	3173	9483	3346	2 989	1 054	3 152	1 2479	30
40	3258	3201	9474	3378	2 960	1 056	3 124	1 2450	20
50	3287	3228	9465	3411	2 932	1 057	3 098	1 2421	10
19°00'	3316	3256	9455	3443	2 904	1 058	3 072	1 2392	71°00'
10	3345	3283	9446	3476	2 877	1 059	3 046	1 2363	50
20	3374	3311	9436	3508	2 850	1 060	3 021	1 2334	40
30	3403	3338	9426	3541	2 824	1 061	2 996	1 2305	30
40	3432	3365	9417	3574	2 798	1 062	2 971	1 2275	20
50	3462	3393	9407	3607	2 773	1 063	2 947	1 2246	10
20°00'	3491	3420	9397	3640	2 747	1 064	2 924	1 2217	70°00'
10	3520	3448	9387	3673	2 723	1 065	2 901	1 2188	50
20	3549	3475	9377	3706	2 699	1 066	2 878	1 2159	40
30	3578	3502	9367	3739	2 675	1 068	2 855	1 2130	30
40	3607	3529	9356	3772	2 651	1 069	2 833	1 2101	20
50	3636	3557	9346	3805	2 628	1 070	2 812	1 2072	10
		cos θ	sin θ	cot θ	tan θ	csc θ	sec θ	θ (radians)	θ (degrees)

Table 3   Trigonometric Functions in Degrees and Radians (continued)

θ (degrees)	θ (radians)	sin θ	cos θ	tan θ	cot θ	sec θ	csc θ	(radians)	θ (degrees)
28°00'	4887	4695	8829	5317	1.881	1.133	2.130	1.0821	62°00'
10	4916	4720	8816	5354	1.868	1.134	2.118	1.0792	50
20	4945	4746	8802	5392	1.855	1.136	2.107	1.0763	40
30	4974	4772	8788	5430	1.842	1.138	2.096	1.0734	30
40	5003	4797	8774	5467	1.829	1.140	2.085	1.0705	20
50	5032	4823	8760	5505	1.816	1.142	2.074	1.0676	10
29°00'	5061	4848	8746	5543	1.804	1.143	2.063	1.0647	61°00'
10	5091	4874	8732	5581	1.792	1.145	2.052	1.0617	50
20	5120	4899	8718	5619	1.780	1.147	2.041	1.0588	40
30	5149	4924	8704	5658	1.767	1.149	2.031	1.0559	30
40	5178	4950	8689	5696	1.756	1.151	2.020	1.0530	20
50	5207	4975	8675	5735	1.744	1.153	2.010	1.0501	10
30°00'	5236	5000	8660	5774	1.732	1.155	2.000	1.0472	60°00'
10	5265	5025	8646	5812	1.720	1.157	1.990	1.0443	50
20	5294	5050	8631	5851	1.709	1.159	1.980	1.0414	40
30	5323	5075	8616	5890	1.698	1.161	1.970	1.0385	30
40	5352	5100	8601	5930	1.686	1.163	1.961	1.0356	20
50	5381	5125	8587	5969	1.675	1.165	1.951	1.0327	10
31°00'	5411	5150	8572	6009	1.664	1.167	1.942	1.0297	59°00'
10	5440	5175	8557	6048	1.653	1.169	1.932	1.0268	50
20	5469	5200	8542	6088	1.643	1.171	1.923	1.0239	40
30	5498	5225	8526	6128	1.632	1.173	1.914	1.0210	30
40	5527	5250	8511	6168	1.621	1.175	1.905	1.0181	20
50	5556	5275	8496	6208	1.611	1.177	1.896	1.0152	10
32°00'	5585	5299	8480	6249	1.600	1.179	1.887	1.0123	58°00'
10	5614	5324	8465	6289	1.590	1.181	1.878	1.0094	50
20	5643	5348	8450	6330	1.580	1.184	1.870	1.0065	40
30	5672	5373	8434	6371	1.570	1.186	1.861	1.0036	30
40	5701	5398	8418	6412	1.560	1.188	1.853	1.0007	20
50	5730	5422	8403	6453	1.550	1.190	1.844	.9977	10
33°00'	5760	5446	8387	6494	1.540	1.192	1.836	.9948	57°00'
10	5789	5471	8371	6536	1.530	1.195	1.828	.9919	50
20	5818	5495	8355	6577	1.520	1.197	1.820	.9890	40
30	5847	5519	8339	6619	1.511	1.199	1.812	.9861	30
40	5876	5544	8323	6661	1.501	1.202	1.804	.9832	20
50	5905	5568	8307	6703	1.492	1.204	1.796	.9803	10
34°00'	5934	5592	8290	6745	1.483	1.206	1.788	.9774	56°00'
10	5963	5616	8274	6787	1.473	1.209	1.781	.9745	50
20	5992	5640	8258	6830	1.464	1.211	1.773	.9716	40
30	6021	5664	8241	6873	1.455	1.213	1.766	.9687	30
40	6050	5688	8225	6916	1.446	1.216	1.758	.9657	20
50	6080	5712	8208	6959	1.437	1.218	1.751	.9628	10
	cos θ	sin θ	cot θ	tan θ	csc θ	sec θ	(radians)		θ (degrees)

θ (degrees)	θ (radians)	sin θ	cos θ	tan θ	cot θ	sec θ	csc θ	(radians)	θ (degrees)
35°00'	6109	5736	8192	7002	1.428	1.221	1.743	9599	55°00'
10	6138	5760	8175	7046	1.419	1.223	1.736	9570	50
20	6167	5783	8158	7089	1.411	1.226	1.729	9541	40
30	6196	5807	8141	7133	1.402	1.228	1.722	9512	30
40	6225	5831	8124	7177	1.393	1.231	1.715	9483	20
50	6254	5854	8107	7221	1.385	1.233	1.708	9454	10
36°00'	6283	5878	8090	7265	1.376	1.236	1.701	9425	54°00'
10	6312	5901	8073	7310	1.368	1.239	1.695	9396	50
20	6341	5925	8056	7355	1.360	1.241	1.688	9367	40
30	6370	5948	8039	7400	1.351	1.244	1.681	9338	30
40	6400	5972	8021	7445	1.343	1.247	1.675	9308	20
50	6429	5995	8004	7490	1.335	1.249	1.668	9279	10
37°00'	6458	6018	7986	7536	1.327	1.252	1.662	9250	53°00'
10	6487	6041	7969	7581	1.319	1.255	1.655	9221	50
20	6516	6065	7951	7627	1.311	1.258	1.649	9192	40
30	6545	6088	7934	7673	1.303	1.260	1.643	9163	30
40	6574	6111	7916	7720	1.295	1.263	1.636	9134	20
50	6603	6134	7898	7766	1.288	1.266	1.630	9105	10
38°00'	6632	6157	7880	7813	1.280	1.269	1.624	9076	52°00'
10	6661	6180	7862	7860	1.272	1.272	1.618	9047	50
20	6690	6202	7844	7907	1.265	1.275	1.612	9018	40
30	6720	6225	7826	7954	1.257	1.278	1.606	8988	30
40	6749	6248	7808	8002	1.250	1.281	1.601	8959	20
50	6778	6271	7790	8050	1.242	1.284	1.595	8930	10
39°00'	6807	6293	7771	8098	1.235	1.287	1.589	8901	51°00'
10	6836	6316	7753	8146	1.228	1.290	1.583	8872	50
20	6865	6338	7735	8195	1.220	1.293	1.578	8843	40
30	6894	6361	7716	8243	1.213	1.296	1.572	8814	30
40	6923	6383	7698	8292	1.206	1.299	1.567	8785	20
50	6952	6406	7679	8342	1.199	1.302	1.561	8756	10
40°00'	6981	6428	7660	8391	1.192	1.305	1.556	8727	50°00'
10	7010	6450	7642	8441	1.185	1.309	1.550	8698	50
20	7039	6472	7623	8491	1.178	1.312	1.545	8668	40
30	7069	6494	7604	8541	1.171	1.315	1.540	8639	30
40	7098	6517	7585	8591	1.164	1.318	1.535	8610	20
50	7127	6539	7566	8642	1.157	1.322	1.529	8581	10
41°00'	7156	6561	7547	8693	1.150	1.325	1.524	8552	49°00'
10	7185	6583	7528	8744	1.144	1.328	1.519	8523	50
20	7214	6604	7509	8796	1.137	1.332	1.514	8494	40
30	7243	6626	7490	8847	1.130	1.335	1.509	8465	30
40	7272	6648	7470	8899	1.124	1.339	1.504	8436	20
50	7301	6670	7451	8952	1.117	1.342	1.499	8407	10
	cos θ	sin θ	cot θ	tan θ	csc θ	sec θ	(radians)		θ (degrees)

Table 3   355

Table 3   Trigonometric Functions in Degrees and Radians (continued)

θ (degrees)	θ (radians)	sin θ	cos θ	tan θ	cot θ	sec θ	csc θ		
42°00′	.7330	.6691	.7431	.9004	1.111	1.346	1.494	.8378	**48°00′**
10	.7359	.6713	.7412	.9057	1.104	1.349	1.490	.8348	50
20	.7389	.6734	.7392	.9110	1.098	1.353	1.485	.8319	40
30	.7418	.6756	.7373	.9163	1.091	1.356	1.480	.8290	30
40	.7447	.6777	.7353	.9217	1.085	1.360	1.476	.8261	20
50	.7476	.6799	.7333	.9271	1.079	1.364	1.471	.8232	10
43°00′	.7505	.6820	.7314	.9325	1.072	1.367	1.466	.8203	**47°00′**
10	.7534	.6841	.7294	.9380	1.066	1.371	1.462	.8174	50
20	.7563	.6862	.7274	.9435	1.060	1.375	1.457	.8145	40
30	.7592	.6884	.7254	.9490	1.054	1.379	1.453	.8116	30
40	.7621	.6905	.7234	.9545	1.048	1.382	1.448	.8087	20
50	.7560	.6926	.7214	.9601	1.042	1.386	1.444	.8058	10
44°00′	.7679	.6947	.7193	.9657	1.036	1.390	1.440	.8029	**46°00′**
10	.7709	.6967	.7173	.9713	1.030	1.394	1.435	.7999	50
20	.7738	.6988	.7153	.9770	1.024	1.398	1.431	.7970	40
30	.7767	.7009	.7133	.9827	1.018	1.402	1.427	.7941	30
40	.7796	.7030	.7112	.9884	1.012	1.406	1.423	.7912	20
50	.7825	.7050	.7092	.9942	1.006	1.410	1.418	.7883	10
45°00′	.7854	.7071	.7071	1.000	1.000	1.414	1.414		
		cos θ	sin θ	cot θ	tan θ	csc θ	sec θ	θ (radians)	θ (degrees)

Answers To Selected Exercises

Chapter 1

Section 1.1 (page 6)
1. $-9, -4, -2, 3, 8$ 3. $-|9|, -|-6|, |-8|$ 5. $-5, -4, -2, -\sqrt{3}, \sqrt{6}, \sqrt{8}, 3$ 7. 3/4, 7/5, $\sqrt{2}$, 22/15, 8/5
9. $-|-8| - |-6|, -|-2| + (-3), -|3|, -|-2|, |-8 + 2|$ 11. 8   13. 6   15. 4   17. 16   19. 8   21. -16
23. 6   25. -6 27. $8 - \sqrt{50}$ 29. $5 - \sqrt{7}$ 31. $\pi - 3$ 33. $x - 4$ 35. $8 - 2k$ 37. $56 - 7m$
39. $8 + 4m$ or $4m + 8$ 41. $y - x$ 43. $3 + x^2$ 45. $1 + p^2$ 47. $6 - \pi$ 49. $\sqrt{7} - 1$ 51. 1   53. (a) 1
(b) 1   (c) 13   (d) 14   (e) 2   55. (a) 2   (b) 7   (c) 2   (d) 0   (e) 9   57. first part of multiplication property   59. second
part of multiplication property   61. first part of multiplication property   63. triangle inequality, $|a + b| \le |a| + |b|$
65. property of absolute value, $|a| \cdot |b| = |ab|$ 67. property of absolute value, $|a/b| = |a|/|b|$ 69. trichotomy property
71. if $x = y$ or $x = -y$ 73. if $y = 0$ or if $|x| \ge |y|$ and x and y have opposite signs   75. if $y = 0$ or $|x| \ge |y|$ and x and y are
either both positive or both negative   77. -1 if $x < 0$ and 1 if $x > 0$ 79. 1   91. x must satisfy $-9 \le x \le 9$
93. $-1 < x < 0$ or $x > 1$

Section 1.2 (page 13)
1–5. 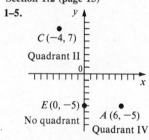   7. 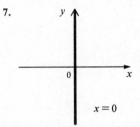   9. 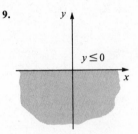   11.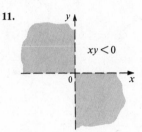

13.   15. 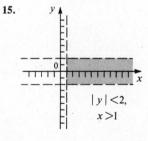   17.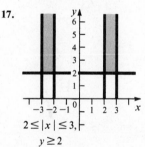

19. $8\sqrt{2}$; $(9, 3)$ 21. $\sqrt{34}$; $(-11/2, -7/2)$ 23. $\sqrt{133}$; $(2\sqrt{2}, 3\sqrt{5}/2)$ 25. 7.616   27. 27.203
29. $(13, 10)$ 31. $(-10, 11)$ 33. yes   35. no   37. no   39. yes   41. no   43. $5, -1$
45. $9 + \sqrt{119}, 9 - \sqrt{119}$ 47. $2\sqrt{2}, -2\sqrt{2}$
51. $x^2 + y^2 = 25$

53. $(2 + \sqrt{7}, 2 + \sqrt{7}), (2 - \sqrt{7}, 2 - \sqrt{7})$
55. $4x + 5y = -41$
63. Starting with the smallest triangle, the hypotenuses have
lengths $\sqrt{2}, \sqrt{3}, \sqrt{4} = 2, \sqrt{5}, \sqrt{6}, \sqrt{7}, \sqrt{8}, \sqrt{9} = 3$,
$\sqrt{10}, \sqrt{11}, \sqrt{12}, \sqrt{13}, \sqrt{14}$, etc.   65. $x = 8$

Section 1.3 (page 21)

1.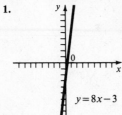
$y = 8x - 3$

3.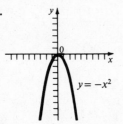
$y = 3x$

5.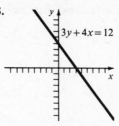
$3y + 4x = 12$

7.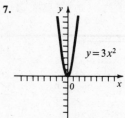
$y = 3x^2$

9.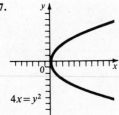
$y = -x^2$

11.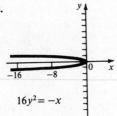
$y = x^2 - 8$

13.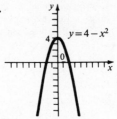
$y = 4 - x^2$

15.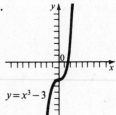
$xy = -9$

17.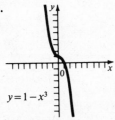
$4x = y^2$

19.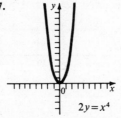
$16y^2 = -x$

21.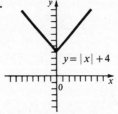
$y^2 = x + 2$

23.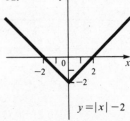
$y = x^3 - 3$

25.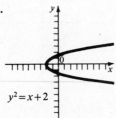
$y = 1 - x^3$

27.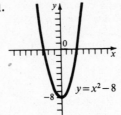
$2y = x^4$

29.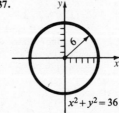
$y = |x| + 4$

31.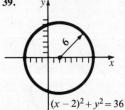
$y = |x| - 2$

33.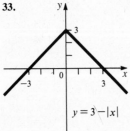
$y = 3 - |x|$

35.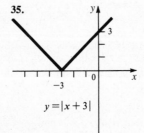
$y = |x + 3|$

37.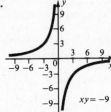
$x^2 + y^2 = 36$

39.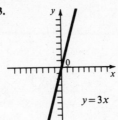
$(x - 2)^2 + y^2 = 36$

41.
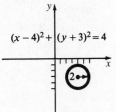
$(x - 4)^2 + (y + 3)^2 = 4$

43.
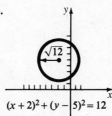
$\sqrt{12}$
$(x + 2)^2 + (y - 5)^2 = 12$

45. $(x - 1)^2 + (y - 4)^2 = 9$ **47.** $(x + 8)^2 + (y - 6)^2 = 25$ **49.** $(x + 1)^2 + (y - 2)^2 = 25$
51. $(x + 3)^2 + (y + 2)^2 = 4$ **53.** $(-3, -4); r = 4$ **55.** $(6, -5); r = 6$ **57.** $(-4, 7); r = 0$ (a point)
59. $(0, 1); r = 7$ **61.** $(1.42, -.7); r = .8$ **63.** $(x + 2)^2 + (y + 3/2)^2 = 149/4$ **65.** yes   **67.** $C = 10\pi, A = 25\pi$
69. $C = 4\pi, A = 4\pi$ **77.** (a) inside (b) outside (c) on (d) outside

Section 1.4 (page 31)

1. (a) 0 (b) 4 (c) 2 (d) 4   **3.** (a) -3 (b) -2 (c) 0 (d) 2   **5.** -1 **7.** -8 **9.** 4   **11.** $3a^2 - 1$ **13.** 9
15. 289   **17.** $-6m - 1$ **19.** $15a - 7$ **21.** $|25p^2 - 20p - 4|$ **23.** $3p - 1 + |4p^2 - 8|$ **25.** $|m^2 - 8| \cdot (3m - 1)$
27. $(3 - m - p)/(m + p)$ **29.** -66.8 **31.** -42.464 **33.** $(-\infty, +\infty); (-\infty, +\infty)$ **35.** $(-\infty, +\infty); [0, +\infty)$
37. $[-8, +\infty); [0, +\infty)$ **39.** $[-4, 4]; [0, 4]$ **41.** $[4, +\infty); [0, +\infty)$ **43.** $(-\infty, +\infty); (0, 1/5]$
45. $(-\infty, 1) \cup (1, 2) \cup (2, +\infty)$ **47.** $(-\infty, -1] \cup [5, +\infty); [0, +\infty)$ **49.** $(-\infty, +\infty); [0, +\infty)$
51. $[-5/2, +\infty); (-\infty, 0]$ **53.** $[-6, 6]; [0, 6]$ **55.** $(-\infty, -3] \cup [3, +\infty); (-\infty, 0]$ **57.** $[-5, 4]; [-2, 6]$
59. $(-\infty, +\infty); (-\infty, 12]$ **61.** $[-3, 4]; [-6, 8]$ **63.** (a) $x^2 + 2xh + h^2 - 4$ (b) $2xh + h^2$ (c) $2x + h$
65. (a) $6x + 6h + 2$ (b) $6h$ (c) 6   **67.** (a) $2x^3 + 6x^2h + 6xh^2 + 2h^3 + x^2 + 2xh + h^2$
(b) $6x^2h + 6xh^2 + 2h^3 + 2xh + h^2$ (c) $6x^2 + 6xh + 2h^2 + 2x + h$ **69.** increasing on $(-\infty, -3]$ and $[3, +\infty)$; decreasing
on $[-3, 3]$ **71.** decreasing on $(-\infty, 3/2]$; increasing on $[3/2, +\infty)$ **73.** increasing on $(-\infty, +\infty)$; never decreasing
75. never increasing; decreasing on $(-\infty, +\infty)$ **77.** increasing on $[0, +\infty)$; decreasing on $(-\infty, 0]$ **79.** increasing on
$(-\infty, -2]$; decreasing on $[-2, +\infty)$ **81.** increasing on $[0, +\infty)$; never decreasing   **83.** even   **85.** even   **87.** neither
89. neither

91.
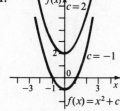
$f(x)$ $c = 2$
2
0
-3 -1 1 3 x
$c = -1$
$f(x) = x^2 + c$

93.
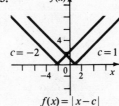
$f(x)$
4
$c = -2$ $c = 1$
-4 0 2 x
$f(x) = |x - c|$

95.
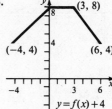
y
8 (3, 8)
$(-4, 4)$ 4 $(6, 4)$
-4 3 6 x
$y = f(x) + 4$

97.
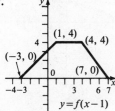
y
4 (1, 4) (4, 4)
$(-3, 0)$
0 (7, 0)
-4 -3 3 6 7 x
$y = f(x - 1)$

99.
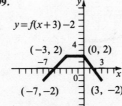
y
$y = f(x + 3) - 2$
$(-3, 2)$ 4 (0, 2)
-7 3
0 x
$(-7, -2)$ $(3, -2)$

101. volume $= x(16 - 2x)(12 - 2x)$ **103.** $t = 1, t = -2$ **105.** does not hold   **107.** does not hold
109. If r is the radius of the circle, then $A = \pi r^2$ and $C = 2\pi r$.   **111.** If x is the length of one side of the rectangle, then the
area of the rectangle is $A = 2x\sqrt{r^2 - x^2/4}$.

Section 1.5 (page 42)

1. 1122 **3.** 97 **5.** $256k^2 + 48k + 2$ **7.** $24x + 4$; $24x + 35$ **9.** $-5x^2 + 20x + 18$; $-25x^2 - 10x + 6$
11. $-64x^3 + 2$; $-4x^3 + 8$ **13.** $1/x^2$; $1/x^2$ **15.** $\sqrt{8x^2 - 4}$; $8x + 10$ **17.** $x/(2 - 5x)$; $2(x - 5)$ **19.** $\sqrt{(x - 1)/x}$;
$-1\sqrt{x + 1}$ **21.** 1 **23.** 2 **31.** $18a^2 + 24a + 9$ **33.** $16\pi t^2$ **35.** one-to-one **37.** one-to-one **39.** one-to-one
41. not one-to-one **43.** not one-to-one **45.** one-to-one **47.** one-to-one **49.** one-to-one **51.** inverses **53.** inverses
55. inverses **57.** not inverses **59.** inverses

61.

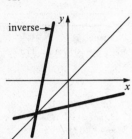

63.

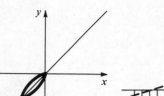

65. $f^{-1}(x) = (x + 5)/4$

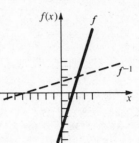

67. $f^{-1}(x) = -\dfrac{5}{2} x$

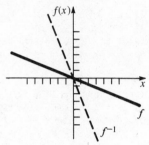

69. $f^{-1}(x) = \sqrt[3]{-x - 2}$

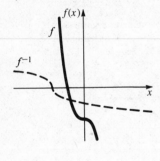

71. $f^{-1}(x) = (9 - x)/3$

73. not one-to-one

75. $f^{-1}(x) = 4/x$

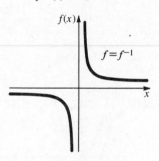

77. $f^{-1}(x) = (x + 5)/(3x + 6)$
domain f:
$(-\infty, 1/3) \cup (1/3, +\infty)$
domain f^{-1}:
$(-\infty, -2) \cup (-2, +\infty)$

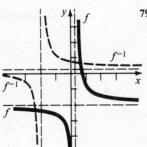

79. $f^{-1}(x) = x^2 - 6$
domain: $[0, +\infty)$

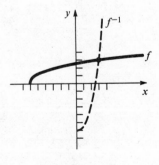

81. 4 **83.** 2 **85.** -2 **87.** 1.14 **89.** $(f \circ g)^{-1}(x) = (\sqrt[3]{19 - x})/2$ **91.** $(g^{-1} \circ f^{-1})(x) = (\sqrt[3]{19 - x})/2$

Chapter 1 Review Exercises (page 46)

1. $-|3 - (-2)|$, $-|-2|$, $|6 - 4|$, $|8 + 1|$ **3.** $-|-8 + \sqrt{13}|$, $-\sqrt{2}$, $\sqrt{15}$, $\sqrt{169}$ **5.** $3 - \sqrt{7}$ **7.** $m - 3$
9. (a) 1 (b) 12 **11.** if $x \geq 0$ **13.** if $x = 0$ and $y = 0$ **15.** if A and B are the same point

17.

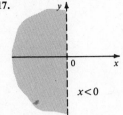

$x < 0$

19.

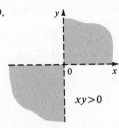

$xy > 0$

21. $d(P, Q) = \sqrt{85}$; $(-1/2, 2)$
23. $(7, -13)$
25. $-7, -1, 8, 23$
27. no such points exist

31. domain: $(-\infty, +\infty)$

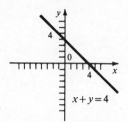

$x + y = 4$

33. domain: $(-\infty, +\infty)$

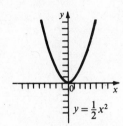

$y = \frac{1}{2}x^2$

35. domain: $(-\infty, 0) \cup (0, +\infty)$ **37.** domain: $[7, +\infty)$

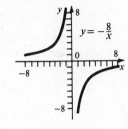

$y = -\frac{8}{x}$

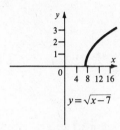

$y = \sqrt{x - 7}$

39. $(x + 2)^2 + (y - 3)^2 = 25$ **41.** $(x + 8)^2 + (y - 1)^2 = 289$ **43.** $(2, -3)$; $r = 1$ **45.** $(-7/2, -3/2)$; $r = 3\sqrt{6}/2$
47. $(-\infty, +\infty)$ **49.** $(-\infty, +\infty)$ **51.** $(-\infty, 8) \cup (8, +\infty)$ **53.** $[-7, 7]$

55.

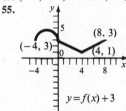

$(-4, 3)$ $(8, 3)$ $(4, 1)$

$y = f(x) + 3$

57.

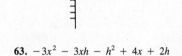

$y = f(x - 2)$
$(-2, 0)$ $(2, 0)$ $(10, 0)$
$(6, -2)$

59.

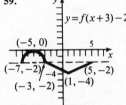

$y = f(x + 3) - 2$
$(-5, 0)$
$(-7, -2)$ $(5, -2)$
$(-3, -2)$ $(1, -4)$

61.

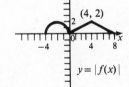

$(4, 2)$

$y = |f(x)|$

63. $-3x^2 - 3xh - h^2 + 4x + 2h$ **65.** $\sqrt{x^2 - 2}$ **67.** $\sqrt{34}$ **69.** 1
71. increasing on $(-\infty, -1]$, decreasing on $[-1, +\infty)$ **73.** never increasing or
decreasing **75.** not one-to-one **77.** not one-to-one **79.** not one-to-one
81. not one-to-one

83. $f^{-1}(x) = (x - 3)/12$

85. $f^{-1}(x) = \sqrt[3]{x + 3}$

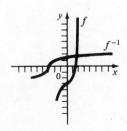

87. $f^{-1}(x) = x^2 + 3$; domain: $(-\infty, 0)$

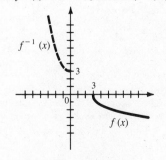

89.
$$(f \circ g)(x) = \begin{cases} 2 \text{ if } x < 0 \\ x \text{ if } 0 \le x \le 1 \\ 2 \text{ if } x > 1 \end{cases}$$

91.
$$(f \circ g)(x) = \begin{cases} 1 \text{ if } x < 0 \\ 4x^2 \text{ if } 0 \le x \le 1/2 \\ 0 \text{ if } 1/2 < x \le 1 \\ 1 \text{ if } x > 1 \end{cases}$$

93. $k = -1$

Chapter 2

Section 2.1 (page 56)
1. $(-\sqrt{2}/2, -\sqrt{2}/2)$ **3.** $(-1, 0)$ **5.** $(1, 0)$ **7.** $(1, 0)$ **9.** $(0, 1)$ **11.** $(-\sqrt{2}/2, \sqrt{2}/2)$ **13.** $(\sqrt{2}/2, \sqrt{2}/2)$
15. $(-\sqrt{2}/2, \sqrt{2}/2)$ **17.** (a) $(2/3, -\sqrt{5}/3)$ (b) $(2/3, \sqrt{5}/3)$ (c) $(-2/3, -\sqrt{5}/3)$ (d) $(-2/3, \sqrt{5}/3)$
19. (a) $(4/5, -3/5)$ (b) $(4/5, 3/5)$ (c) $(-4/5, -3/5)$ (d) $(-4/5, 3/5)$ **21.** (a) $(-1/2, -\sqrt{3}/2)$ (b) $(-1/2, \sqrt{3}/2)$
(c) $(1/2, -\sqrt{3}/2)$ (d) $(1/2, \sqrt{3}/2)$ **23.** (a) $(-2/5, \sqrt{21}/5)$ (b) $(-2/5, -\sqrt{21}/5)$ (c) $(2/5, \sqrt{21}/5)$
(d) $(2/5, -\sqrt{21}/5)$ **25.** 0; 1 **27.** $\sqrt{2}/2$; $\sqrt{2}/2$ **29.** -1; 0 **31.** 0; -1 **33.** $-\sqrt{2}/2$; $\sqrt{2}/2$
35. $-\sqrt{2}/2$; $-\sqrt{2}/2$ **37.** III **39.** III **41.** II **45.** (b) $(1/2, \sqrt{3}/2)$ (c) $(1/2, -\sqrt{3}/2)$; $(-1/2, \sqrt{3}/2)$;
$(-1/2, -\sqrt{3}/2)$ **47.** $\sqrt{3}/2$ **49.** $-1/2$ **51.** $-\sqrt{3}/2$ **53.** $1/2$ **55.** $\sin s = p^2$; $\cos s = m$
57. $\sin s = z$; $\cos s = 3/8$ **59.** $\sin s = 3b$; $\cos s = 5a$ **61.** $\sin s = 4/5$; $\cos s = 3/5$ **63.** $\sin s = 5/8$;
$\cos s = -\sqrt{39}/8$ **65.** $\sin s = 2\sqrt{3}/\sqrt{13}$ or $2\sqrt{39}/13$; $\cos s = -1/\sqrt{13}$ or $-\sqrt{13}/13$ **67.** $\sin s = -\sqrt{2}/\sqrt{11}$ or
$-\sqrt{22}/11$; $\cos s = 3/\sqrt{11}$ or $3\sqrt{11}/11$ **69.** $\sin s = b/\sqrt{a^2 + b^2}$; $\cos s = a/\sqrt{a^2 + b^2}$

Section 2.2 (page 62)
1. $\tan s = \sqrt{3}/3$; $\cot s = \sqrt{3}$; $\sec s = 2\sqrt{3}/3$; $\csc s = 2$ **3.** $\tan s = -4/3$; $\cot s = -3/4$; $\sec s = -5/3$; $\csc s = 5/4$
5. $\tan s = -\sqrt{3}$; $\cot s = -\sqrt{3}/3$; $\sec s = 2$; $\csc s = -2\sqrt{3}/3$
In Exercises 7–17 and 23–33 answers are given in the order sine, cosine, tangent, cotangent, secant, cosecant. **7.** 0; -1; 0;
undefined; -1; undefined **9.** $\sqrt{2}/2$; $-\sqrt{2}/2$; -1; -1; $-\sqrt{2}$; $\sqrt{2}$ **11.** $-\sqrt{2}/2$; $\sqrt{2}/2$; -1; -1; $\sqrt{2}$; $-\sqrt{2}$
13. $1/2$; $\sqrt{3}/2$; $\sqrt{3}/3$; $\sqrt{3}$; $2\sqrt{3}/3$; 2 **15.** $\sqrt{3}/2$; $-1/2$; $-\sqrt{3}$; $-\sqrt{3}/3$; -2; $2\sqrt{3}/3$
17. $-1/2$; $-\sqrt{3}/2$; $\sqrt{3}/3$; $\sqrt{3}$; $-2\sqrt{3}/3$; -2 **19.** $+$, $+$ **21.** $-$, $-$, $+$, $+$, $-$, $-$
23. $1/4$; $\sqrt{15}/4$; $1/\sqrt{15}$ or $\sqrt{15}/15$; $\sqrt{15}$; $4/\sqrt{15}$ or $4\sqrt{15}/15$; 4
25. $2\sqrt{10}/7$; $-3/7$; $-2\sqrt{10}/3$; $-3/(2\sqrt{10})$ or $-3\sqrt{10}/20$; $-7/3$; $7/(2\sqrt{10})$ or $7\sqrt{10}/20$
27. $-\sqrt{3}/2$; $1/2$; $-\sqrt{3}$; $-1/\sqrt{3}$ or $-\sqrt{3}/3$; 2; $-2/\sqrt{3}$ or $-2\sqrt{3}/3$
29. $\sqrt{6}/\sqrt{7}$ or $\sqrt{42}/7$; $-1/\sqrt{7}$ or $-\sqrt{7}/7$; $-\sqrt{6}$; $-1/\sqrt{6}$ or $-\sqrt{6}/6$; $-\sqrt{7}$; $\sqrt{7}/\sqrt{6}$ or $\sqrt{42}/6$
31. $a/\sqrt{a^2 + b^2}$; $b/\sqrt{a^2 + b^2}$; a/b; b/a; $\sqrt{a^2 + b^2}/b$; $\sqrt{a^2 + b^2}/a$ **33.** q; $-\sqrt{1 - q^2}$; $-q/\sqrt{1 - q^2}$; $-\sqrt{1 - q^2}/q$;
$-1/\sqrt{1 - q^2}$; $1/q$ **35.** impossible **37.** possible **39.** possible **41.** possible **43.** impossible **45.** impossible
47. II **49.** III **51.** II or III **53.** III or IV **55.** $\sin s = \sqrt{65}/9$; $\tan s = -\sqrt{65}/4$; $\cot s = -4\sqrt{65}/65$;
$\sec s = -9/4$; $\csc s = 9\sqrt{65}/65$ **57.** $\sin s = 3\sqrt{13}/13$; $\cos s = 2\sqrt{13}/13$; $\cot s = 2/3$; $\sec s = \sqrt{13}/2$
59. $\cos t = -2\sqrt{5}/5$; $\tan t = -1/2$; $\cot t = -2$; $\sec t = -\sqrt{5}/2$; $\csc t = \sqrt{5}$
61. $\sin s = \sqrt{21}/7$; $\cos s = -2\sqrt{7}/7$; $\cot s = 2\sqrt{3}/3$; $\csc s = -\sqrt{21}/3$
In Exercises 63–71, answers may differ slightly in the last digits, depending on the method used to get the number.
63. $\cos t = -.986425$; $\tan t = -.166475$; $\cot t = -6.00691$; $\sec t = -1.01376$; $\csc t = 6.08958$ **65.** $\sin t = -.540362$;
$\cos t = -.841433$; $\cot t = 1.55716$; $\sec t = -1.18845$; $\csc t = -1.85061$ **67.** a, $\sqrt{1 - a^2}$, $a\sqrt{1 - a^2}/(1 - a^2)$,
$\sqrt{1 - a^2}/a$, $\sqrt{1 - a^2}/(1 - a^2)$, $1/a$ **69.** $\sin s = .903687$, $\tan s = -2.11047$, $\cot s = -.473829$,
$\sec s = -2.33540$, $\csc s = 1.10658$ **71.** $\sin t = -.095832$, $\cos t = .995397$, $\tan t = -.096275$, $\cot t = -10.3869$,
$\sec t = 1.00462$ **77.** $\cos s$ **79.** $\csc s$ **81.** $\tan s$ **83.** s

Section 2.3 (page 72)
1. 320° **3.** 235° **5.** 90° **7.** 179° **9.** $\pi/3$ **11.** $\pi/2$ **13.** $3\pi/4$ **15.** $5\pi/3$ **17.** $9\pi/4$ **19.** $7\pi/9$
21. 60° **23.** 315° **25.** 330° **27.** $-30°$ **29.** 900° **31.** 63° **33.** 2.4289 **35.** 1.1221 **37.** $-.5185$
39. -3.6612 **41.** 114.5916° ; 114° 35′ **43.** 99.6947° ; 99° 42′ **45.** 5.2254° ; 5° 14′ **47.** 564.2276° ; 564° 14′
49. $3/2$ **51.** 1.23 **53.** 3.09309 **55.** 980 mi **57.** 3700 mi **59.** 25.1 in **61.** 25.8 cm **63.** 5.05 m
65. 53.429 m **67.** (a) 11.6 in (b) 37° 05′ **69.** $5\pi/4$ radians **71.** $\pi/25$ radians/sec **73.** 9 min **75.** $72\pi/5$ cm/sec
77. 6 radians/sec **79.** 9.29755 cm/sec **81.** 18π cm **83.** 12 sec **85.** $3\pi/32$ radians/sec **87.** 24.8647 cm **89.** $\pi/6$
radians per hour **91.** $200\pi/3$ radians per minute **93.** $\pi/18$ radians per second **95.** (a) 2π radians/day; $\pi/12$ radians/hr
(b) 0 (c) $12,800\pi$ km/day or 533π km/hr (d) 9050π km/day or 377π km/hr

Section 2.4 (page 81)

In Exercises 1–15, answers are given in the order sine, cosine, tangent, cotangent, secant, and cosecant.
1. $\sqrt{3}/2$, $-1/2$, $-\sqrt{3}$, $-\sqrt{3}/3$, -2, $2\sqrt{3}/3$ **3.** $1/2$, $-\sqrt{3}/2$, $-\sqrt{3}/3$, $-\sqrt{3}$, $-2\sqrt{3}/3$, 2
5. $-\sqrt{3}/2$, $-1/2$, $\sqrt{3}$, $\sqrt{3}/3$, -2, $-2\sqrt{3}/3$ **7.** $-1/2$, $\sqrt{3}/2$, $-\sqrt{3}/3$, $-\sqrt{3}$, $2\sqrt{3}/3$, -2 **9.** $\sqrt{3}/2$, $1/2$, $\sqrt{3}$, $\sqrt{3}/3$,
2, $2\sqrt{3}/3$ **11.** $1/2$, $-\sqrt{3}/2$, $-\sqrt{3}/3$, $-\sqrt{3}$, $-2\sqrt{3}/3$, 2 **13.** 0, -1, 0, undefined, -1, undefined   **15.** -1, 0,
undefined, 0, undefined, -1 **17.** $\sqrt{3}/3$, $\sqrt{3}$ **19.** $\sqrt{3}/2$, $\sqrt{3}/3$, $2\sqrt{3}/3$ **21.** -1, -1 **23.** $-\sqrt{3}/2$, $-2\sqrt{3}/3$
In Exercises 25–51 answers are given in the order sine, cosine, tangent, cotangent, secant, and cosecant.
25. $4/5$, $-3/5$, $-4/3$, $-3/4$, $-5/3$, $5/4$ **27.** $7/25$, $24/25$, $7/24$, $24/7$, $25/24$, $25/7$ **29.** $-5/13$, $-12/13$, $5/12$, $12/5$, $-13/12$,
$-13/5$ **31.** $-4/5$, $-3/5$, $4/3$, $3/4$, $-5/3$, $-5/4$ **33.** $-\sqrt{2}/2$; $\sqrt{2}/2$; -1; -1; $\sqrt{2}$; $-\sqrt{2}$ **35.** $-2/3$; $\sqrt{5}/3$; $-2\sqrt{5}/5$;
$-\sqrt{5}/2$; $3\sqrt{5}/5$; $-3/2$ **37.** $\sqrt{3}/4$; $-\sqrt{13}/4$; $-\sqrt{39}/13$; $-\sqrt{39}/3$; $-4\sqrt{13}/13$; $4\sqrt{3}/3$ **39.** $-\sqrt{10}/5$; $\sqrt{15}/5$; $-\sqrt{6}/3$;
$-\sqrt{6}/2$; $\sqrt{15}/3$; $-\sqrt{10}/2$ **41.** $-.34727$; $.93777$; $-.37031$; -2.7004; 1.0664; -2.8796 **43.** $-.5638$; $-.8259$; $.6826$;
1.465; -1.211; -1.774 **45.** $-.633$; $.774$; $-.818$; 1.22; 1.29; -1.58 **47.** $-5/\sqrt{26}$ or $-5\sqrt{26}/26$; $-1/\sqrt{26}$ or
$-\sqrt{26}/26$; 5; $1/5$; $-\sqrt{26}$; $-\sqrt{26}/5$ **49.** $3/5$, $4/5$, $3/4$, $4/3$, $5/4$, $5/3$ **51.** $21/29$, $20/29$, $21/20$, $20/21$, $29/20$, $29/21$
53. $.759260$, $.650787$, 1.16668 **55.** 1 **57.** $23/4$ **59.** $1/2 + \sqrt{3}$ **61.** $-29/12$ **63.** $-\sqrt{3}/3$ **65.** false
67. true   **69.** false   **71.** true   **73.** true   **75.** $30°$; $150°$ **77.** $60°$; $240°$ **79.** $120°$; $240°$ **81.** $240°$; $300°$
83. $90°$; $270°$ **85.** $0°$; $180°$ **89.** $a = 12$; $b = 12\sqrt{3}$; $d = 12\sqrt{3}$; $c = 12\sqrt{6}$ **91.** $m = 7\sqrt{3}/3$; $a = 14\sqrt{3}/3$;
$n = 14\sqrt{3}/3$; $q = 14\sqrt{6}/3$

Section 2.5 (page 91)

1. $35°$ **3.** $37°$ **5.** $69° 50'$ **7.** $.0684$ **9.** $.3142$ **11.** $.8814$ **13.** $.6338$ **15.** $-.6248$ **17.** $-.5712$
19. -3.072 **21.** $.9636$ **23.** $-.6361$ **25.** $.5704$ **27.** 1.162 **29.** $.1628$ **31.** $-.3131$ **33.** $58° 00'$
35. $30° 30'$ **37.** $46° 10'$ **39.** $81° 10'$ **41.** $.2095$ **43.** 1.4426 **45.** 1.4748 **47.** 1.0180 **49.** $.4188$
51. $.3462$ **53.** (a) $30°$ (b) $60°$ (c) $75°$ (d) $86°$ (e) $86°$ (f) $60°$ **55.** $.8417$ **57.** $.1801$ **59.** $.0100$
61. $a = 571$ m; $b = 777$ m   **63.** $n = 154$ m; $p = 198$ m   **65.** $c = 84.816$ cm; $a = 62.942$ cm   **67.** 50.7 m
69. 28.94 in   **71.** $31° 20'$ **73.** 3.3 ft   **75.** 84.7 ft   **77.** 2×10^8 m per sec   **79.** $19°$ **81.** $48.7°$

Chapter 2 Review Exercises (page 96)

1. $(\sqrt{2}/2, -\sqrt{2}/2)$ **3.** $(-1, 0)$ **5.** $-\sqrt{2}/2$; $-\sqrt{2}/2$; 1 **7.** $-\sqrt{3}/2$; $1/2$; $-\sqrt{3}$ **9.** $(-2/3, -\sqrt{5}/3)$
11. $(2/3, \sqrt{5}/3)$ **13.** II   **15.** III   **17.** $\cos s = \sqrt{5}/3$, $\tan s = 2\sqrt{5}/5$, $\cot s = \sqrt{5}/2$, $\sec s = 3\sqrt{5}/5$, $\csc s = 3/2$
19. $\sin s = -2\sqrt{5}/5$, $\cos s = \sqrt{5}/5$, $\tan s = -2$, $\csc s = -\sqrt{5}/2$ **21.** $1280°$ **23.** $47.420°$ **25.** $74° 17' 54''$
27. $183° 05' 50''$ **29.** $135°$ **31.** $1116°$ **33.** $3\pi/2$ **35.** $17\pi/3$ **37.** $\sqrt{3}/2$ **39.** $\sqrt{3}$ **41.** -1 **43.** Sine is
$-\sqrt{2}/2$; cosine is $-\sqrt{2}/2$; tangent is 1.   **45.** 1.428 **47.** $-.9216$ **49.** $.5289$ **51.** $55° 40'$ **53.** $12° 40'$
55. $.3898$ **57.** $.5148$ **59.** $a = 638$; $b = 391$ **61.** $a = 32.38$ m; $c = 50.66$ m   **63.** $15/32$ sec   **65.** $\pi/20$ rad/sec
67. 285.3 cm

Chapter 3

Section 3.1 (page 106)

1. 2

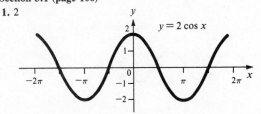

$y = 2 \cos x$

3. $2/3$

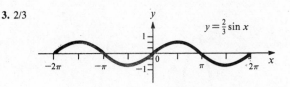

$y = \frac{2}{3} \sin x$

5. 1

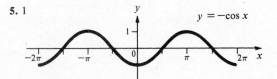

$y = -\cos x$

7. 2

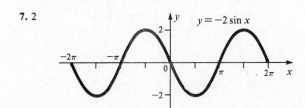

$y = -2 \sin x$

9. 4π; 1; none

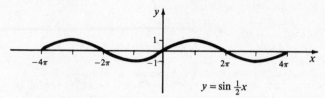

$$y = \sin \tfrac{1}{2}x$$

11. 6π; 1; none

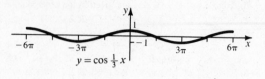

$$y = \cos \tfrac{1}{3}x$$

13. $2\pi/3$; 1; none

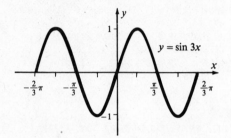

$$y = \sin 3x$$

15. π; 1; none

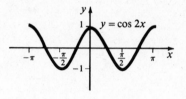

$$y = \cos 2x$$

17. $\pi/2$; 1; none

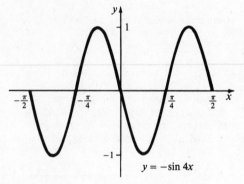

$$y = -\sin 4x$$

19. 8π; 2; none

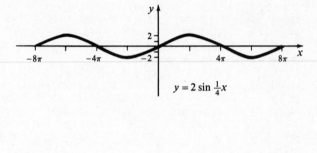

$$y = 2 \sin \tfrac{1}{4}x$$

21. $2\pi/3$; 2; none

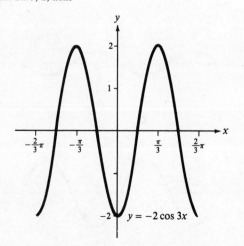

$$y = -2 \cos 3x$$

23. $2\pi/3$; 1/2; none

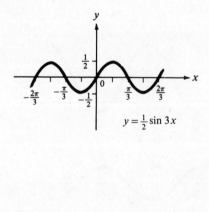

$$y = \tfrac{1}{2} \sin 3x$$

25. $8\pi/3$; $2/3$; none

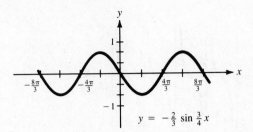

$$y = -\tfrac{2}{3}\sin\tfrac{3}{4}x$$

27. 2π; 1; 2

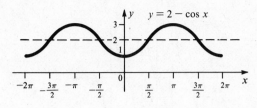

$$y = 2 - \cos x$$

29. 2π; 2; -3

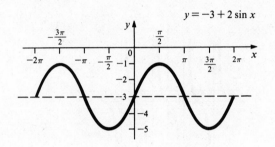

$$y = -3 + 2\sin x$$

31. 4π; 2; 1

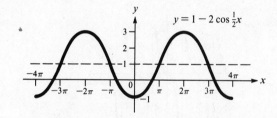

$$y = 1 - 2\cos\tfrac{1}{2}x$$

33. 2; 1; none

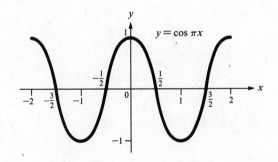

$$y = \cos \pi x$$

35.

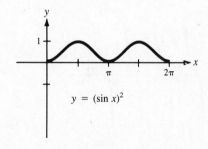

$$y = (\sin x)^2$$

37.

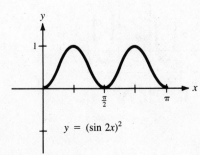

$$y = (\sin 2x)^2$$

39. (a) 20   (b) 75   **41.** (a) about 7/4 hr   (b) 1 yr   **43.** 1; $4\pi/3$
45. (a) 5; 1/60   (b) 60   (c) 5; 1.545; -4.045; -4.045; 1.545
(d)

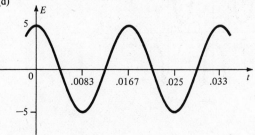

Section 3.2 (page 113)

1.

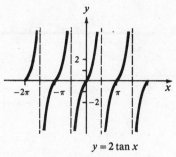

$$y = 2 \tan x$$

3.

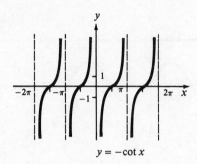

$$y = -\cot x$$

5.

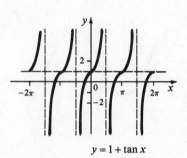

$$y = 1 + \tan x$$

7.

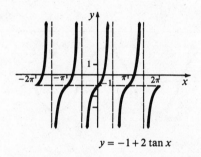

$$y = -1 + 2 \tan x$$

9.

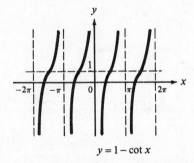

$$y = 1 - \cot x$$

11.

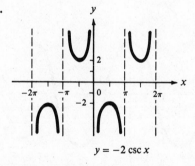

$$y = -2 \csc x$$

13.

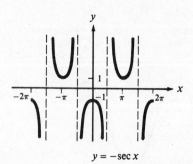

$$y = -\sec x$$

15.

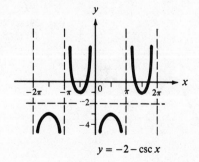

$$y = -2 - \csc x$$

17. $\pi/2$

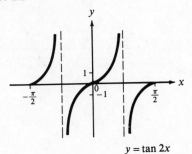

$y = \tan 2x$

19. $\pi/3$

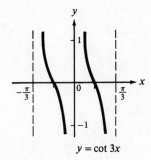

$y = \cot 3x$

21. $\pi/2$

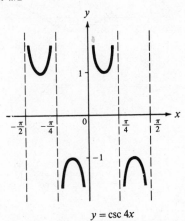

$y = \csc 4x$

23. 4π

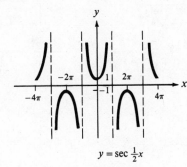

$y = \sec \frac{1}{2}x$

25. 4π

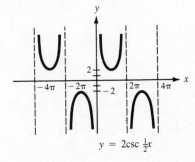

$y = 2\csc \frac{1}{2}x$

27. (a) 0   (b) -2.9 m   (c) -12.3 m   (d) 12.3 m
(e) It leads to tan $\pi/2$, which does not exist.
(f) all t except $.25\,k$, where k is an integer

29. (b)

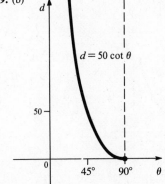

$d = 50 \cot \theta$

Section 3.3 (page 118)

1. 2; 2π; none; none   **3.** 4; 4π; none; none   **5.** not applicable; $\pi/2$; none; $\pi/2$ to the right   **7.** not applicable; $\pi/3$; none;
$\pi/12$ to the left   **9.** 1; $2\pi/3$; up 2; $\pi/15$ to the right   **11.** not applicable; $\pi/4$; down 2; $\pi/4$ to the left

13.

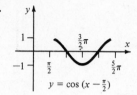

$$y = \cos\left(x - \frac{\pi}{2}\right)$$

15.

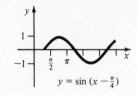

$$y = \sin\left(x - \frac{\pi}{4}\right)$$

17.

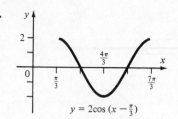

$$y = 2\cos\left(x - \frac{\pi}{3}\right)$$

19.

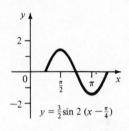

$$y = \frac{3}{2}\sin 2\left(x - \frac{\pi}{4}\right)$$

21.

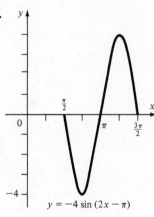

$$y = -4\sin(2x - \pi)$$

23.

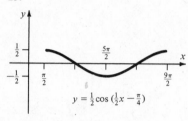

$$y = \frac{1}{2}\cos\left(\frac{1}{2}x - \frac{\pi}{4}\right)$$

25.

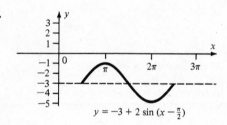

$$y = -3 + 2\sin\left(x - \frac{\pi}{2}\right)$$

27.

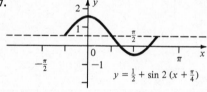

$$y = \frac{1}{2} + \sin 2\left(x + \frac{\pi}{4}\right)$$

29.

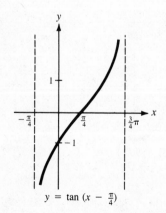

$$y = \tan\left(x - \frac{\pi}{4}\right)$$

31.

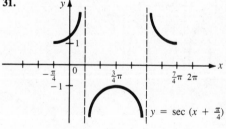

$$y = \sec\left(x + \frac{\pi}{4}\right)$$

33.

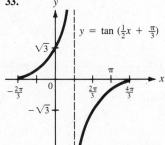

$$y = \tan\left(\tfrac{1}{2}x + \tfrac{\pi}{3}\right)$$

35.

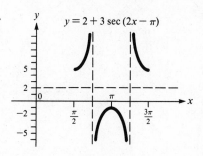

$$y = 2 + 3\sec(2x - \pi)$$

37.

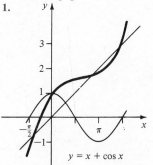

$$y = 1 - \tfrac{1}{2}\csc\left(x - \tfrac{3\pi}{4}\right)$$

39.

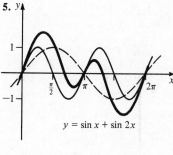

$$y = \tfrac{2}{3}\tan\left(\tfrac{3}{4}x - \pi\right) - 2$$

Section 3.4 (page 121)

1.

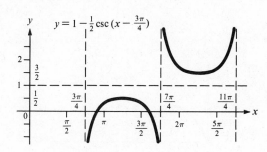

$$y = x + \cos x$$

3.

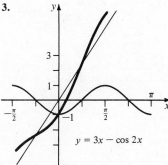

$$y = 3x - \cos 2x$$

5.

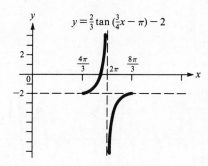

$$y = \sin x + \sin 2x$$

7.

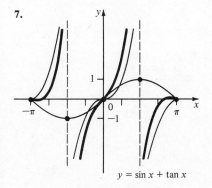

$$y = \sin x + \tan x$$

9.

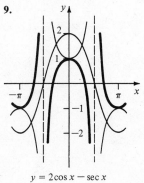

$$y = 2\cos x - \sec x$$

11.

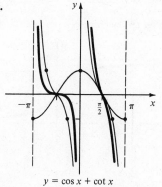

$$y = \cos x + \cot x$$

13.

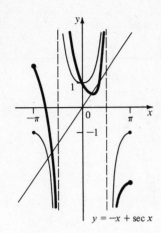

$y = -x + \sec x$

15.

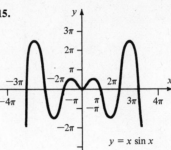

$y = x \sin x$

17.

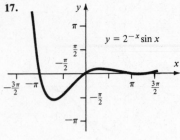

$y = 2^{-x} \sin x$

19.

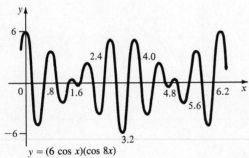

$y = (6 \cos x)(\cos 8x)$

21.

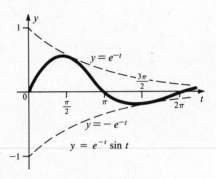

$y = e^{-t}$

$y = -e^{-t}$

$y = e^{-t} \sin t$

Section 3.5 (page 126)
1. (a) $y = \sin 2t$; 1; π; $1/\pi$ (b) $y = \sin 3t$; 1; $2\pi/3$; $3/2\pi$ (c) $y = \sin 4t$; 1; $\pi/2$; $2/\pi$
3. $y = 2 \sin(t + \pi/3)$; 2; 2π; $1/2\pi$ **5.** $8/\pi^2$ **7.** (a) period, $2\pi\sqrt{m/k}$; frequency, $(1/2\pi)\sqrt{k/m}$ (b) $1/\pi^2$

Section 3.6 (page 131)
1. $-\pi/3$ **3.** $\pi/4$ **5.** $-\pi/2$ **7.** $\pi/3$ **9.** $3\pi/4$ **11.** $-\pi/3$ **13.** $-7°\,40'$ **15.** $113°\,30'$ **17.** $22°\,00'$
19. $48°\,00'$ **21.** $-42°\,40'$ **23.** $.8058$ **25.** -1.332 **29.** $-26°\,21'\,29''$ **31.** $23°\,52'\,41''$ **33.** $\sqrt{7}/3$
35. $\sqrt{5}/5$ **37.** $-\sqrt{5}/2$ **39.** $\sqrt{34}/3$ **41.** $1/2$ **43.** -1 **45.** 2   **47.** $\pi/4$ **49.** $\pi/3$ **51.** 0
53. $3\pi/4$ **55.** 0   **57.** $.957826$ **59.** $.123430$ **61.** $x = \sin(y/4)$ **63.** $x = (1/2)\tan 2y$ **65.** $x = -2 + \sin y$
67. $\sqrt{1 - u^2}$ **69.** $\sqrt{u^2 + 1}/u$ **71.** $\sqrt{1 - u^2}/u$ **73.** $\sqrt{u^2 - 4}/u$ **75.** $u/\sqrt{2}$ or $\sqrt{2u}/2$
77. $\pm 2/\sqrt{4 - u^2}$ or $\pm 2\sqrt{4 - u^2}/(4 - u^2)$
79. real numbers; $0 < y < \pi$ **81.** $x \le -1$ or $x \ge 1$; $0 \le y < \pi/2$ or $\pi \le y < 3\pi/2$

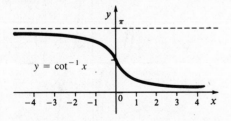

$y = \cot^{-1} x$

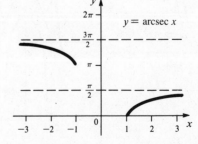

$y = \text{arcsec } x$

83. $-1 \le x \le 1$; $0 \le y \le 2\pi$ **89.** false   **91.** false   **93.** false   **95.** true   **97.** $113°$ **99.** $60°$

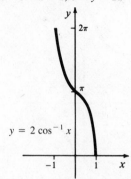

$y = 2\cos^{-1} x$

Chapter 3 Review Exercises (page 133)
1. 2; 2π; none; none   **3.** 1/2; $2\pi/3$; none; none   **5.** 2; 8π; 1; none   **7.** 3; 2π; none; $\pi/2$ to the left
9. none; π; none; $\pi/8$ to the right   **11.** 1/3; $2\pi/3$; none; $\pi/9$ to the right

13.
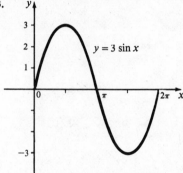
$y = 3\sin x$

15.
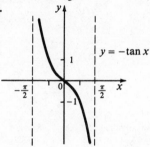
$y = -\tan x$

17.
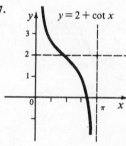
$y = 2 + \cot x$

19.
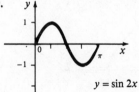
$y = \sin 2x$

21.
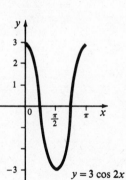
$y = 3\cos 2x$

23.
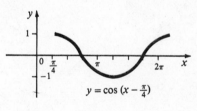
$y = \cos\left(x - \frac{\pi}{4}\right)$

25.
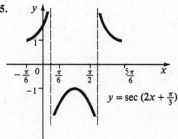
$y = \sec\left(2x + \frac{\pi}{3}\right)$

27.
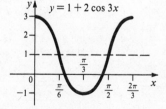
$y = 1 + 2\cos 3x$

29.

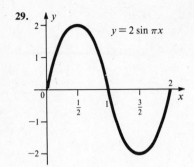

$y = 2 \sin \pi x$

31.

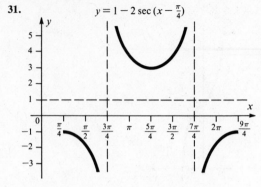

$y = 1 - 2 \sec (x - \frac{\pi}{4})$

33.

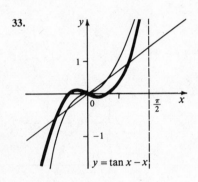

$y = \tan x - x$

35.

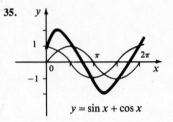

$y = \sin x + \cos x$

37. (a) about 20 years   (b) from about 10,000 to about 150,000   **39.** $\pi/4$ **41.** $-\pi/3$ **43.** -0.7214 **45.** $1/2$
47. -1 **49.** $3\pi/4$ **51.** $u/\sqrt{1 + u^2}$ or $u\sqrt{1 + u^2}/(1 + u^2)$ **53.** $1/u$ **55.** $-1 \le x \le 1; 0 \le y \le \pi$
57. (a) $t = [1/(2\pi f)][\arcsin (e/E_{\max})]$ (b) .0007

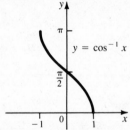

$y = \cos^{-1} x$

Chapter 4

Section 4.1 (page 140)
1. $\sqrt{7}/4$ **3.** $-2\sqrt{5}/5$ **5.** $\sqrt{21}/2$ **7.** $\cos \theta = -\sqrt{5}/3$; $\tan \theta = -2\sqrt{5}/5$; $\cot \theta = -\sqrt{5}/2$; $\sec \theta = -3\sqrt{5}/5$;
$\csc \theta = 3/2$ **9.** $\sin \theta = -\sqrt{17}/17$; $\cos \theta = 4\sqrt{17}/17$; $\cot \theta = -4$; $\sec \theta = \sqrt{17}/4$; $\csc \theta = -\sqrt{17}$
11. $\sin \theta = 2\sqrt{2}/3$; $\cos \theta = -1/3$; $\tan \theta = -2\sqrt{2}$; $\cot \theta = -\sqrt{2}/4$; $\csc \theta = 3\sqrt{2}/4$ **13.** $\sin \theta = 3/5$; $\cos \theta = 4/5$;
$\tan \theta = 3/4$; $\sec \theta = 5/4$; $\csc \theta = 5/3$ **15.** $\sin \theta = -\sqrt{7}/4$; $\cos \theta = 3/4$; $\tan \theta = -\sqrt{7}/3$; $\cot \theta = -3\sqrt{7}/7$;
$\csc \theta = -4\sqrt{7}/7$ **17.** (b)   **19.** (e)   **21.** (a)   **23.** (a)   **25.** (d)   **27.** 1   **29.** $-\sin \alpha$ **31.** 0
33. $(1 + \sin \theta)/\cos \theta$ **35.** 1   **37.** -1 **39.** $\sin^2 \theta/\cos^4 \theta$ **41.** 1   **43.** $(\cos^2 \alpha + 1)/(\sin^2 \alpha \cos^2 \alpha)$
45. $(\sin^2 s - \cos^2 s)/\sin^4 s$ **47.** $\pm\sqrt{1 + \cot^2 \theta}/(1 + \cot^2 \theta)$; $\pm\sqrt{\sec^2 \theta - 1}/\sec \theta$ **49.** $\pm\sin \theta\sqrt{1 - \sin^2 \theta}/(1 - \sin^2 \theta)$;
$\pm\sqrt{1 - \cos^2 \theta}/\cos \theta$; $\pm\sqrt{\sec^2 \theta - 1}$; $\pm\sqrt{\csc^2 \theta - 1}/(\csc^2\theta - 1)$ **51.** $\pm\sqrt{1 - \sin^2 \theta}/(1 - \sin^2 \theta)$; $\pm\sqrt{\tan^2 \theta + 1}$;
$\pm\sqrt{1 + \cot^2 \theta}/\cot \theta$; $\pm\csc \theta\sqrt{\csc^2 \theta - 1}/(\csc^2 \theta - 1)$ **53.** $\sin \theta = \pm\sqrt{2x + 1}/(x + 1)$
61. $4 \sec \theta$; $3x/\sqrt{16 + 9x^2}$; $4/\sqrt{16 + 9x^2}$ **63.** $\sin^3 \theta$; $\sqrt{1 - x^2}$; $\sqrt{1 - x^2}/x$
65. $(\tan^2 \theta \sec \theta)/16$; $4x/\sqrt{1 + 16x^2}$; $1/\sqrt{1 + 16x^2}$ **67.** $(25\sqrt{6} - 60)/12$; $-(25\sqrt{6} + 60)/12$

Section 4.2 (page 145)
1. $1/(\sin \theta \cos \theta)$ **3.** $1 + \cos s$ **5.** 1   **7.** 1   **9.** $2 + 2 \sin t$ **11.** $(\sin \gamma + 1)(\sin \gamma - 1)$ **13.** $4 \sin x$
15. $(2 \sin x + 1)(\sin x + 1)$ **17.** $(4 \sec x - 1)(\sec x + 1)$ **19.** $(\cos^2 x + 1)^2$ **21.** $\sin \theta$ **23.** 1   **25.** $\tan^2 \beta$
27. $\tan^2 x$ **29.** $\sec^2 x$ **75.** identity   **77.** not an identity   **79.** not an identity   **81.** not an identity

Section 4.3 (page 150)
1. $\cot 3°$ **3.** $\sin 5\pi/12$ **5.** $\sec 104° 24'$ **7.** $\cos (-\pi/8)$ **9.** $\csc (-56° 42')$ **11.** $\tan (-86.9814°)$ **13.** $\tan$
15. $\cos$ **17.** $\csc$ **19.** true   **21.** false   **23.** true   **25.** true   **27.** 15°   **29.** $(140/3)°$ **31.** 20°
33. $(\sqrt{6} - \sqrt{2})/4$ **35.** $(\sqrt{2} - \sqrt{6})/4$ **37.** $(\sqrt{2} - \sqrt{6})/4$ **39.** 0   **41.** $\sqrt{2}/2$ **43.** 0   **45.** 0
47. $(\sqrt{3} \cos \theta - \sin \theta)/2$ **49.** $(\cos \theta - \sqrt{3} \sin \theta)/2$ **51.** $-\sin x$ **53.** $(4 - 6\sqrt{6})/25; (4 + 6\sqrt{6})/25$
55. $16/65; -56/65$ **57.** $-77/85; -13/85$ **59.** $(2\sqrt{638} - \sqrt{30})/56; (2\sqrt{638} + \sqrt{30})/56$ **71.** $\cos (0 - t) = \cos t;$
$\cos (0 + t) = \cos t; \cos (\pi/2 - t) = \sin t; \cos (\pi/2 + t) = -\sin t; \cos (\pi - t) = -\cos t; \cos (\pi + t) = -\cos t;$
$\cos (3\pi/2 - t) = -\sin t; \cos (3\pi/2 + t) = \sin t$ **75.** $36/85$ **77.** $(8\sqrt{89} - 5\sqrt{1335})/356$

Section 4.4 (page 157)
1. $(\sqrt{6} - \sqrt{2})/4$ **3.** $2 - \sqrt{3}$ **5.** $-(\sqrt{6} + \sqrt{2})/4$ **7.** $(\sqrt{6} + \sqrt{2})/4$ **9.** $\sqrt{2}/2$ **11.** -1 **13.** 0   **15.** 1
17. 0   **19.** 0   **21.** $\sqrt{2}(\sin \theta + \cos \theta)/2$ **23.** $(\sqrt{3} \tan \theta + 1)/(\sqrt{3} - \tan \theta)$ **25.** $(1 + \tan s)/(1 - \tan s)$ **27.** $\sin \theta$
29. $\tan \theta$ **31.** $-\sin \theta$ **33.** $63/65; 33/65; 63/16; 33/56$ **35.** $(4\sqrt{2} + \sqrt{5})/9; (4\sqrt{2} - \sqrt{5})/9;$
$(-8\sqrt{5} - 5\sqrt{2})/(20 - 2\sqrt{10}); (-8\sqrt{5} + 5\sqrt{2})/(20 + 2\sqrt{10})$ **37.** $77/85; 13/85; -77/36; 13/84$
39. $-33/65; -63/65; 33/56; 63/16$ **41.** $1; -161/289;$ undefined; $-161/240$ **55.** $-.996728$ **57.** $-.959882$
59. $2 \sin (x + 330°)$ **61.** $13 \sin (A + 23°)$ **63.** $17 \sin (B + 332°)$ **65.** $25 \sin (t + 106°)$ **67.** $5 \sin (x + 143°)$
69. $y = \sin x - \sqrt{3} \cos x$ or   **71.** $y = -\sin x - \cos x$ or
$y = 2 \sin (x + 5\pi/3)$ $y = \sqrt{2} \sin (x + 5\pi/4)$

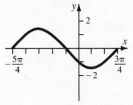

75. $\cos A \cos B \cos C - \cos A \sin B \sin C - \sin A \cos B \sin C - \sin A \sin B \cos C$ **79.** $80.8°$ **81.** $(8\sqrt{17} + \sqrt{85})/51$
83. $23/80$

Section 4.5 (page 164)
1. $\sin 20°$ **3.** $\tan 73.5°$ **5.** $\tan 29.87°$ **7.** $\cos 9x$ **9.** $\tan 4\theta$ **11.** $\cos x/8$ **13.** $\cos 30°$ or $\sqrt{3}/2$
15. $\sin 2\pi/3$ or $\sqrt{3}/2$ **17.** $(1/2) \sin \pi/4$ or $\sqrt{2}/4$ **19.** $(1/2) \tan 102°$ **21.** $(1/16) \sin 59°$ **23.** $\cos 4\alpha$
25. $-$ **27.** $+$ **29.** $\sin 22.5° = \sqrt{2 - \sqrt{2}}/2, \cos 22.5° = \sqrt{2 + \sqrt{2}}/2, \tan 22.5° = \sqrt{3 - 2\sqrt{2}}$ or $\sqrt{2} - 1$
31. $\sin 195° = -\sqrt{2 - \sqrt{3}}/2, \cos 195° = -\sqrt{2 + \sqrt{3}}/2, \tan 195° = \sqrt{7 - 4\sqrt{3}}$
33. $\sin 5\pi/2 = 1, \cos 5\pi/2 = 0, \tan 5\pi/2$ does not exist   **35.** $k = 4; t = 6$ **37.** $k = 6; t = 2$ **39.** $-\sqrt{42}/12$
41. $-\sqrt{6}/4$ **43.** $\cos x = -\sqrt{42}/12, \sin x = \sqrt{102}/12, \tan x = -\sqrt{119}/7, \sec x = -2\sqrt{42}/7, \csc x = 2\sqrt{102}/17, \cot x$
$= -\sqrt{119}/17$ **45.** $\cos 2\theta = 17/25, \sin 2\theta = -4\sqrt{21}/25, \tan 2\theta = -4\sqrt{21}/17, \sec 2\theta = 25/17, \csc 2\theta = -25\sqrt{21}/84,$
$\cot 2\theta = -17\sqrt{21}/84$ **47.** $\tan 2x = -4/3, \sec 2x = -5/3, \cos 2x = -3/5, \cot 2x = -3/4, \sin 2x = 4/5, \csc 2x = 5/4$
49. $\sin \alpha/2 = \sqrt{3}/3, \cos \alpha/2 = -\sqrt{6}/3, \tan \alpha/2 = -\sqrt{2}/2, \cot \alpha/2 = -\sqrt{2}, \sec \alpha/2 = -\sqrt{6}/2, \csc \alpha/2 = \sqrt{3}$
73. $4 \tan^2 x/(1 - 2 \tan^2 x + \tan^4 x)$ **75.** $4 \cos^3 x - 3 \cos x$ **77.** $8 \cos^4 x - 8 \cos^2 x + 1$ **79.** $84°$ **81.** $60°$
83. 3.9 **85.** -0.843580 **87.** 0.537003 **89.** -1.570905 **91.** $.892230$ **93.** -1.97579 **95.** (a) $\sqrt{3 - 2\sqrt{2}}$
97. 1   **99.** $\sec^2 4x$

Section 4.6 (page 170)
1. $(1/2)[\sin 60° - \sin 10°] = (1/2)[\sqrt{3}/2 - \sin 10°]$ **3.** $(3/2)[\cos 8x + \cos 2x]$
5. $(1/2)[\cos 2\theta - \cos(-4\theta)] = (1/2)[\cos 2\theta - \cos 4\theta]$ **7.** $-4[\cos 9y + \cos(-y)] = -4(\cos 9y + \cos y)$
9. $2 \cos 45° \sin 15°$ **11.** $2 \cos 95° \cos (-53°) = 2 \cos 95° \cos 53°$ **13.** $2 \cos 15\beta/2 \sin 9\beta/2$
15. $-6 \cos (7x/2) \sin (-3x/2) = 6 \cos (7x/2) \sin (3x/2)$

Section 4.7 (page 177)
1. $\{3\pi/4, 7\pi/4\}$ **3.** $\{\pi/3, 5\pi/3\}$ **5.** $\{\pi/6, 7\pi/6, 4\pi/3, 5\pi/3\}$ **7.** $\{\pi/3, 5\pi/3, \pi/6, 11\pi/6\}$ **9.** $\{\pi\}$ **11.** $\{3\pi/2, 7\pi/6, 11\pi/6\}$ **13.** $\{\pi/4, 3\pi/4, 5\pi/4, 7\pi/4\}$ **15.** $\{0, \pi/2, \pi, 3\pi/2\}$ **17.** $\{\pi/12, 7\pi/12, 13\pi/12, 19\pi/12\}$ **19.** $\{3\pi/8, 5\pi/8, 11\pi/8, 13\pi/8\}$ **21.** $\{\pi/2, 3\pi/2\}$ **23.** $\emptyset$ **25.** $\{71.6°, 251.6°, 63.4°, 243.4°\}$ **27.** $\{135°, 315°, 71.6°, 251.6°\}$
29. $\{33.6°, 326.4°\}$ **31.** $\{45°, 225°\}$ **33.** $\{0°, 90°, 180°, 270°\}$ **35.** $\{30°, 60°, 210°, 240°\}$ **37.** $\{0°\}$ **39.** $\{120°, 240°\}$
41. $\{270°, 30°, 150°\}$ **43.** $\{90°, 270°, 45°, 225°\}$ **45.** $\{70.5°, 289.5°\}$ **47.** $\{90°, 270°, 30°, 150°, 210°, 330°\}$
49. $\{x | x = 90° + 360° \cdot n, 180° + 360° \cdot n, n \text{ any integer}\}$ **51.** $\{x | x = 30° + 360° \cdot n, 150° + 360° \cdot n, 90° + 360° \cdot n, 270° + 360° \cdot n, n \text{ any integer}\}$ **53.** $\{x | x = 45° + 180° \cdot n, 108.5° + 180° \cdot n, n \text{ any integer}\}$
55. $\{x | x = 22\frac{1}{2}° + 180° \cdot n, 112\frac{1}{2}° + 180° \cdot n, n \text{ any integer}\}$ **57.** $\{53.6°, 126.4°, 187.9°, 352.1°\}$
59. $\{149.6°, 329.6°, 106.3°, 286.3°\}$ **61.** $\emptyset$ **63.** $\{339.2°, 237.7°\}$
65. $\{\pi/12, \pi/2, 5\pi/12, 3\pi/2, 13\pi/12, 17\pi/12\}$ **67.** $\{0, \pi/4, 3\pi/4, \pi, 5\pi/4, 7\pi/4\}$ **69.** $\{\pi/6, \pi/2, 3\pi/2, 5\pi/6\}$ **71.** 1/4 sec
73. 2 sec   **75.** 14°   **77.** (a) 42.2°   (b) 90°   (c) 48.0°   **79.** $\{3/5\}$
81. $\{4/5\}$ **83.** $\{0\}$ **85.** $\{1/2\}$ **87.** $\{-1/2\}$ **89.** $\{0\}$

Chapter 4 Review Exercises (page 180)
1. $\sin x = -4/5$, $\tan x = -4/3$, $\sec x = 5/3$, $\csc x = -5/4$, $\cot x = -3/4$ **3.** $\sin (x + y) = (4 + 3\sqrt{15})/20$, $\cos (x - y) = (4\sqrt{15} + 3)/20$ **5.** $\sin x = \sqrt{2 - \sqrt{2}}/2$, $\cos x = \sqrt{2 + \sqrt{2}}/2$, $\tan x = \sqrt{3 - 2\sqrt{2}}$ **7.** j
9. c   **11.** d   **13.** a   **15.** f   **17.** e   **19.** 1   **21.** $1/\cos^2 \theta$ or $\sec^2 \theta$ **23.** $1/\sin^2 \theta \cos^2 \theta$ **53.** $\{\pi/2, 3\pi/2\}$
55. $\{\pi/6, 5\pi/6\}$ **57.** $\{\pi/8, 3\pi/8, 5\pi/8, 7\pi/8, 9\pi/8, 11\pi/8, 13\pi/8, 15\pi/8\}$ **59.** $\{\pi/2\}$ **61.** $\{\pi/3, 5\pi/3, \pi\}$
63. $\{60°, 300°\}$ **65.** $\{270°\}$ **67.** $\{0°, 45°, 180°, 225°\}$ **69.** (a) 48.8°   (b) the light beam is completely under water
79. $2 - \sqrt{3}$

Chapter 5

Section 5.1 (page 190)
1. $A = 17° 00'$; $a = 39.1$ in; $c = 134$ in   **3.** $c = 85.9$ yd; $A = 62° 50'$; $B = 27° 10'$ **5.** $b = 42.3$ cm; $A = 24° 10'$; $B = 65° 50'$ **7.** $B = 36° 36'$; $a = 310.8$ ft; $b = 230.8$ ft   **9.** $A = 50° 51'$; $a = .4832$ m; $b = .3934$ m
11. $A = 71° 36'$; $B = 18° 24'$; $a = 7.413$ m   **13.** $A = 47.568°$; $b = 143.97$ m; $c = 213.38$ m
15. $B = 32.791°$; $a = 156.77$ cm; $b = 101.00$ cm   **17.** $a = 115.072$ m; $A = 33.4901°$; $B = 56.5099°$ **19.** 26.6 m
21. 52° 30'   **23.** 59.8 m   **25.** 11 ft   **27.** 35° 50'   **29.** 26° 20'   **31.** 1580 ft   **33.** 51.4 m   **35.** 8200 ft
37. 446 ft   **39.** 114 ft   **41.** 5.18 m   **43.** $h = k(\tan B - \tan A)$ **45.** 156 mi   **47.** 120 mi   **49.** 38°
51. $m(1 - \cos \alpha)$ **53.** 54° 40'   **55.** 6.993752×10^9 mi   **57.** $\sqrt{15}/4$ **59.** $\sqrt{5}/3$

Section 5.2 (page 202)
1. $C = 80° 40'$, $a = 79.5$ mm, $c = 108$ mm   **3.** $b = 37.3°$, $a = 38.5$ ft, $b = 51.0$ ft   **5.** $C = 57.36°$, $b = 11.13$ ft, $c = 11.55$ ft   **7.** $b = 18.5°$, $a = 239$ yd, $c = 230$ yd   **9.** $A = 56° 00'$, $c = 361$ ft, $a = 308$ ft   **11.** $B = 110.0°$, $a = 27.01$ m, $c = 21.36$ m   **13.** $A = 34.72°$, $a = 3326$ ft, $c = 5704$ ft   **15.** $C = 97° 34'$, $b = 283.2$ m, $c = 415.2$ m
17. $B_1 = 49.1°$, $C_1 = 101.2°$, $B_2 = 130.9°$, $C_2 = 19.4°$ **19.** $B = 26° 30'$, $A = 112° 10'$ **21.** no such triangle
23. $B = 27.19°$, $C = 10.68°$ **25.** $A = 43° 50'$, $B = 6° 52'$ **27.** $B = 20.6°$, $C = 116.9°$, $c = 20.6$ ft   **29.** no such triangle   **31.** $B_1 = 49° 20'$, $C_1 = 92° 00'$, $c_1 = 15.5$ km, $B_2 = 130° 40'$, $C_2 = 10° 40'$, $c_2 = 2.88$ km   **33.** $A_1 = 52° 10'$, $C_1 = 95° 00'$, $c_1 = 9520$ cm, $A_2 = 127° 50'$, $C_2 = 19° 20'$, $c_2 = 3160$ cm   **35.** $B = 37.77°$, $C = 45.43°$, $c = 4.174$ ft
37. $A_1 = 53.23°$, $C_1 = 87.09°$, $c_1 = 37.16$ m, $A_2 = 126.77°$, $C_2 = 13.55°$, $c_2 = 8.721$ m   **39.** 118 m   **41.** 1.93 mi
43. 10.4 in   **45.** 111°   **47.** does not exist   **49.** 46.4 m²   **51.** 356 cm²   **53.** 722.9 in²   **55.** 1071 cm²   **57.** 100

Section 5.3 (page 208)
1. $c = 2.83$ in, $A = 44.9°$, $B = 107°$ **3.** $c = 6.46$ m, $A = 53.1°$, $B = 81.3°$ **5.** $a = 156$ cm, $B = 64° 50'$, $C = 34° 30'$ **7.** $b = 9.529$ in, $A = 64.59°$, $C = 40.61°$ **9.** $a = 15.7$ m, $B = 21.6°$, $C = 45.6°$ **11.** $c = 139.0$ m, $A = 49° 20'$, $B = 105° 51'$ **13.** $A = 29.9°$, $B = 56.3°$, $C = 93.8°$ **15.** $A = 81.4°$, $B = 37.5°$, $C = 61.1°$
17. $A = 42.0°$, $B = 35.9°$, $C = 102.1°$ **19.** $A = 47.7°$, $B = 44.9°$, $C = 87.4°$ **21.** $A = 35° 22'$, $B = 50° 58'$, $C = 93° 40'$ **23.** $A = 28° 10'$, $B = 21° 56'$, $C = 129° 54'$ **25.** 257 m   **27.** 281 km   **29.** 22 ft
31. 18 ft   **33.** 1470 m   **35.** 6.01 km   **37.** 25.24983 mi   **39.** The angle at Mackinac West Base is 16.42821°; the angle at Green Island is 123.13624°; the angle at St. Ignace West Base is 40.43555°.   **41.** 140 in²   **43.** 12,600 cm²
45. 3650 ft²   **47.** 1921 ft²   **49.** 33 cans

Section 5.4 (page 221)

1. m and p, n and r **3.** m and p equal $2t$ or t is one-half m or p; also $m = 1p$ and $n = 1r$; $m = p = -q$; $r = -s$

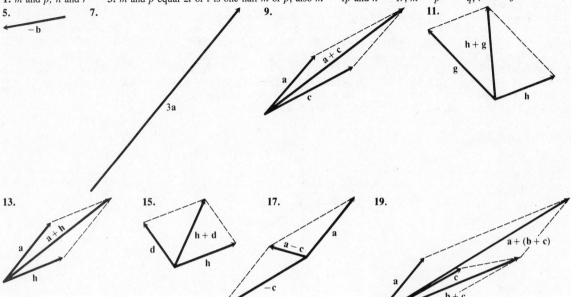

5.   **7.**   **9.**   **11.**

$-\mathbf{b}$ $3\mathbf{a}$ $\mathbf{a} + \mathbf{c}$, $\mathbf{a}$, $\mathbf{c}$ $\mathbf{h} + \mathbf{g}$, $\mathbf{g}$, $\mathbf{h}$

13.   **15.**   **17.**   **19.**

$\mathbf{a} + \mathbf{h}$, $\mathbf{a}$, $\mathbf{h}$ $\mathbf{h} + \mathbf{d}$, $\mathbf{d}$, $\mathbf{h}$ $\mathbf{a} - \mathbf{c}$, $\mathbf{a}$, $-\mathbf{c}$ $\mathbf{a} + (\mathbf{b} + \mathbf{c})$, $\mathbf{a}$, $\mathbf{c}$, $\mathbf{b} + \mathbf{c}$, $\mathbf{b}$

21. <1, 3>   **23.** <22, −22>   **25.** $10\sqrt{2}$, $10\sqrt{2}$ **27.** 14.2, 24.8   **29.** −123, 155   **31.** −22.3, −65.4
33. $\sqrt{2}$, 45°   **35.** 16, 315°   **37.** 17, 332°   **39.** 6, 180°   **41.** $-5\mathbf{i} + 8\mathbf{j}$ **43.** $2\mathbf{i}$ **45.** $4\sqrt{2}\mathbf{i} + 4\sqrt{2}\mathbf{j}$
47. $-.2536\mathbf{i} + .5438\mathbf{j}$ **49.** 530 newtons   **51.** 27.2 lb   **53.** 88.2 lb   **55.** 94° 00′   **57.** 17 lb
59. 18°   **61.** magnitude 2.86, equilibrant makes an angle of 55.3° with the 4.72 lb force   **63.** weight 64.8 lb, tension 61.9 lb
65. 190, 283 pounds respectively   **67.** 173.1°   **69.** 39.2 km   **71.** 237°, 470 mph   **73.** 358°, 170 mph   **75.** The ship
traveled 55.9 mi on its modified course, for 55.9 − 50 = 5.9 additional mi.   **83.** −22   **85.** −50   **89.** 151° 00′

Chapter 5 Review Exercises (page 225)

1. $B = 42° 40′$, $c = 58.4$ cm   **3.** $A = 56° 00′$, $B = 34° 00′$ **5.** $a = 11.7$ ft, $c = 402$ ft   **7.** 23.9 ft   **9.** 4040 ft
11. 70.8 m   **13.** triangle does not exist   **15.** 17° 10′   **17.** 1300 ft   **19.** 13 m   **21.** 25° 00′   **23.** 19° 50′
25. 14.8 m   **27.** $B = 17° 10′$, $C = 137° 40′$, $c = 11.0$ yd   **29.** 153,000 m²   **31.** 185 cm²
33.   **35.**

$\mathbf{a}$, $\mathbf{a} + \mathbf{b}$, $\mathbf{b}$ 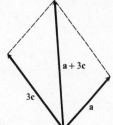 $\mathbf{a} + 3\mathbf{c}$, $3\mathbf{c}$, $\mathbf{a}$

37. horizontal 17.9, vertical 66.8   **39.** 2, 2   **41.** $-4\sqrt{3}$, −4   **43.** $\sqrt{40}$, 161° 30′   **45.** 2, 270°   **47.** $2\mathbf{i} - \mathbf{j}$
49. $10\mathbf{i}\sqrt{3} + 10\mathbf{j}$ **51.** 6   **53.** 10   **55.** 52° 10′   **57.** 30°   **59.** 28 lb   **61.** 135 newtons   **63.** 270 lb, 56° 20′
65. 122 lb   **67.** 306°, 524 mph

Chapter 6

Section 6.1 (page 235)
1. imaginary and complex   **3.** real and complex   **5.** imaginary and complex   **7.** complex   **9.** $0 + 10i$ **11.** $0 - 20i$
13. $0 - i\sqrt{39}$ **15.** $5 + 2i$ **17.** $-6 - 14i$ **19.** $9 - 5i\sqrt{2}$ **21.** -5 **23.** $-4 + 0i$ **25.** $2 + 0i$
27. $-2 + 0i$ **29.** $7 - i$ **31.** $2 + 0i$ **33.** $1 - 10i$ **35.** $-10 + 5i$ **37.** $8 - i$ **39.** $-14 + 2i$ **41.** $5 - 12i$
43. $-8 - 6i$ **45.** $13 + 0i$ **47.** $7 + 0i$ **49.** $0 + 25i$ **51.** $12 + 9i$ **53.** $0 + i$ **55.** $7/25 - (24/25)i$
57. $26/29 + (7/29)i$ **59.** $-2 + i$ **61.** $-2i$ **63.** $[(3 - 2\sqrt{5}) + (-2 - 3\sqrt{5})i]/13$ **65.** i **67.** i **69.** 1
71. $-i$ **73.** 1 **75.** i **77.** $5/2 + i$ **79.** $-16/65 - (37/65)i$ **81.** $27/10 + (11/10)i$ **83.** $17/10 - (11/10)i$
85. $37/34 + (165/34)i$ **87.** $a = 23; b = 5$ **89.** $a = 18; b = -3$ **91.** $a = -5; b = 1$
93. $a = -3/4; b = 3$ **95.** $-18 + 19i$ **109.** 25 **111.** $a = 0$ or $b = 0$

Section 6.2 (page 242)
1. 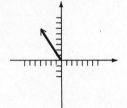   **3.** 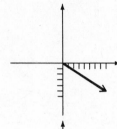   **5.**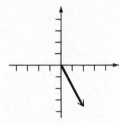

7.    **9.**

11. $1 + i$ **13.** $-2 + 2i$ **15.** $-2 + 4i$ **17.** $2 + 4i$ **19.** $7 + 9i$ **21.** $\sqrt{2} + i\sqrt{2}$ **23.** $0 + 10i$
25. $-2 - 2i\sqrt{3}$ **27.** $\sqrt{3}/2 + i/2$ **29.** $5/2 - 5i\sqrt{3}/2$ **31.** $-\sqrt{2} + 0i$ **33.** $\sqrt{13}(\cos 56° 20' + i \sin 56° 20')$
35. $-1.0179 - 2.8221i$ **37.** $2(\cos 160° 00' + i \sin 160° 00')$ **39.** $\sqrt{34}(\cos 59° 00' + i \sin 59° 00')$
41. $3\sqrt{2}(\cos 315° + i \sin 315°)$ **43.** $6(\cos 240° + i \sin 240°)$ **45.** $2(\cos 330° + i \sin 330°)$
47. $5\sqrt{2}(\cos 225° + i \sin 225°)$ **49.** $2\sqrt{2}(\cos 45° + i \sin 45°)$ **51.** $4(\cos 180° + i \sin 180°)$ **53.** $-3\sqrt{3} + 3i$
55. $0 - 4i$ **57.** $12\sqrt{3} + 12i$ **59.** $-15\sqrt{2}/2 + 15i\sqrt{2}/2$ **61.** $0 - 3i$ **63.** $\sqrt{3} - i$ **65.** $-1 - i\sqrt{3}$
67. $-1/6 - i\sqrt{3}/6$ **69.** $2\sqrt{3} - 2i$ **71.** $-1/2 - i/2$ **73.** $\sqrt{3} + i$ **75.** $20 + 15i$ **77.** $1.2 - .14i$
79. $2.39 + 15.0i$ **81.** $.378 + 3.52i$
83. 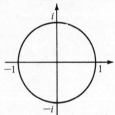   **85.** 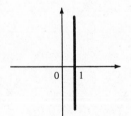   **87.**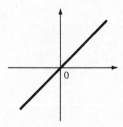

Section 6.3 (page 249)
1. $0 + 27i$ **3.** $1 + 0i$ **5.** $27/2 - 27i\sqrt{3}/2$ **7.** $-16\sqrt{3} + 16i$ **9.** $-128 + 128i\sqrt{3}$ **11.** $128 + 128i$
13. $-.1892 + .0745i$ **15.** $5520 + 9550i$

17. (cos 0° + *i* sin 0°),
(cos 120° + *i* sin 120°),
(cos 240° + *i* sin 240°)

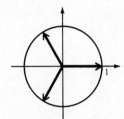

19. 2(cos 20° + *i* sin 20°),
2(cos 140° + *i* sin 140°),
2(cos 260° + *i* sin 260°)

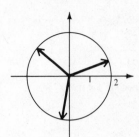

21. 2(cos 90° + *i* sin 90°)
2(cos 210° + *i* sin 210°),
2(cos 330° + *i* sin 330°)

23. 4(cos 60° + *i* sin 60°),
4(cos 180° + *i* sin 180°),
4(cos 300° + *i* sin 300°)

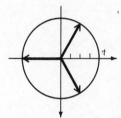

25. $\sqrt[3]{2}$(cos 20° + *i* sin 20°),
$\sqrt[3]{2}$(cos 140° + *i* sin 140°),
$\sqrt[3]{2}$(cos 260° + *i* sin 260°)

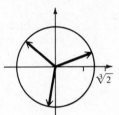

27. $\sqrt[3]{4}$(cos 50° + *i* sin 50°),
$\sqrt[3]{4}$(cos 170° + *i* sin 170°),
$\sqrt[3]{4}$(cos 290° + *i* sin 290°)

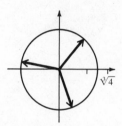

29. (cos 0° + *i* sin 0°),
(cos 180° + *i* sin 180°)

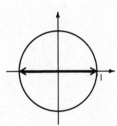

31. (cos 0° + *i* sin 0°), (cos 60° + *i* sin 60°),
(cos 120° + *i* sin 120°), (cos 180° + *i* sin 180°),
(cos 240° + *i* sin 240°), (cos 300° + *i* sin 300°)

33. (cos 45° + *i* sin 45°),
(cos 225° + *i* sin 225°)

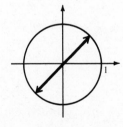

35. (cos 0° + *i* sin 0°), (cos 120° + *i* sin 120°), (cos 240° + *i* sin 240°)   **37.** (cos 90° + *i* sin 90°), (cos 210° + *i* sin 210°),
(cos 330° + *i* sin 330°)   **39.** 2(cos 0° + *i* sin 0°), 2(cos 120° + *i* sin 120°), 2(cos 240° + *i* sin 240°)
41. (cos 45° + *i* sin 45°), (cos 135° + *i* sin 135°), (cos 225° + *i* sin 225°), (cos 315° + *i* sin 315°)
43. (cos 22 1/2° + *i* sin 22 1/2°), (cos 112 1/2° + *i* sin 112 1/2°), (cos 202 1/2° + *i* sin 202 1/2°), (cos 292 1/2° + *i* sin 292 1/2°)

45. $2(\cos 20° + i \sin 20°), 2(\cos 140° + i \sin 140°), 2(\cos 260° + i \sin 260°)$
47. $1.3606 + 1.2637i, -1.7747 + .5464i, .4141 - 1.8102i$ **49.** $1.6309 - 2.5259i, -1.6309 + 2.5259i$
51. $\sqrt[4]{2}(\cos 22\ 1/2° + i \sin 22\ 1/2°), \sqrt[4]{2}(\cos 202\ 1/2° + i \sin 202\ 1/2°)$
53. $\sqrt[4]{18}(\cos 157\ 1/2° + i \sin 157\ 1/2°), \sqrt[4]{18}(\cos 337\ 1/2° + i \sin 337\ 1/2°)$

Section 6.4 (page 256)

1–11.

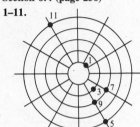

13.

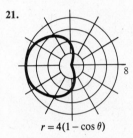

$r = 2 + 2 \cos \theta$

15.

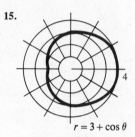

$r = 3 + \cos \theta$

17.

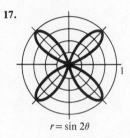

$r = \sin 2\theta$

19.

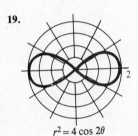

$r^2 = 4 \cos 2\theta$

21.

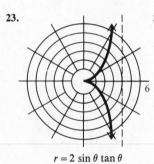

$r = 4(1 - \cos \theta)$

23.

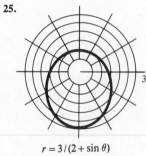

$r = 2 \sin \theta \tan \theta$

25.

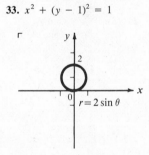

$r = 3/(2 + \sin \theta)$

27.

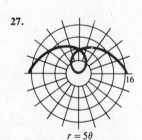

$r = 5\theta$

29.

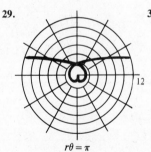

$r\theta = \pi$

31.

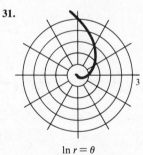

$\ln r = \theta$

33. $x^2 + (y - 1)^2 = 1$

$r = 2 \sin \theta$

35. $y^2 = 4(x + 1)$

$r = 2/(1 - \cos \theta)$

37. $x^2 + y^2 + 2x + 2y = 0$ or
$(x + 1)^2 + (y + 1)^2 = 2$

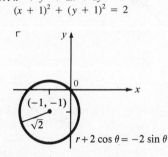

$(-1, -1)$
$\sqrt{2}$
$r + 2 \cos \theta = -2 \sin \theta$

39. $x = 2$

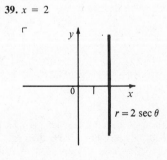

$r = 2 \sec \theta$

41. $x + y = 2$

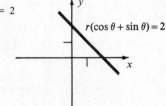

$r(\cos \theta + \sin \theta) = 2$

43. $y = -2$

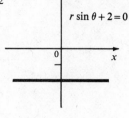

$r \sin \theta + 2 = 0$

45. $r(\cos \theta + \sin \theta) = 4$
47. $r = 4$
49. $r = 2 \csc \theta$ or $r \sin \theta = 2$
51. $r \sin^2 \theta = 25 \cos \theta$
53. $r^2(\cos^2 \theta + 9 \sin^2 \theta) = 36$

Chapter 6 Review Exercises (page 257)
1. $-2 - 3i$ **3.** $5 + 4i$ **5.** $29 + 37i$ **7.** $-32 + 24i$ **9.** $-2 - 2i$ **11.** $0 + i$ **13.** $8/5 + (6/5)i$
15. $-5/26 + (1/26)i$ **17.** $7/2 - (11/4)i$ **19.** $0 + 2i\sqrt{3}$ **21.** $5 + 4i$
23.

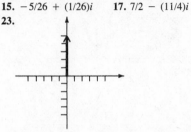

25.

27. $2\sqrt{2}(\cos 135° + i \sin 135°)$ **29.** $-\sqrt{2} - i\sqrt{2}$ **31.** $\sqrt{2}(\cos 315° + i \sin 315°)$ **33.** $4(\cos 270° + i \sin 270°)$
35. $0 - 30i$ **37.** $-1/8 + i\sqrt{3}/8$ **39.** $0 + 8i$ **41.** $-1/2 - i\sqrt{3}/2$ **43.** $8^{1/10}(\cos 27° + i \sin 27°)$;
$8^{1/10}(\cos 99° + i \sin 99°)$; $8^{1/10}(\cos 171° + i \sin 171°)$; $8^{1/10}(\cos 243° + i \sin 243°)$; $8^{1/10}(\cos 315° + i \sin 315°)$
45. $\cos 0° + i \sin 0°$; $\cos 60° + i \sin 60°$; $\cos 120° + i \sin 120°$; $\cos 180° + i \sin 180°$; $\cos 240° + i \sin 240°$;
$\cos 300° + i \sin 300°$ **47.** $5(\cos 60° + i \sin 60°)$; $5(\cos 180° + i \sin 180°)$; $5(\cos 300° + i \sin 300°)$
49.

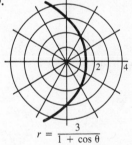

$r = \dfrac{3}{1 + \cos \theta}$

51.

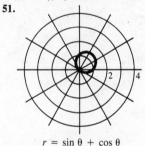

$r = \sin \theta + \cos \theta$

53. $r \cos \theta = -3$ **55.** $r = \tan \theta \sec \theta$ **63.** $(3, 0°)$, $(3, 72°)$, $(3, 144°)$, $(3, 216°)$, $(3, 288°)$

Chapter 7

Section 7.1 (page 267)
1. yes   **3.** no   **5.** 4/9   **7.** $\sqrt{6}/3$
9. (a)

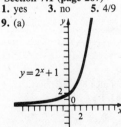

$y = 2^x + 1$

(b)

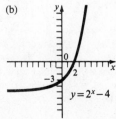

$y = 2^x - 4$

(c)

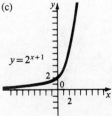

$y = 2^{x+1}$

(d)

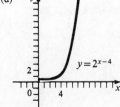

$y = 2^{x-4}$

11.

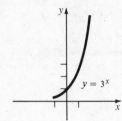

$y = 3^x$

13.

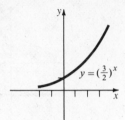

$y = \left(\frac{3}{2}\right)^x$

15.

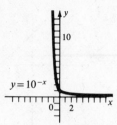

$y = 10^{-x}$

17.

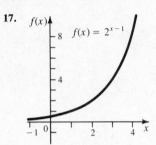

$f(x) = 2^{x-1}$

19.

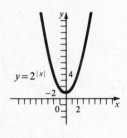

$y = 2^{|x|}$

21.

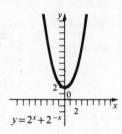

$y = 2^x + 2^{-x}$

23.

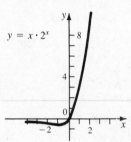

$y = x \cdot 2^x$

25.

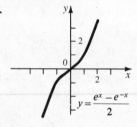

$y = \dfrac{e^x - e^{-x}}{2}$

27.

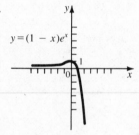

$y = (1 - x)e^x$

29. {1/2}   **31.** {−2}   **33.** {0}   **35.** {3}   **37.** {8}   **39.** {1/5}   **41.** {0}   **43.** {3/5}   **45.** {−2/3}   **47.** {−3, 3}
49. {−1, 5}   **51.** {0}   **55.** \$3,516.56   **57.** \$6,799.21   **59.** \$8,158.66   **61.** (a) \$12,075.32   (b) \$12,348.76
(c) \$12,650.78   **63.** (a) 500 g   (b) 409 g   (c) 335 g   (d) 184 g   (e) $Q(t)$

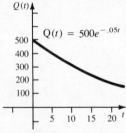

$Q(t) = 500e^{-.05t}$

65. (a) about 2,690,000   (b) about 3,020,000   (c) about 9,600,000   (d) about 38,400,000   **67.** 11.6 pounds per square inch
71. 0   **73.** 2   **75.** neither   **79.** 2.718   **83.** $4/(e^{-x} - e^x)^2$

Section 7.2 (page 276)
1. $\log_3 81 = 4$ **3.** $\log_{10} 10,000 = 4$ **5.** $\log_{1/2} 16 = -4$ **7.** $\log_{10} .0001 = -4$ **9.** $6^2 = 36$ **11.** $(\sqrt{3})^8 = 81$
13. $10^{-4} = .0001$ **15.** $m^n = k$ **17.** 2   **19.** 1   **21.** −3   **23.** 1/2   **25.** −1/6   **27.** 9   **29.** 8   **31.** 4
33. 1/2   **35.** 2   **37.** x **39.** 9   **41.** {1/5}   **43.** {3/2}   **45.** {16}   **47.** {1}   **49.** $\log_3 2 - \log_3 5$
51. $\log_2 6 + \log_2 x - \log_2 y$ **53.** $1 + (1/2) \log_5 7 - \log_5 3$ **55.** cannot be simplified using the properties of logarithms
57. $\log_k p + 2 \cdot \log_k q - \log_k m$ **59.** $(1/2)(\log_m 5 + 3 \cdot \log_m r - 5 \cdot \log_m z)$ **61.** $\log_a (xy)/m$ **63.** $\log_m (a^2/b^6)$
65. $\log_x (a^{3/2}b^{-4})$ or $\log_x (a^{3/2}/b^4)$ **67.** $\log_b [7x(x + 2)/8]$ **69.** $\log_a [(z + 1)^2(3z + 2)]$
71. $\log_5 (5^{1/3}m^{-1/3})$ or $\log_5(5^{1/3}/m^{1/3})$

73.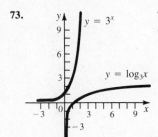

75. (a)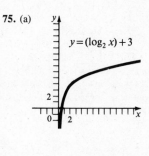

(b)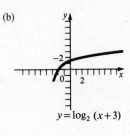

(c)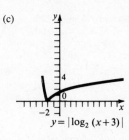

77.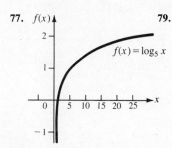

79.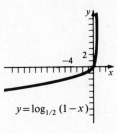

81.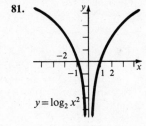

83.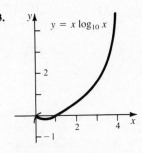

85. .7781   **87.** .9542   **89.** 1.4771   **91. (a)** about 239   **(b)** about 477   **(c)** about 759   **(d)**

93. (a) 21   **(b)** 70   **(c)** 91   **(d)** 120   **(e)** 140   **95.** 398,000,000 I_0

99.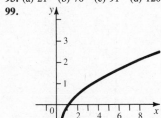

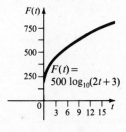

Section 7.3 (page 282)
1. 1   **3.** 4   **5.** -3 **7.** 1.2   **9.** 1.3863   **11.** 2.8332   **13.** 5.8579   **15.** 3.8918   **17.** 3.7377   **19.** 11.3022
21. 1.43   **23.** .59   **25.** -1.58 **27.** .96   **29.** 1.89   **31.** -5.05
33.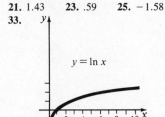

35.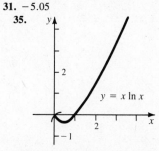

37. 1   **39. (a)** 2.03   **(b)** 2.28   **(c)** 2.17   **(d)** 1.21   **41. (a)** about 495   **(b)** about 6   **43.** 378 feet
45. (a) 7,298,000; 77,338,000; 168,192,000; 175,139,000   **(b)** According to the given equation, the population can never increase to
197,273,000.   **47.** .86

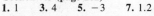

Section 7.4 (page 289)

1. {1.631}   **3.** {1.069}   **5.** {−.535}   **7.** {−.080}   **9.** {3.797}   **11.** {2.386}   **13.** {−.123}   **15.** 0   **17.** {1.104}
19. {17.475}   **21.** {−2.141, 2.141}   **23.** {1.386}   **25.** {11}   **27.** {5}   **29.** {10}   **31.** {5}   **33.** {4}   **35.** 0
37. {11}   **39.** {3}   **41.** {0}   **43.** {−2, 2}   **45.** {1, 10}   **47.** {2.423}   **49.** {−.204}   **51.** 15 seconds   **53.** 46 days
55. {0}   **57.** {0, ln 5}   **59.** $x = \ln[y/(y + 1)]$ **61.** $x = \ln (y + \sqrt{y^2 + 1})$ **63.** $t = (−2/R) \ln (1 − RI/E)$
65. $n = \ln[A/(A − Pi)]/\ln (1 + i)$ **67.** $x = (1 − y)/2$ **69.** $(1/3, + \infty)$ **71.** $(0, 1) \cup (1, 10)$ **73.** $(0, 1)$
75. $[−\sqrt[6]{3}, −1) \cup (−1, 1) \cup (1, \sqrt[6]{3}]$ **77.** 89 decibels is about twice as loud as 86 decibels, which is a 100% increase.
79. $n = [\ln (A/P)]/[m \ln (1 + i/m)]$ **81.** $t = −[\ln (M/434)]/.08$

Section 7.5 (page 296)

1. about 14 months   **3.** about 283 grams   **5.** about 11.3°   **7.** 5600 years   **11.** about 15,000 years   **13.** $2,166.57
15. $17,143.21   **17.** $12,652.54   **19.** about 13.9 years   **21.** $i = [\ln (A/P)]/n$ **23.** 81.25°C   **25.** 117.5°C
27. about 1611   **29.** in about 1.8 decades or 18 years

Section 7.6 (page 302)

1. 2   **3.** 6   **5.** −4   **7.** −5   **9.** −3.0491 or .9509 − 4   **11.** 4.8338   **13.** .8825   **15.** 4.6331   **17.** −2.0918 or
.9082 − 3   **19.** 34.4   **21.** .0747   **23.** 3120   **25.** .00637   **27.** 4.83   **29.** −.02   **31.** 5.62   **33.** 3.32
35. 4.73   **37.** 1.79   **39.** {.40}   **41.** {−.19}   **43.** {−1.47}   **45.** {.38}   **47.** 2.01   **49.** 10.6   **51.** 21.5
53. (a) about 350 years   (b) about 4000 years   (c) about 2300 years   **55.** 3.2   **57.** 1.8   **59.** 2.0×10^{-3}
61. 1.6×10^{-5} **63.** 8706   **65.** (a) about 78   (b) about 38   (c) about 2800

Chapter 7 Appendix (page 305)

1. 3.3701   **3.** 1.6836   **5.** .7975 − 2   **7.** 1.4322   **9.** 62.26   **11.** .004534   **13.** 493,200   **15.** .0001667
17. 1.396   **19.** 1.550   **21.** .9308   **23.** 15.57   **25.** .1405

Chapter 7 Review Exercises (page 306)

1.   **3.**   **5.**   **7.**

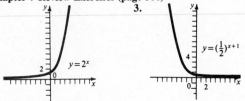

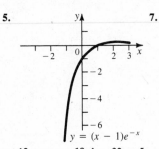

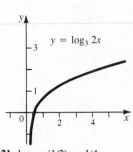

9. {5/3}   **11.** {3/2}   **13.** 800 g   **15.** $1790.19   **17.** about 12 years   **19.** $\log_2 32 = 5$ **21.** $\log_{1/16} (1/2) = 1/4$
23. $\log 3 = .4771$ **25.** $\ln 1.1052 = .1$ **27.** $10^{-3} = .001$ **29.** $10^{.537819} = 3.45$ **31.** $\log_3 m + \log_3 n − \log_3 p$
33. $2 \log_5 x + 4 \log_5 y + (3/5) \log_5 m + (1/5) \log_5 p$ **35.** .9279   **37.** 3.0212   **39.** 3150   **41.** .0446   **43.** 5.97
45. 1.44   **47.** .8 m   **49.** 1.5 m   **51.** (a) 207   (b) 235   (c) 249   (d)

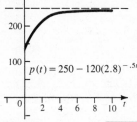

53. {1.490}   **55.** {1.303}   **57.** {.747}   **59.** {4}   **61.** {2}   **63.** {2}   **65.** $x = [\ln(y + \sqrt{y^2 + 1})]/\ln 5$
67. $x = (4 + a \pm \sqrt{a^2 + 12a})/2$ **69.** $c = de^{(N − a)/b}$ **71.** −5.3   **73.** 4.4886   **75.** 2.0794   **77.** −3.2403
79. 1.2785   **81.** −2.0082 or .9918 − 3   **83.** $7,182.87   **85.** $1,241.99   **87.** about 28 quarters or 7 years
89. about 5.3 years   **91.** $16,095.74   **93.** $580,792.63   **95.** (a) about 286 grams   (b) about 28.9 days
97. (a) −.223; .182   (b) .405 + .288 = .693   (c) .405 + .693 = 1.098; 3(.693) = 2.079
99. $100! \approx 9.32 \times 10^{157}$; $200! \approx 7.88 \times 10^{374}$

Chapter 8

Section 8.1 (page 318)

1. 3/5   **3.** undefined   **5.** 2   **7.** 5/9   **9.** undefined   **11.** 2   **13.** .5785   **15.** 45°   **17.** 60°   **19.** 108°
21. 115°   **23.** $2x + y = 5$ **25.** $y = 1$ **27.** $x + 3y = 10$ **29.** $3x + 4y = 12$ **31.** $2x - 3y = 6$ **33.** $x = -6$
35. $5.081x + y = -4.69256$

37.

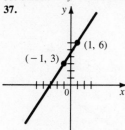

39.

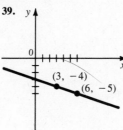

41.

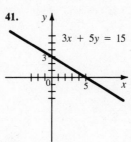

43.

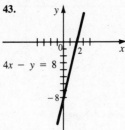

45.

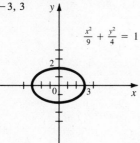

47.

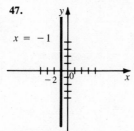

49.

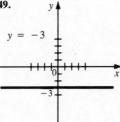

51. $x + 3y = 11$ **53.** $x - y = 7$ **55.** $2x - y = -4$ **57.** $x - 11y = 9$ **59.** no   **61.** (a) $-1/2$ (b) $-7/2$
63. 81.9°   **65.** 10.3°   **67.** 45°   **69.** 70°, 81°, 29°

Section 8.2 (page 328)

1. $-3, 3$

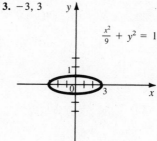

3. $-3, 3$

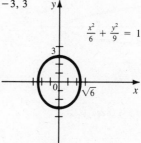

5. $-3, 3$

7. $-4, 4$

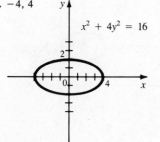

9. $y = \pm x$

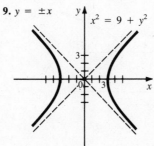

11. $-2\sqrt{2}, 2\sqrt{2}$

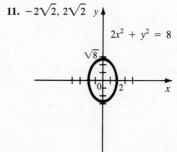

13. $y = \pm 5x/2$

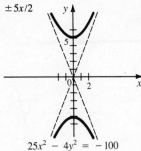

$$25x^2 - 4y^2 = -100$$

15. $x^2/16 + y^2/12 = 1$ **17.** $x^2/36 + y^2/20 = 1$ **19.** $x^2/16 + y^2/25 = 1$
21. $x^2/9 - y^2/7 = 1$ **23.** $x^2/9 - y^2/7 = 1$ **25.** $y^2/9 - x^2/25 = 1$
27. about 141.6 million miles

Section 8.3 (page 336)

1.

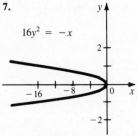

$$3y = x^2$$

3.

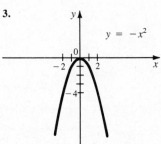

$$y = -x^2$$

5.

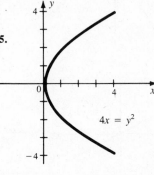

$$4x = y^2$$

7.

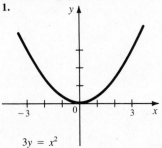

$$16y^2 = -x$$

9. $(0, 4)$; $y = -4$; y-axis   **11.** $(0, -1/8)$; $y = 1/8$; y-axis   **13.** $(1/64, 0)$;
$x = -1/64$; x-axis   **15.** $(-4, 0)$; $x = 4$; x-axis   **17.** $x^2 = -8y$
19. $y^2 = -2x$ **21.** $y^2 = 4x$ **23.** $x^2 = -2y$ **25.** $x^2 = -y$ **27.** $1/2$
29. $\sqrt{2}$ **31.** $\sqrt{21/7}$ **33.** $\sqrt{10/3}$ **35.** $\sqrt{2}/2$ **37.** $x^2 = 32y$
39. $x^2/36 + y^2/27 = 1$ **41.** $x^2/36 - y^2/108 = 1$ **43.** $x^2 = -4y$
45. $25x^2/81 + y^2/9 = 1$ **49.** 0

Section 8.4 (page 341)

1. ellipse   **3.** parabola   **5.** circle   **7.** hyperbola
9.

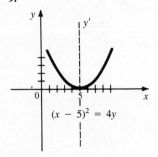

$$(x - 5)^2 = 4y$$

11.

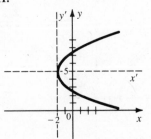

$$(y - 5)^2 = 3(x + 2)$$

13.

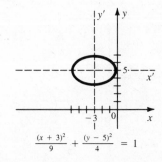

$$\frac{(x + 3)^2}{9} + \frac{(y - 5)^2}{4} = 1$$

15.

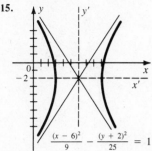

$$\frac{(x-6)^2}{9} - \frac{(y+2)^2}{25} = 1$$

17.

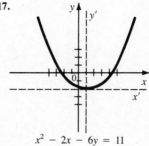

$$x^2 - 2x - 6y = 11$$

19.

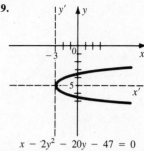

$$x - 2y^2 - 20y - 47 = 0$$

21.

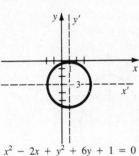

$$x^2 - 2x + y^2 + 6y + 1 = 0$$

23. no graph

25.

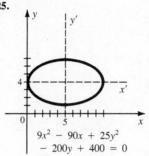

$$9x^2 - 90x + 25y^2$$
$$- 200y + 400 = 0$$

27.

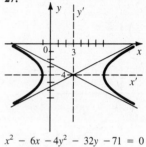

$$x^2 - 6x - 4y^2 - 32y - 71 = 0$$

29. the point $(-4, 2)$

Section 8.5 (page 346)

1. circle or ellipse   **3.** hyperbola   **5.** parabola   **7.** $30°$ **9.** $60°$ **11.** $22.5°$

13.

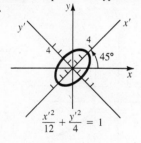

$$\frac{x'^2}{12} + \frac{y'^2}{4} = 1$$

15.

$$\frac{x'^2}{9} + \frac{y'^2}{4} = 1$$

17.

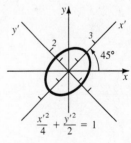

$$\frac{x'^2}{4} + \frac{y'^2}{2} = 1$$

19.

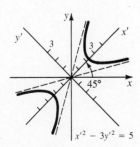

$$x'^2 - 3y'^2 = 5$$

21.

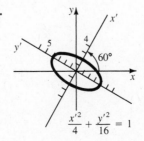

$$\frac{x'^2}{4} + \frac{y'^2}{16} = 1$$

23.

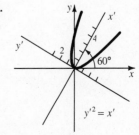

$$y'^2 = x'$$

25.

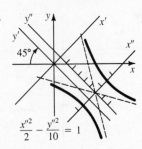

$$\frac{x''^2}{2} - \frac{y''^2}{10} = 1$$

27.

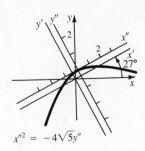

$$x''^2 = -4\sqrt{5}y''$$

29.

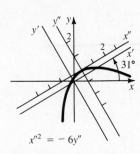

$$x''^2 = -6y''$$

Chapter 8 Review Exercises (page 347)
1. undefined slope; 90.0°   **3.** 9/4; 66.0°   **5.** 1/5; 11.3°   **7.** $x + 3y = 10$ **9.** $2x + y = 1$ **11.** $5x - 3y = -15$
13. $3x - y = 7$ **15.** 75.0°   **17.** 85.6°
19. (3/4, 0); $x = -3/4$; $y = 0$ **21.** (0, 5/4); $y = -5/4$; $x = 0$ **23.** vertices (0, 18),
(0, −18); foci $(0, 6\sqrt{5})$,
$(0, -6\sqrt{5})$

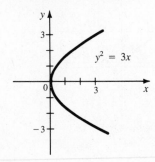

$y^2 = 3x$

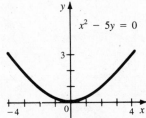

$x^2 - 5y = 0$

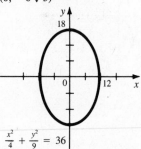

$$\frac{x^2}{4} + \frac{y^2}{9} = 36$$

25. $y = \pm 5x/3$
vertices (3, 0), (−3, 0);
foci $(\sqrt{34}, 0)$, $(-\sqrt{34}, 0)$

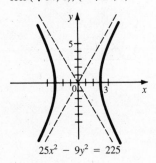

$25x^2 - 9y^2 = 225$

27. $x^2 = -16y$ **29.** $y = 7x^2$ **31.** $x^2/16 + y^2/7 = 1$
33. $y^2/4 - 5x^2/16 = 1$ **35.** hyperbola; $e = \sqrt{13/9}$ **37.** parabola; $e = 1$
39. circle; $e = 0$ **41.** $4x^2/245 + 4y^2/441 = 1$ **43.** $y^2 = 4x$
45. $y^2/16 - x^2/20 = 1$

47.

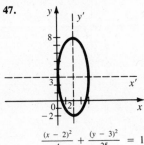

$$\frac{(x - 2)^2}{4} + \frac{(y - 3)^2}{25} = 1$$

49.

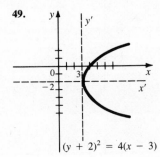

$(y + 2)^2 = 4(x - 3)$

51.

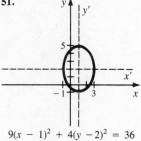

$$9(x - 1)^2 + 4(y - 2)^2 = 36$$

53.

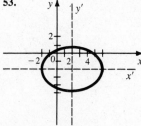

$$x^2 + 2y^2 - 4x + 8y - 4 = 0$$

55.

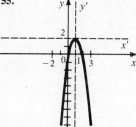

$$3x^2 - 6x + y + 1 = 0$$

57. hyperbola
59. hyperbola
61. 15°

63.

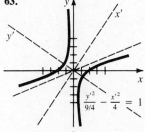

$$\frac{y'^2}{9/4} - \frac{x'^2}{4} = 1$$

$$24xy - 7y^2 + 36 = 0$$

65.

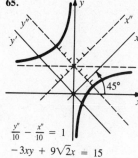

$$\frac{y''^2}{10} - \frac{x''^2}{10} = 1$$

$$-3xy + 9\sqrt{2}x = 15$$

Index

45°–45° Right Triangle

In a 45°–45° right triangle, the hypotenuse has a length that is $\sqrt{2}$ times as long as the length of either of the shorter sides.

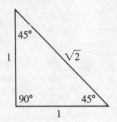

30°–60° Right Triangle

In a 30°–60° right triangle, the hypotenuse is always twice as long as the shortest side, and the medium side has a length that is $\sqrt{3}$ times as long as that of the shortest side. Also, the shortest side is opposite the 30° angle, and the medium side is opposite the 60° angle.

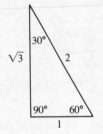

Fundamental Identities

$$\tan A = \frac{\sin A}{\cos A} \qquad \cot A = \frac{\cos A}{\sin A}$$

$$\cot A = \frac{1}{\tan A} \qquad \csc A = \frac{1}{\sin A} \qquad \sec A = \frac{1}{\cos A}$$

$$\sin^2 A + \cos^2 A = 1$$

$$\sin^2 A = 1 - \cos^2 A \qquad \cos^2 A = 1 - \sin^2 A$$

$$\tan^2 A + 1 = \sec^2 A \qquad 1 + \cot^2 A = \csc^2 A$$

$$\sin (-A) = -\sin A \qquad \cos (-A) = \cos A$$

$$\tan (-A) = -\tan A$$

Sum and Difference Identities

$$\cos (A - B) = \cos A \cos B + \sin A \sin B$$

$$\cos (A + B) = \cos A \cos B - \sin A \sin B$$

$$\sin (A - B) = \sin A \cos B - \cos A \sin B$$

$$\sin (A + B) = \sin A \cos B + \cos A \sin B$$

$$\tan (A - B) = \frac{\tan A - \tan B}{1 + \tan A \tan B}$$

$$\tan (A + B) = \frac{\tan A + \tan B}{1 - \tan A \tan B}$$

Cofunction Identities

$$\sin \left(\frac{\pi}{2} - A \right) = \cos A$$

$$\cos \left(\frac{\pi}{2} - A \right) = \sin A$$

$$\tan \left(\frac{\pi}{2} - A \right) = \cot A$$

Multiple-Angle and Half-Angle Identities

$$\cos 2A = \cos^2 A - \sin^2 A$$

$$\cos 2A = 1 - 2 \sin^2 A$$

$$\cos 2A = 2 \cos^2 A - 1$$

$$\sin 2A = 2 \sin A \cos A$$

$$\tan 2A = \frac{2 \tan A}{1 - \tan^2 A}$$

$$\cos \frac{A}{2} = \pm \sqrt{\frac{1 + \cos A}{2}}$$

$$\sin \frac{A}{2} = \pm \sqrt{\frac{1 - \cos A}{2}}$$

$$\tan \frac{A}{2} = \pm \sqrt{\frac{1 - \cos A}{1 + \cos A}}$$

$$\tan \frac{A}{2} = \frac{\sin A}{1 + \cos A}$$

$$\tan \frac{A}{2} = \frac{1 - \cos A}{\sin A}$$